# The Theory of Canonical
# Moments with Applications
# in Statistics, Probability,
# and Analysis

# The Theory of Canonical Moments with Applications in Statistics, Probability, and Analysis

HOLGER DETTE
Ruhr-Universität Bochum
Bochum, Germany

WILLIAM J. STUDDEN
Purdue University
Lafayette, Indiana

A Wiley-Interscience Publication
JOHN WILEY & SONS, INC.
New York · Chichester · Weinheim · Brisbane · Singapore · Toronto

*Library of Congress Cataloging-in-Publication Data*

Dette, Holger, 1961–
    The theory of canonical moments with applications in statistics,
  probability, and analysis / Holger Dette, William J. Studden.
       p.  cm. -- (Wiley series in probability and statistics.
  Applied probability and statistics)
    "A Wiley-Interscience publication."
    Includes bibliographical references (p. 311–19 ) and index.
    ISBN 0-471-10991-6 (alk. paper)
    1. Chebyshev systems.  2. Probability measures.  I. Studden.
  William J.  II. Title.  III. Series.
  QA355.D47 1997
  519.2--dc21                        96-53188

*To Gabi and Alina*
*and*
*To Myrna*

# Preface

This monograph presents a viewpoint on moment theory and its application in statistics, probability, and analysis that is somewhat different from that which is usually found in the literature on moment theory. The material is concerned with the theory and application of canonical moments of probability measures on intervals of the real line and, to a lesser extent, measures on the circle. For a given probability measure $\mu$ on the interval $[a, b]$ the ordinary moments are given by

$$c_k = c_k(\mu) = \int_a^b x^k d\mu(x), \qquad k \geq 1.$$

The canonical moments may be defined in the following geometrical manner: For a given measure $\mu$ with moments $c_k(\mu)$, $k \geq 1$, let

$$c_k^- = c_k^-(\mu) = \min_\eta c_k(\eta), \quad c_k^+ = c_k^+(\mu) = \max_\eta c_k(\eta),$$

where the minimum and maximum are taken over those probability measures $\eta$ for which the moments up to degree $k - 1$ coincide with the moments of $\mu$, that is, $c_i(\eta) = c_i(\mu)$ for $i \leq k - 1$. The canonical moments are given by

$$p_k = \frac{c_k - c_k^-}{c_k^+ - c_k^-}, \qquad k \geq 1.$$

Here it is assumed that $c_k^- < c_k^+$; otherwise, the sequence $p_k$ is appropriately terminated. The $k$th canonical moment depends on $c_1, \ldots, c_k$. If $M$ denotes the convex set of all ordinary moment sequences and $M_k$ the projection onto the first $k$ coordinates, then this definition yields a one-to-one mapping between the interior of $M_k$ and the cube $(0, 1)^k$ for each $k \in \mathbb{N}$. Roughly speaking, the canonical moments successively define the "relative position" of $c_k$, given $c_1, \ldots, c_{k-1}$, in $M_k$ or $M$.

The canonical moments have a number of simple interesting properties. For example, they are invariant under linear transformations of the measure $\mu$ on $[a, b]$ onto any other interval $[c, d]$, and measures symmetric about the midpoint of the underlying interval have the odd canonical moments $p_{2i-1} \equiv \frac{1}{2}$. The "middle" of the moment space $M$ where $p_i \equiv \frac{1}{2}$ for all $i$ corresponds, relative to the interval $[-1, 1]$, to the arc-sine distribution with density

$$\frac{1}{\pi\sqrt{1 - x^2}}, \qquad x \in (-1, 1),$$

and to the classical Chebyshev polynomials $T_n(x) = \cos n\theta$, $x = \cos\theta$.

The canonical moments are also related historically to orthogonal polynomials and continued fractions. A sequence $\{P_k\}_{k \geq 0}$ of polynomials, with leading coefficients one, is orthogonal with respect to a measure $\mu$ on the real line with infinite support if and only if

$$\int_{-\infty}^{\infty} P_k(x)P_\ell(x)d\mu(x) = 0$$

whenever $k \neq \ell$. It can be shown that this property is equivalent to the fact that the polynomials satisfy a three term recurrence relation

$$P_{k+1}(x) = (x - \alpha_{k+1})P_k(x) - \beta_{k+1}P_{k-1}(x), \qquad k = 0, 1, 2, \ldots,$$

where $P_{-1}(x) = 0$, $P_0(x) \equiv 1$ and $\{\alpha_k\}_{k \geq 1}$ and $\{\beta_k\}_{k \geq 1}$ are real sequences such that $\beta_k > 0$ for all $k \geq 1$. This was discovered by Favard (1935). The corresponding measure of orthogonality $\mu$ is supported on the nonnegative real axis if there exist a sequence $\{\zeta_k\}_{k \geq 1}$ of positive numbers such that the sequences $\{\alpha_k\}$ and $\{\beta_k\}$ satisfy

$$\beta_{k+1} = \zeta_{2k-1}\zeta_{2k} \quad \text{and} \quad \alpha_{k+1} = \zeta_{2k} + \zeta_{2k+1}.$$

It was further discovered by Wall (1940) that $\mu$ is concentrated on the interval $[0, 1]$ if the sequence $\{\zeta_k\}_{k \geq 1}$ is a chain sequence; that is, it can be further decomposed as

$$\zeta_k = q_{k-1}p_k,$$

where $q_k = 1 - p_k$ ($q_0 = 1$) and $0 < p_k < 1$. In this case the $p_k$ turn out to be the canonical moments of the measure $\mu$.

The sequences $\{\alpha_k\}$, $\{\beta_k\}$, and $\{\zeta_k\}$ and the orthogonal polynomials are related through the Stieltjes transform of the measure $\mu$ and its corresponding continued fraction expansion. For a probability measure $\mu$ on the interval $[0, 1]$ this takes the following form:

$$\int_0^1 \frac{d\mu(x)}{z - x} = \frac{1}{|z|} - \frac{p_1}{|1|} - \frac{q_1 p_2}{|z|} - \frac{q_2 p_3}{|1|} - \cdots$$

$$:= \cfrac{1}{z - \cfrac{p_1}{1 - \cfrac{q_1 p_2}{z - \cdots}}}, \qquad z \notin [0, 1].$$

The theory of general Chebyshev systems and their corresponding moment spaces was first described by Krein (1951) and further expounded by Karlin and Studden (1966) and Krein and Nudelman (1977). The work of both Karlin and Krein, in fields too numerous to mention, has been a source of inspiration for the present volume. The geometrical aspects of the classical moment spaces $M_k$ were delineated by Karlin and Shapley in (1953) where the geometric properties of the arc-sine law and the associated Chebyshev polynomials are discussed. The general geometric description of the canonical moments seems to have first appeared in a paper of Seall and Wetzel (1959) where the associated continued fractions are also considered. Our interest in canonical moments began with our basic previous interest in moment spaces in general and the paper by Morris Skibinsky (1967). In this paper Skibinsky rediscovers and defines the canonical moments or coordinates while investigating the range of the $(k + 1)$th moment $c_{k+1}$ for given moments $c_1, \ldots, c_k$ which he showed to be $\prod_{i=1}^k p_i q_i$. In a number of further excellent papers Skibinsky proves numerous other interesting properties of the canonical moments.

Our main interest stems from the discovery that the canonical moments appear to be more intimately or intrinsically related to the measure $\mu$ than the ordinary moments. Moreover, the canonical moments turn out to be very useful in many areas of statistics, probability, and analysis including problems in the design of experiments, simple random walks or birth and death chains, and in approximation theory.

The general outline of the book is as follows: Chapters 1 to 3 provide the theoretical development of the theory of canonical moments for measures on intervals $[a, b]$, and Chapters 4, 5, 6, 7, 8, and 10 contain various applications. The canonical moments for the circle or trigonometric functions, both the theory and applications, are in Chapter 9.

Chapter 1 discusses the basic theory of the classical moment spaces and carefully defines the canonical moments both geometrically and as ratios of Hankel determinants. Elementary properties and simple examples are provided in order to indicate that it is often easier to describe measures in terms of canonical moments instead of ordinary moments. Later sections of Chapter 1 attempt to describe some simple motivating applications to design of experiments, to random walks, and indicate the relation between canonical moments, orthogonal polynomials, and continued fractions. These aspects are discussed more fully in the subsequent chapters.

Chapter 2 is devoted to orthogonal polynomials. We prove some basic properties and present some of the classical examples of the theory. The orthogonal polynomials are viewed geometrically as hyperplanes to the moment spaces $M_k$. These are used to provide further insight into the canonical moments and provide an explicit description of the (inverse) mapping of the set of canonical moments onto the set of ordinary moments.

Chapter 3 gives a very brief discussion of continued fractions which provide a connecting link between the ordinary moments and canonical moments. The ordinary moments are the coefficients in the Taylor expansion of the Stieltjes transform of a measure $\mu$, while the corresponding continued fraction expansion uses the canonical moments. These results are used for identifying measures from their canonical moment sequences in Chapter 4 which deals with special sequences of canonical moments. The corresponding probability measures turn out to be important for the application of canonical moments in the following chapters.

One of the major applications of canonical moments, namely the determination of optimal designs for polynomial regression, is presented in Chapters 5 and 6. After a short introduction to the basic concepts of optimal design in linear models, it is demonstrated that canonical moments are a very powerful tool for the calculation of optimal designs in such models.

Applications that give some interesting extensions of some classical results in analysis are presented in Chapter 7. Here canonical moments are used for deriving the asymptotic zero distribution of classical orthogonal polynomials, identities for orthogonal polynomials, and for solving nonlinear extremal problems in approximation theory.

Chapter 8 deals with the relationships between canonical moments, random walks, and orthogonal polynomials. It is shown that to every Hausdorff moment sequence there exists a random walk on the nonnegative integers such that the given moment sequence coincides with the sequence of return probabilities to the origin of the random walk. The one-step transition probabilities of this random walk are precisely the canonical moments of the given moment sequence.

The canonical moments for the circle or trigonometric functions are discussed in Chapter 9. The theory corresponding to much of the earlier chapters has a corresponding analog for the trigonometric functions and in some aspects the theory here is even richer than for the powers. Because of space limitations many applications of this theory in analysis and stochastic processes are omitted. Finally two other applications of canonical moments are considered in Chapter 10. We study the problem of Bayesian estimation of a binomial probability and the asymptotic distribution of "random" moment sequences.

The authors would like to thank T. Franke, A. Munk, M. Sahm, and T. Wagner for constructive comments after reading the early drafts of this monograph. It was a pleasure to collaborate with F. C. Chang, J. Fill, J. H. B. Kemperman, T. S. Lau, Y. B. Lim, J. Pitman, I. Röder, and

W. K. Wong. Many of the results obtained with our coauthors are included in this book. Special thanks go to T. S. Lau whose Ph.D. dissertation at Purdue University helped to organize the material in its preliminary stages. We are also grateful to Morris Skibinsky for some personal communication regarding canonical moments and to J. H. B. Kemperman for sending us copious unpublished notes on moment theory, random walks, orthogonal polynomials, and canonical moments. Finally we convey our deepest appreciation to Teena Seele who did most of the typing of this monograph with considerable technical expertise.

During the preparation of this monograph H. Dette was working at different places, and this author would like to thank several institutions and departments. Special thanks go to the Institut für Mathematische Stochastik, Technische Universität Dresden, for a stimulating environment during 1993–1995, the Department of Statistics, Purdue University, and the Institut für Mathematische Stochastik, Universität Göttingen, for the warm hospitality during numerous visits at these places. Travel support provided by the Deutsche Forschungsgemeinschaft made several visits to the United States possible.

Finally we are most grateful to the publisher who gave us the opportunity to present our personal point of view on these—at a first glance—very different topics.

HOLGER DETTE
WILLIAM J. STUDDEN

# Contents

# Interdependence of Chapters

Chapters 1, 2, and 3 form the core theoretical results of the book. Chapters 5 and 6 contain applications to design theory and rely on Chapter 4 only in identifying certain distributions from their canonical moments. The approximation theory in Chapter 7 depends on the design theory and illustrates the important idea that essentially all design theory problems have interesting dual formulations in approximation theory. Chapter 8 on random walks depends considerably on canonical moments and continued fractions. Chapter 9 is to some extent independent of all the other chapters. However, it uses results on continued fractions in Chapter 3 and contains analogues of the results in Chapters 1 to 3 and a small section on design theory.

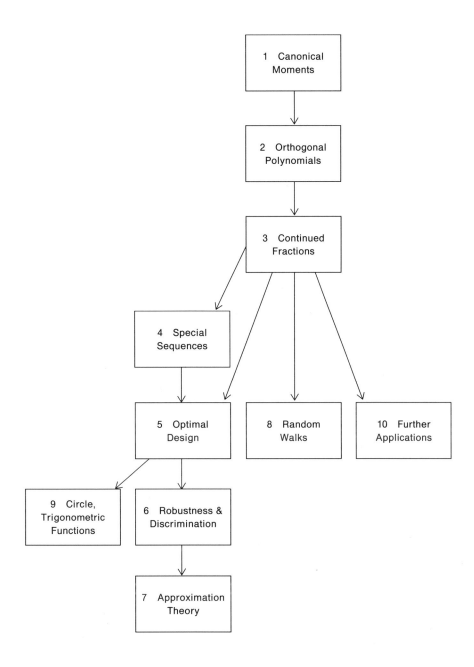

The Theory of Canonical
Moments with Applications
in Statistics, Probability,
and Analysis

CHAPTER 1

# Canonical Moments

## 1.1 INTRODUCTION

This book is concerned with some investigations and results involving moments of probability measures on intervals of the real line and on the circle. For the moment we will confine our discussion to the real line. For a given probability measure $\mu$ on a finite interval $[a, b]$ the (ordinary) *moments* of $\mu$ are given by

$$c_k = c_k(\mu) = \int_a^b x^k \, d\mu(x), \qquad k = 1, 2, \ldots .. \tag{1.1.1}$$

Problems involving moments and the relationship between the sequence $\{c_k\}_{k \geq 0}$ and the measure $\mu$ have a long history. This history involves numerous subject areas, including continued fractions, orthogonal polynomials, and approximation theory. The reader is referred to the recent book edited by H. J. Landau (1987), *Moments in Mathematics*, for a number of excellent papers describing some modern aspects of the theory and numerous references.

It is well known that the moments in (1.1.1) are somewhat difficult to work with and various related quantities are introduced. Although numerical questions are not our primary concern, the sequence of functions $1, x, x^2, x^3, \ldots$ and the moments in (1.1.1) are numerically unstable. In addition qualitative properties of the measure $\mu$ are difficult to discern from the moments $\{c_k\}$. For example, whether a measure $\mu$ is singular, discrete, or has a smooth density is difficult to discern from the moments of $\mu$. Other simple properties such as symmetry are also difficult to see except on symmetric intervals. For example, the uniform measure on $[-1, 1]$ has moments $c_0 = 1, 0, 1/3, 0, 1/5, 0, \ldots$; however, for the uniform measure on $[0, 1]$ the symmetry is not transparent from the moments $c_0 = 1, 1/2, 1/3, 1/4, \ldots .. $ A similar problem is to establish that two measures are related by a linear transformation from their sequences of moments. As an example consider the sequence of moments $c_0 = 1, 3/2, 7/3, 15/4, \ldots$ which corresponds to the uniform distribution on the interval $[1, 2]$. If only this sequence is given, it is difficult to recognize that the corresponding probability measure is only a "shift" of the uniform distribution on the interval $[0, 1]$.

Throughout the text fundamental use will be made of the moment spaces $M_n$. To define them, we let $\mathcal{P} = \mathcal{P}(a, b)$ denote the set of all probability measures on the Borel sets of the interval $[a, b]$, and for $\mu \in \mathcal{P}$ we let

$$c = c(\mu) = (c_1(\mu), c_2(\mu), \ldots) = (c_1, c_2, \ldots)$$

denote the sequence of moments of $\mu$ where $c_k(\mu)$ is given by (1.1.1). Finally, if $\mu_x$ is a Dirac measure concentrated at $x$, we define

$$c(x) = c(\mu_x) = (x, x^2, \ldots),$$

$$c_n(x) = c_n(\mu_x) = (x, \ldots, x^n).$$

**Definition 1.1.1**    *The moment space $M = M(a, b)$ is given by*

$$M = \{c(\mu) | \mu \in \mathcal{P}\}.$$

*The nth-moment space $M_n$ is defined by truncating the sequence $c(\mu)$ to $c_n(\mu) = (c_1(\mu), \ldots, c_n(\mu))$ or by projecting $M$ onto its first n coordinates.*
    *For $c_n \in M_n$, let*

$$\mathcal{P}(c_n) = \{\mu \in \mathcal{P} | c_n = \int_a^b c_n(x) d\mu(x)\} \qquad (n \geq 1)$$

*denote the set of all measures $\mu \in \mathcal{P}$ with moments up to the order n equal to $c_n = (c_1, \ldots, c_n)$. Finally define $\mathcal{P}(c_0) = \mathcal{P}$.*

In our studies in the optimal design of statistical experiments many functions

$$h(c_n) = h(c_1, \ldots, c_n)$$

are encountered which we wish to maximize over the moment space $M_n$. For example, we may have

$$h(c_1, \ldots, c_{2m}) = \begin{vmatrix} c_0 & \cdots & c_m \\ \vdots & & \vdots \\ c_m & \cdots & c_{2m} \end{vmatrix}$$

where it is required to maximize the above determinant over $(c_1, \ldots, c_{2m}) \in M_{2m}$. Many problems of this sort will be discussed extensively in Chapters 5 and 6. The main difficulty in maximizing functions over the moment space is that the set $M_n$ of all such moment sequences $(c_1, \ldots, c_n)$ is, in some sense, cumbersome to write down. This is primarily due to the fact that the range of $c_k$ is a complicated function of $c_1, \ldots, c_{k-1}$. This same difficulty appears in such seemingly simple problems as finding the Lebesgue measure of $M_n$, that is,

$$\int_{M_n} \cdots \int dc_1 dc_2 \ldots dc_n. \tag{1.1.2}$$

The quantities that circumvent these difficulties are some parameters that appear to be more intrinsically related to the measure $\mu$ than the sequence $\{c_k\}_{k \geq 0}$. These parameters will be called the canonical moments of the measure $\mu$. They appear in recursively defining the polynomials orthogonal with respect to the measure $\mu$ and in the continued fraction expansion of the Stieltjes transform of $\mu$. They can also be defined using certain Hankel determinants involving $\{c_k\}_{k \geq 0}$ and also have a very interesting geometrical definition.

## 1.2  MOMENT SPACES

In order to define the canonical moments of a probability measure $\mu$, a basic familiarity with the moment spaces $M(a, b)$ and $M_n(a, b)$ is needed. Some simple properties of these spaces are introduced here; further details are given in Chapter 2. For a discussion of moment spaces corresponding to more general systems of functions, the reader is referred to the monograph of Karlin and Studden (1966). It will usually be assumed that the interval $[a, b] = [0, 1]$ and therefore the dependence on $[a, b]$ will be omitted so that $M$ and $M_n$ will refer to $[0,1]$. In places where $[-1, 1]$ or more general intervals are under discussion, these will be explicitly mentioned.

Recall the definition of $c_n(x) = (x, x^2, \ldots, x^n)$, and let

$$C_n = Co\{c_n(x)|0 \leq x \leq 1\}$$

where $Co(A)$ denotes the smallest convex set containing $A$. Since $C_n$ consists of all points of the form

$$\sum_{i=1}^{r} \lambda_i \, c_n(x_i) \qquad \left(\lambda_i > 0, \sum_{i=1}^{r} \lambda_i = 1\right)$$

for arbitrary $r$, it follows from *Caratheodory's theorem* that each such point can be represented using at most $r = n + 1$ terms. In this case simple convergence arguments show that $C_n$ is closed. It is also bounded, and hence compact, and, of course, convex. Furthermore, since $1, x, x^2, \ldots, x^n$ are linearly independent, $C_n$ is an $n$-dimensional body, i.e. it has a nonempty $n$-dimensional interior. The set $C_n$ is depicted in Figure 1.1 for $n = 2$, and the following result shows that this set coincides with the $n$th moment space $M_n$:

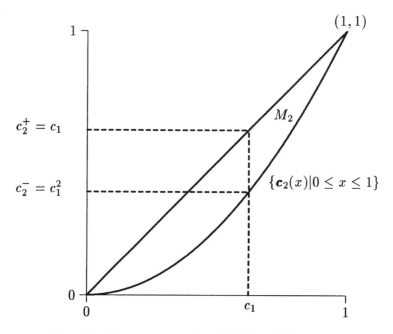

**Figure 1.1** The moment space $M_n$ and definition of $c_n^+$ and $c_n^-$ for $n = 2$.

**Theorem 1.2.1** $M_n = C_n$

*Proof* Clearly $C_n \subset M_n$. If there exists a point $c_n^0 = (c_1^0, \ldots, c_n^0) \in M_n$ such that $c_n^0 \notin C_n$, then, since $C_n$ is closed, there is a hyperplane strictly separating $c_n^0$ and $C_n$. In other words, there exists a vector $a_n = (a_1, a_2, \ldots, a_n)$ with $\sum_{i=1}^n a_i^2 > 0$ and a constant $b$ such that

$$\sum_{i=1}^n a_i c_i^0 < b,$$

$$\sum_{i=1}^n a_i c_i \geq b,$$

for all $c_n = (c_1, \ldots, c_n) \in C_n$. For $c_n = c_n(x) \in C_n$ we obtain $\sum_{i=1}^n a_i x^i \geq b$ whenever $x \in [0, 1]$. Now, if $\mu_0 \in \mathcal{P}$ is a probability measure such that $c_n^0 = c_n(\mu_0)$, it follows that

$$\int_0^1 \left( \sum_{i=1}^n a_i x^i \right) d\mu_0(x) = \sum_{i=1}^n a_i c_i^0 \geq b,$$

which is a contradiction.                                                              ∎

**Corollary 1.2.2** *Each $c_n \in M_n$ can be written as*

$$c_n = \sum_{i=1}^{r} \lambda_i c_n(x_i), \qquad (1.2.1)$$

*where $x_i \in [0, 1]$, $\lambda_i > 0$ $(i = 1, \ldots, r)$, $\sum_{i=1}^{r} \lambda_i = 1$, and $r \le n + 1$.*

In the following discussion frequent use will be made of the fact that

$$\sum_{i=1}^{n} a_i c_i \ge b \text{ for all } c_n = (c_1, \ldots, c_n) \in M_n \qquad (1.2.2)$$

$$\Leftrightarrow \sum_{i=1}^{n} a_i x^i \ge b \qquad \text{for all } x \in [0, 1]$$

which follows readily from the proof of Theorem 1.2.1.

**Definition 1.2.3** *We refer to Equation (1.2.1) as a representation of $c_n$. The corresponding $\{x_i\}$ are called the roots or support (points) of the representation, and $\{\lambda_i\}$ are called the weights or masses. The index of a representation is the number of support points in such a representation with the convention that the $x_i$ in the interior $(0, 1)$ are counted as one and the $x_i$ equal to an end point 0 or 1 are counted as one-half.*

*The index $I(c_n)$ of a point $c_n \in M_n$ is the minimal index over all representations of the point $c_n$.*

Note that if a representation has index $m$, then the set of roots consists of $m$ interior points or $m - 1$ interior points and both end points. If the index is equal to $m + \frac{1}{2}$, then it has $m$ interior points and one of the end points. Schematically this is represented by

$$\text{index} = m \qquad \begin{array}{l} 0 = s_0 < s_1 < \ldots < s_{m-1} < s_m = 1 \\ 0 < t_1 < t_2 < \ldots < t_m < 1 \end{array}$$

$$\qquad (1.2.3)$$

$$\text{index} = m + \frac{1}{2} \qquad \begin{array}{l} 0 < s_1 < s_2 < \ldots < s_m < s_{m+1} = 1 \\ 0 = t_0 < t_1 < \ldots < t_m < 1 \end{array}$$

**EXAMPLE 1.2.4** As an illustration of the preceding definition, we discuss the second-moment space $M_2$ which is depicted in Figure 1.2. Consider the point $c_2 = (1/2, 5/16)$ which is an interior point of $M_2$ and the three

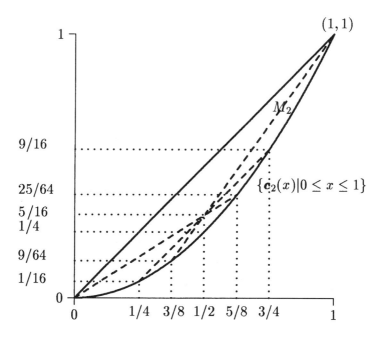

**Figure 1.2**   Different representation of the point (1/2, 5/16).

probability measures

> $\mu_1$   with equal masses at the two points $\frac{1}{4}$ and $\frac{3}{4}$
>
> $\mu_2$   with masses $\frac{1}{5}$ and $\frac{4}{5}$ at the two points 0 and $\frac{5}{8}$
>
> $\mu_3$   with masses $\frac{4}{5}$ and $\frac{1}{5}$ at the two points $\frac{3}{8}$ and 1

It is easy to see that $c_2(\mu_i) = c_2 = (1/2, 5/16)$ for $i = 1, 2, 3$ (see also Figure 1.2). The representation $\mu_1$ involves two interior points and has index 2, while the representations $\mu_2$ and $\mu_3$ have index 3/2 because we count the boundary points 0 and 1 as one-half. On the other hand, there are at least two points of the set $C_n$ needed in order to represent $c_2$ and this cannot be done by chosing the points $(0, 0)$ and $(1, 1)$ simultaneously (see Figure 1.2). Therefore the index of the point $c_2 = (1/2, 5/16)$ is given by 3/2.

The boundary and interior of the $n$th-moment space $M_n$ are denoted by $\partial M_n$ and Int $M_n$. The interior of the moment space $M$ is defined by

$$\text{Int } M = \{c \in M | c_n \in \text{Int } M_n \, \forall n \in \mathbb{N}\}$$

These sets play a very important role in the subsequent discussion.

**Theorem 1.2.5**   *The following statements are equivalent:*

1. $c_n \in \partial M_n$.

2. $I(c_n) \leq \dfrac{n}{2}$.

3. $\mathcal{P}(c_n)$ is a singleton, in which case the representation is unique.

*Proof*  We first show the equivalence between parts 1 and 2. The point $c_n = (c_1, \ldots, c_n)$ is in $\partial M_n$ if and only if there exists a *supporting hyperplane* to $M_n$ at $c_n$. By (1.2.2) this means that there exists a vector $(a_1, \ldots, a_n)$ with $\sum_{i=1}^{n} a_i^2 > 0$ and a constant $b$ such that

$$\sum_{i=1}^{n} a_i x^i \geq b \qquad \text{for all } x \in [0, 1]$$

and

$$\sum_{i=1}^{n} a_i c_i = b.$$

Therefore, if $\mu$ corresponds to $c_n$ (i.e., $c_n = c_n(\mu)$), then the polynomial $P_n(x) = \sum_{i=1}^{n} a_i x^i - b$ is nonnegative on the interval $[0, 1]$ and satisfies

$$\int_0^1 P_n(x) d\mu(x) = \sum_{i=1}^{n} a_i c_i - b = 0.$$

Hence $P_n(x) = 0$ on the support of any representation of $c_n$. A careful counting argument shows that this implies that $I(c_n) \leq n/2$, since the polynomial $P_n(x)$ is not constant. For example, if $n = 2m$, then $P_n(x) \geq 0$ for $x \in [0, 1]$, and $P_n$ can have at most $m$ interior zeros or $m - 1$ interior zeros and possibly either one or both end points. In both cases we obtain $I(c_n) \leq m$. A similar argument holds for $n = 2m + 1$.

The converse also holds since one can find a nonnegative, nontrivial polynomial of degree at most $n$ vanishing on a given set of index at most $n/2$. The coefficients of this polynomial determine a supporting hyperplane to $M_n$ at $c_n$ which implies that $c_n \in \partial M_n$.

For the equivalence between parts 1 and 3 we will show now that if $c_n \in \partial M_n$ or equivalently $I(c_n) \leq n/2$, then $\mathcal{P}(c_n)$ is a singleton, or there is a unique measure representing $c_n$. The reverse implication will follow from the next theorem which states that $\mathcal{P}(c_n)$ is infinite for $c_n \in \text{Int } M_n$.

To show uniqueness, we suppose that there are two such representations:

$$\sum_i \alpha_i c_n(u_i) = c_n = \sum_i \beta_i c_n(v_i),$$

both of index $\leq n/2$. A simple counting argument shows that the total number of distinct points $w_1, \ldots, w_t$ among the roots of both representations can be at most $n + 1$. Therefore the difference of the two representations gives

$$\sum_{i=1}^{t} \gamma_i c_n(w_i) = 0 \qquad \text{where } t \leq n + 1. \tag{1.2.4}$$

Since also $\sum_{i=1}^{t} \gamma_i = 0$ and the matrix $(w_j^i)_{i,j=0}^{t}$ is nonsingular, this can happen only if $\gamma_i \equiv 0$. ∎

**EXAMPLE 1.2.6** Consider the case where $n = 2$. It is shown in Example 1.2.4 that the point $c_2 = (1/2, 5/16)$ has index $I(c_2) = 3/2$, and therefore it is an interior point of $M_2$. The points $\{(x, x^2)|x \in [0, 1]\}$ have index $1/2$ or $1$ according to $x \in \{0, 1\}$ or $x \in (0, 1)$. These points form the "lower" boundary of the moment space $M_2$ (see Figure 1.2). The "upper" boundary of $M_2$ is given by the points $\{(x, x)|x \in (0, 1)\}$, which always have index $1$. The nomenclature "lower" and "upper" boundary is motivated by Figure 1.2. The points corresponding to $x = 0$ or $1$ could be included in either the upper or lower boundary. For higher dimensional moment spaces, a justification for these phrases is given in the subsequent chapters.

Note that if $c_n \in \partial M_n$, then the corresponding measure $\mu$ or the representation is unique; in which case the moments of higher order $c_{n+\ell}(\mu)$ for $\ell \geq 1$ are uniquely determined. By the proof of Theorem 1.2.5, the support must be the zero set of some nonzero polynomial $P_n(x) = \sum_{i=0}^{n} a_i x^i \geq 0$ of degree at most $n$. One can recursively obtain $c_{n+\ell}$ for $\ell \geq 0$ from the equations

$$0 = \int_0^1 x^\ell P_n(x) d\mu(x) = \sum_{i=0}^{n} a_i c_{i+\ell}. \tag{1.2.5}$$

If $c_{n-1} \in \text{Int } M_{n-1}$, then $P_n$ must actually be of degree exactly $n$; otherwise, the polynomial would show $c_{n-1} \in \partial M_{n-1}$.

Recalling the definition of $\mathcal{P}(c_n)$ in Definition 1.1.1, we now define

$$c_{n+1}^+ = \max_{\mu \in \mathcal{P}(c_n)} c_{n+1}(\mu) \quad \text{and} \quad c_{n+1}^- = \min_{\mu \in \mathcal{P}(c_n)} c_{n+1}(\mu) \tag{1.2.6}$$

as the maximum and minimum of the $(n + 1)$th moment over the set of all probability measures $\mu$ whose moments up to the order $n$ coincide with the given moments $c_n = (c_1, \ldots, c_n)$. We remark here that the minimum and maximum in (1.2.6) exist because the moment space $M_{n+1}$ is compact.

**Theorem 1.2.7** *If $c_n \in \text{Int } M_n$, then $c_{n+1}^+ > c_{n+1}^-$ and $\mathcal{P}(c_n)$ is infinite.*

*Proof* The fact that $c_{n+1}^+ > c_{n+1}^-$ follows, since $M_{n+1}$ is convex and has an

$(n + 1)$-dimensional interior. Each $c_{n+1} \in [c_{n+1}^-, c_{n+1}^+]$ gives rise to a distinct $\mu \in \mathcal{P}(c_n)$ so that $\mathcal{P}(c_n)$ is infinite. ∎

**EXAMPLE 1.2.8** If $n = 0$, then $\mathcal{P}(c_0)$ is the set of all probability measures on the interval $[0, 1]$, and we obtain $c_1^+ = 1, c_1^- = 0$ by considering Dirac measures at the points 1 and 0, respectively. The moment space $M_2$ is depicted in Figure 1.1, and the set of all second moments of measures $\mu$ with first moment equal to $c_1$ is given by the line connecting the two points $(c_1, c_1^2)$ and $(c_1, c_1)$. Consequently the maximum and minimum second moments are $c_2^+ = c_1$ and $c_2^- = c_1^2$, respectively. For higher dimensional moment spaces the calculation of $c_{n+1}^+$ and $c_{n+1}^-$ is more complicated and deferred to Section 1.4.

In the following we would like to show that each $c_n \in \text{Int } M_n$ has precisely two representations of index $(n + 1)/2$. Since each of the points $(c_n, c_{n+1}^+)$ and $(c_n, c_{n+1}^-)$ lie in $\partial M_{n+1}$, they both have index at most $(n + 1)/2$. However, $I(c_n) > n/2$ if $c_n \in \text{Int } M_n$ so that the index of both $(c_n, c_{n+1}^+)$ and $(c_n, c_{n+1}^-)$ must be equal to $(n + 1)/2$. Further any representation $\mu$ of $c_n$ of index $(n + 1)/2$ must be such that $(c_n, c_{n+1}(\mu)) \in \partial M_{n+1}$; thus $c_{n+1}(\mu)$ must be either $c_{n+1}^+$ or $c_{n+1}^-$.

**Theorem 1.2.9** *Each $c_n \in \text{Int } M_n$ has precisely two representations of index $(n + 1)/2$. These correspond to the unique representations $\sigma_n^+$ of $(c_n, c_{n+1}^+)$ and $\sigma_n^-$ of $(c_n, c_{n+1}^-)$. The roots of these two representations must strictly interlace, $\sigma_n^+$ has the end point one in its support while $\sigma_n^-$ does not.*

*Proof* The only part remaining to prove is the last sentence. The proof is very similar to the second part in the proof of Theorem 1.2.5. First, the combined set of roots of the two representations must contain $n + 2$ points; otherwise (1.2.4) would show these two representations are actually the same. If the combined set contains $n + 2$ points, then the sets of roots of $\sigma_n^+$ and $\sigma_n^-$ must be disjoint. Note further that the two end points 0 and 1 must be included. We write the combined set of ordered roots as $\{w_i\}_0^{n+1}$ and write the difference of the moment points $(c_n, c_{n+1}^+)$ and $(c_n, c_{n+1}^-)$ as

$$\int_0^1 c_{n+1}(x) d\sigma^+(x) - \int_0^1 c_{n+1}(x) d\sigma_n^-(x)$$

$$= \sum_{i=0}^{n+1} \gamma_i c_{n+1}(w_i) = (0, \ldots, 0, c_{n+1}^+ - c_{n+1}^-).$$

(1.2.7)

Note also that $\sum_{i=0}^{n+1} \gamma_i = 0$. Since the $w_i$ are ordered and $c_{n+1}^+ - c_{n+1}^- > 0$, we obtain that none of the $\gamma_i$ vanish, and these must alternate in sign, starting with a plus sign for the largest root $w_{n+1} = 1$. This follows by a straightforward application of Cramer's rule to the matrix $(w_j^i)_{i,j=0}^{n+1}$. Therefore the roots alternate and $\sigma_n^+$ has 1 as a root. ∎

**Definition 1.2.10** *The representations $\sigma_n^+$ and $\sigma_n^-$ of $c_n \in Int\ M_n$ of index $(n+1)/2$ are called principal representations; the representation $\sigma_n^+$ of $(c_n, c_{n+1}^+)$ is referred to as the upper principal representation, while the representation $\sigma_n^-$ of $(c_n, c_{n+1}^-)$ is the lower principal representation.*

The roots of $\sigma_n^+$ and $\sigma_n^-$ follow the pattern in the display (1.2.3). The roots of $\sigma_n^+$ and $\sigma_n^-$ are denoted by $\{s_i\}$ and $\{t_i\}$, respectively. If $n = 2m - 1$, we have by Theorem 1.2.9,

$$0 = s_0 < t_1 < s_1 < t_2 < \ldots < s_{m-1} < t_m < s_m = 1 \qquad (1.2.8)$$

and if $n = 2m$,

$$0 = t_0 < s_1 < t_1 < s_2 < \ldots < s_m < t_m < s_{m+1} = 1 \qquad (1.2.9)$$

Note that in either case there are precisely $n$ interior roots from the combined set of $\{s_i\}$ and $\{t_i\}$.

**EXAMPLE 1.2.11** Consider the moment space $M_2$ and the interior point $c_2 = (1/2, 5/16)$ (see Figure 1.2). The lower principal representation of $c_2$ has masses $1/5$ and $4/5$ at the points $0$ and $5/8$, and the upper principal representation has masses $4/5$ and $1/5$ at the points $3/8$ and $1$ (see Example 1.2.4). If $n$ is odd, the number of roots of the upper and lower principal representation differs by 1 (but the index of the representations is the same). For example, consider the moment space $M_3$ and the point $c_3 = (1/2, 5/16, 7/32)$. A representation of $c_3$ of index $\leq 3/2$ has to involve at least one of the end points $0$ or $1$, and it is easy to see that such a representation does not exist. For this reason $c_3$ is an interior point of $M_3$. The lower principal representation has equal masses at the two points $1/4$ and $3/4$, while the upper principal representation has masses $1/8, 3/4$, and $1/8$ at the three points $0, 1/2$, and $1$, respectively.

**Lemma 1.2.12** *If $c_n \in Int\ M_n$ then for any given point $x_0 \in [0, 1]$ there is a representation of $c_n$ with positive mass at $x_0$.*

*Proof* To provide the representation, we draw the line segment from $c_n(x_0) = (x_0, x_0^2, \ldots, x_0^n)$ through $c_n$ and protruding beyond to reach the boundary of $M_n$. That is, we take the value of $\alpha > 1$ such that

$$(1 - \alpha)c_n(x_0) + \alpha c_n = b_n \in \partial M_n.$$

This gives the required representation of $c_n$:

$$c_n = \frac{1}{\alpha}b_n + \frac{\alpha - 1}{\alpha}c_n(x_0). \qquad \blacksquare$$

## 1.3  CANONICAL MOMENTS

For each $c \in M$ let

$$N = N(c) = \min\{n \in \mathbb{N} | c_n \in \partial M_n\} \tag{1.3.1}$$

denote the minimum integer such that $c_N$ is on the boundary of the $N$th moment space $M_N$. If $c_n \in \operatorname{Int} M_n$ for all $n \geq 1$, define $N(c) = \infty$ while $N(c) = 1$ if $c_1 \in \partial M_1$. Thus $c_k \in \operatorname{Int} M_k$ for $k < N(c)$, $c_N \in \partial M_N$, and of course $c_k = \partial M_k$ for $k \geq N + 1$ (see Theorem 1.2.5). The *canonical moment sequence* is then defined for $k \leq N(c)$ by

$$p_k = p_k(c) = \frac{c_k - c_k^-}{c_k^+ - c_k^-}, \tag{1.3.2}$$

where $c_k^-$ and $c_k^+$ are defined in (1.2.6). Note that the canonical moments vary in the interval $[0, 1]$. Moreover, if $N = N(c) < \infty$, then $p_j \in (0, 1)$, $1 \leq j < N$, and $p_N$ is either 0 or 1 corresponding to a lower or upper principal representation of the interior point $c_{N-1} \in \operatorname{Int} M_{N-1}$, respectively (see Theorem 1.2.9).

**EXAMPLE 1.3.1**  In this example we discuss the calculation of the first two canonical moments. For the first canonical moment we observe from Example 1.2.8 that $c_1^+ = 1$, $c_1^- = 0$ and obtain by definition (1.3.2) that $p_1 = c_1$. The calculation of the second canonical moment is slightly more complicated. Note that $c_1 \in \operatorname{Int} M_1$ if and only if $c_1 \in (0, 1)$. By Example 1.2.8 we have (see also Figure 1.1) $c_2^+ = c_1$ and $c_2^- = c_1^2$ which give for $c_1 \in (0, 1)$,

$$p_2 = \frac{c_2 - c_1^2}{c_1(1 - c_1)}.$$

Conversely, we can express the second moment $c_2$ in terms of the first two canonical moments $p_1, p_2$ and obtain

$$c_2 = p_1(p_1 + q_1 p_2) \tag{1.3.3}$$

where $q_1 = 1 - p_1$.

In general, the problem of calculating the canonical moments of a given measure $\mu \in \mathcal{P}$ is very complicated if the geometric definition (1.3.2) is used directly. More efficient methods are needed for this purpose. In Section 1.4 we present a representation of the canonical moments in terms of Hankel determinants. Other methods that are related to the theory of continued fractions and orthogonal polynomials will be discussed in Chapters 2 and 3. The converse problem is the problem of identifying the measure corresponding to a given sequence of canonical moments. It will turn out in the subsequent text that in many special cases the measure corresponding to a given sequence of moments can be identified much easier from its sequence of canonical moments.

Since the moments of a probability measure on a bounded interval uniquely determine the measure (e.g., see Feller, 1966, p. 223), it follows that for each $c \in M$ there corresponds a unique $\mu \in \mathcal{P}$. Thus there is a one-to-one mapping of $M$ onto $\mathcal{P}$. For this reason we simultaneously use the notation $p_k(\mu)$ and $p_k(c)$ for the canonical moments corresponding to a given moment sequence $c = c(\mu) \in M$. Whenever the corresponding sequence $c$ or measure $\mu$ is clear from the context, this dependence will be suppressed.

The above definition (1.3.2) provides a one-to-one map $T$, from the moment space $M$ onto a set $S$ defined by

$$S = \left( \bigcup_{k=0}^{\infty} S_k \right) \bigcup S_\infty, \qquad (1.3.4)$$

where

$$S_\infty = \{(p_1, p_2, \ldots) | 0 < p_i < 1, \text{ for all } i \geq 1\},$$

and for $k \geq 0$,

$$S_k = \{(p_1, \ldots, p_k, p_{k+1}) | 0 < p_i < 1, \quad 1 \leq i \leq k, \quad p_{k+1} = 0 \text{ or } 1\}.$$

Any $c \in \text{Int } M$ corresponds to some point in $S_\infty$, while for any sequence of canonical moments $\boldsymbol{p} = (p_1, p_2, \ldots) \in S_\infty$ the corresponding sequence $c = (c_1, c_2, \ldots)$ can be defined successively from (1.3.2). Therefore it is evident that $T$ maps Int $M$ onto $S_\infty$ in a one-to-one manner. Similarly each $(p_1, \ldots, p_n)$ is uniquely determined by $(c_1, c_2, \ldots, c_n)$. From the definition of $N = N(c)$, it follows that $(c_1, \ldots, c_{N-1}) \in \text{Int } M_{N-1}, (c_1, \ldots, c_N) \in \partial M_N$, and we define

$$T(c) = (p_1, \ldots, p_N) \in S_{N-1}.$$

The geometric definition (1.3.2) allows us to show that the canonical moments have some interesting invariance properties. The first of these relates to the invariance under a linear transformation of the underlying interval $[a, b]$. For the general interval $[a, b]$ the moments are denoted by

$$b_k = \int_a^b y^k d\mu_{ab}(y), \qquad k = 1, 2, \ldots,$$

where $\mu_{ab}$ is a probability measure on the interval $[a, b]$. The set of probability measures on the interval $[a, b]$ can be related in a one-to-one manner with the set of probability measures on $[0, 1]$ through the linear transformation $y = a + (b - a)x$ mapping $x \in [0, 1]$ onto $y \in [a, b]$. The sequences $\{c_k\}_{k \geq 0}$ and

$\{b_k\}_{k\geq 0}$ obtained from $\mu$ on $[0, 1]$ and the linearly transformed measure $\mu_{ab}$ on $[a, b]$ are related for $n \geq 1$ by

$$c_n = \frac{1}{(b-a)^n} \sum_{j=0}^{n} \binom{n}{j} (-a)^{n-j} b_j = \frac{1}{(b-a)^n} b_n + \ldots, \qquad (1.3.5)$$

$$b_n = \sum_{j=0}^{n} \binom{n}{j} (b-a)^j a^{n-j} c_j = (b-a)^n c_n + \ldots.$$

For the moment spaces $M(a, b)$ and $M_n(a, b)$ corresponding to the set $\mathcal{P}(a, b)$ of all probability measures on the interval $[a, b]$, the previous definitions and discussion carry over in precise analogy. The first $(n-1)$ moments $(c_1, \ldots, c_{n-1})$ are in one-to-one correspondence with $(b_1, \ldots, b_{n-1})$. If $b > a$, it is evident from (1.3.5) that

$$b_n - b_n^- = (b-a)^n (c_n - c_n^-)$$

and                                                                                           (1.3.6)

$$b_n^+ - b_n^- = (b-a)^n (c_n^+ - c_n^-).$$

The geometric definition (1.3.2) yields the following interesting *invariance property* for canonical moments of probability measures on different intervals related by a linear transformation. This result, along with Theorem 1.3.3 and Corollary 1.3.4 is contained in Skibinsky (1969).

**Theorem 1.3.2**   *The canonical moments remain invariant under linear transformation of the measure $\mu$, i.e.*

$$p_k(\mu) = p_k(\mu_{ab}), \qquad k = 1, 2, \ldots, N(c),$$

*where $\mu_{ab}$ denotes the measure induced by $\mu$ through the linear transformation $y = a + (b-a)x$, $b > a$, of $[0, 1]$ onto $[a, b]$.*

A similar analysis applies to the case $y = 1 - x$ where $a = 1$ and $b = 0$, and the interval $[0, 1]$ is reversed. For $n$ even, equations (1.3.6) remain true, but for $n$ odd we have

$$c_n - c_n^- = b_n^+ - b_n$$

and

$$c_n^+ - c_n^- = b_n^+ - b_n^-.$$

This provides the following property for the *reflection* $\mu^{(r)}$ of the measure $\mu$ with respect to the point $\frac{1}{2}$.

**Theorem 1.3.3** *Let $\mu^{(r)}$ denote the measure corresponding to $\mu$ under the transformation $y = 1 - x$ with canonical moments $p_1^{(r)}, p_2^{(r)}, \ldots$ . For $1 \leq k \leq N(c)$,*

$$p_k^{(r)} = \begin{cases} p_k & \text{if } k \text{ is even,} \\ q_k & \text{if } k \text{ is odd,} \end{cases}$$

*where $q_k = 1 - p_k$ $(k \geq 1)$ and $p_k = p_k(\mu)$ denote the canonical moments of $\mu$, $1 \leq k \leq N(c)$.*

**Corollary 1.3.4** *The measure $\mu$ is symmetric; that is, $\mu^{(r)} = \mu$ if and only if $p_{2k-1} = \frac{1}{2}$ for $k \geq 1$ and $2k - 1 \leq N(c)$.*

*Proof* If $\mu^{(r)} = \mu$, then $p_{2k-1}^{(r)} = p_{2k-1} = q_{2k-1}$ or $p_{2k-1} = \frac{1}{2}$. If $p_{2k-1} = \frac{1}{2}$, for $k \geq 1$ then, by Theorem 1.3.3, the measure $\mu^{(r)}$ has the same canonical moments as $\mu$. In this case all of the ordinary moments are also the same and $\mu^{(r)} = \mu$. ∎

One further property that will be extremely useful later is related to symmetric measures on the interval $[-1, 1]$. Consider the mapping $x = y^2$ for $y \in [-1, 1]$ and $x \in [0, 1]$. The symmetric measures $\mu^{(s)}$ on the interval $[-1, 1]$ are in one-to-one correspondence with measures $\mu$ on $[0, 1]$ through the above transformation, that is,

$$\mu^{(s)}([-x, x]) = \mu([0, x^2]). \tag{1.3.7}$$

For the moments of the measure $\mu^{(s)}$ we have for $n \geq 1$

$$b_{2n} = \int_{-1}^{1} y^{2n} d\mu^{(s)}(y) = \int_{0}^{1} x^n d\mu(x) = c_n,$$

$$b_{2n-1} = 0. \tag{1.3.8}$$

**Theorem 1.3.5** *Let $p_k^{(s)}$ and $p_k$ denote the canonical moments of two probability measures $\mu^{(s)}$ and $\mu$ on the intervals $[-1, 1]$ and $[0, 1]$, respectively. The measures $\mu^{(s)}$ and $\mu$ are related by (1.3.7) if and only if $p_{2k-1}^{(s)} = \frac{1}{2}$ and $p_{2k}^{(s)} = p_k$ for $1 \leq k \leq N(c)$.*

*Proof* If $\mu^{(s)}$ and $\mu$ satisfy (1.3.7), then (1.3.8) holds. In this case $b_{2k-1} = 0$, $b_{2k} = c_k$, and hence $p_{2k}^{(s)} = p_k$ because of the one-to-one correspondence between the symmetric measures on the interval $[-1, 1]$ and the probability measures on the interval $[0, 1]$ through (1.3.7) and (1.3.8). Since $\mu^{(s)}$ is symmetric, $p_{2k-1}^{(s)} = \frac{1}{2}$ from Corollary 1.3.4. The reverse implication follows again from the fact that the canonical moments uniquely define the corresponding measure. ∎

We will conclude this section by presenting two examples that give the canonical moments for the Beta and Binomial distributions. The proofs of the following statements are deferred to subsequent sections where more powerful tools for the calculation of canonical moments are available:

**EXAMPLE 1.3.6** *Beta Distribution.* Let $\xi_{\alpha\beta}$ denote the Beta distribution on the interval $(0, 1)$ with density

$$w_{(\alpha,\beta)}(x) = \frac{1}{B(\beta+1, \alpha+1)} x^\beta (1-x)^\alpha, \qquad 0 < x < 1, \tag{1.3.9}$$

where $\alpha, \beta > -1$, and

$$B(p,q) = \int_0^1 x^{p-1}(1-x)^{q-1} dx = \frac{\Gamma(p)\Gamma(q)}{\Gamma(p+q)}, \qquad (p, q > 0) \tag{1.3.10}$$

denotes the *Beta integral* and $\Gamma(\cdot)$ the *Gamma function* (see Johnson and Kotz, 1970). The ordinary moments of $\xi_{\alpha\beta}$ are

$$c_j = \frac{B(\beta+1+j, \alpha+1)}{B(\beta+1, \alpha+1)} = \frac{\Gamma(\beta+j+1)}{\Gamma(\beta+1)} \frac{\Gamma(\alpha+\beta+2)}{\Gamma(\alpha+\beta+2+j)}, \qquad j \geq 1.$$

The canonical moments of the Beta distribution are given in Skibinsky (1969). It will be shown in Example 1.5.3, by a different method, that the canonical moments of $\xi_{\alpha\beta}$ are given by

$$p_{2j} = \frac{j}{2j+1+\alpha+\beta}, \qquad p_{2j-1} = \frac{\beta+j}{2j+\alpha+\beta}, \qquad j \geq 1. \tag{1.3.11}$$

Note that $p_{2j-1} = \frac{1}{2}$ if and only if $\alpha = \beta$, which means, by Corollary 1.3.4, that $\xi_{\alpha\beta}$ is symmetric with respect to the midpoint $x = \frac{1}{2}$. Two special cases should be mentioned. For $\alpha = \beta = 0$, $\xi_{\alpha\beta}$ is the uniform distribution on the interval $[0, 1]$, and the canonical moments are

$$p_{2k} = \frac{k}{2k+1}, \qquad p_{2k-1} = \frac{1}{2}, \qquad k \geq 1.$$

For $\alpha = \beta = -\frac{1}{2}$, $\xi_{\alpha\beta}$ gives the *arc-sine distribution* with canonical moments $p_k \equiv \frac{1}{2}$ for all $k$. This indicates that the arc-sine distribution with density

$$w_{(-1/2,-1/2)}(x) = \frac{1}{\pi\sqrt{x(1-x)}}, \qquad 0 < x < 1,$$

has moments in the *center* of the moment space. It will play an important role in many places.

**EXAMPLE 1.3.7**  *Binomial Distribution.* The Binomial distribution $\xi_B$ is given by the mass distribution

$$b(x; N, p) = \binom{N}{x} p^x (1-p)^{N-x}, \qquad x = 0, 1, \ldots, N,$$

where $p \in (0,1)$ and $N \in \mathbb{N}$. The ordinary moments of $\xi_B$ are somewhat complicated and are given by

$$c_r = \sum_{x=0}^{N} \binom{N}{x} p^x (1-p)^{N-x} x^r = N! \sum_{j=0}^{r} \frac{S(r,j)}{(N-j)!} p^j \quad (r \geq 1),$$

where $S(r,j)$ denote the *Stirling numbers* of the second kind defined by

$$S(r,j) = \frac{1}{j!} \sum_{k=0}^{j} (-1)^{j-k} \binom{j}{k} k^r \qquad (j \leq r)$$

(see Johnson, Kotz, and Kemp, 1992). The canonical moments of the Binomial distribution have a much simpler form and were obtained by Skibinsky (1969) as

$$p_{2j-1} = p, \qquad p_{2j} = \frac{j}{N}, \qquad j = 1, 2, \ldots, N.$$

These will be verified in Example 2.3.8. They are calculated with reference to the interval $[0, N]$ or alternatively on $[0, 1]$ by moving the mass $b(x; N, p)$ to $x/N$ for $x = 0, 1, \ldots, N$. Note that the sequence of canonical moments terminates at $p_{2N} = 1$ which reflects the fact that the Binomial distribution is supported on a finite number of points. According to Definition 1.2.10 this would be an upper principal representation of index $N$.

## 1.4  HANKEL DETERMINANTS

The description of the canonical moments in Section 1.3 was interesting and geometrically intuitive. In order to calculate the $p_k$ from the moments $c_k$, a procedure is needed to calculate $c_k^+$ and $c_k^-$ in terms of $c_1, c_2, \ldots, c_{k-1}$. This can be accomplished with the aid of certain determinants involving the moments $c_k$ providing an explicit expression for $p_k$ in terms of $c_1, c_2, \ldots, c_k$. (The reverse map expressing the ordinary moments $c_k$ in terms of $p_1, \ldots, p_k$ will be discussed in Section 2.4 of Chapter 2.) Two basic results given in Theorems 1.4.1 and 1.4.2 are required. These results relate the classical moment problem to a characterization problem for nonnegative polynomials on the interval $[0, 1]$.

It is clear that if $c_n \in M_n$, then

$$\sum_{i=0}^{n} a_i x^i \geq 0 \quad \text{for } x \in [0,1] \text{ implies } \sum_{i=0}^{n} a_i c_i \geq 0, \tag{1.4.1}$$

where $c_0 = \int_0^1 d\mu(x) = 1$. The first result is that the simple condition (1.4.1) is also a sufficient condition for $c_n$ to be in $M_n$. The set of nonnegative polynomials on $[0, 1]$ of degree at most $n$ is denoted by $\mathcal{P}_{n+1}$. We associate $\mathcal{P}_{n+1}$ with the set of $(n + 1)$-dimensional vectors $\boldsymbol{a} = (a_0, a_1, a_2, \ldots, a_n)$ defined by the coefficients of nonnegative polynomials.

**Theorem 1.4.1**
1. $c_n \in M_n$ if and only if $\sum_{i=0}^{n} a_i c_i \geq 0$ whenever $\boldsymbol{a} \in \mathcal{P}_{n+1}$.
2. $c_n \in \text{Int } M_n$ if and only if $\sum_{i=0}^{n} a_i c_i > 0$ whenever $\boldsymbol{a} \in \mathcal{P}_{n+1}, \boldsymbol{a} \neq 0$

*Proof* (1) If $c_n \notin M_n$, then there exists a hyperplane separating $c_n$ and $M_n$. Thus by Theorem 1.2.1 [or equivalently by (1.2.2)] there exists an $\boldsymbol{a} \in \mathcal{P}_{n+1}$ so that $\sum_{i=0}^{n} a_i x^i \geq 0$ for $x \in [0, 1]$ and $\sum_{i=0}^{n} a_i c_i < 0$. The reverse implication is very easy and was alluded to in (1.4.1).

(2) If $c_n \notin \text{Int } M_n$, the argument is essentially the same as above; that is, there exist an $\boldsymbol{a} \in \mathcal{P}_{n+1}$ such that $\sum_{i=0}^{n} a_i c_i \leq 0$. To prove the converse assume that $c_n \in \text{Int } M_n$ and that $P(x) = \sum_{i=0}^{n} a_i x^i$ is a nonnegative polynomial on the interval $[0, 1]$ with $\boldsymbol{a} \neq 0$. Then $P(x_0) > 0$ for some $x_0 \in [0, 1]$. By Lemma 1.2.12 there exist a representation $\mu$ of $c_n$ with mass at $x_0$. In this case

$$\sum_{i=0}^{n} a_i c_i = \int_0^1 P(x) d\mu(x) > 0. \qquad \blacksquare$$

**Theorem 1.4.2** *P is a nonnegative polynomial of degree $\leq n$ on the interval $[0, 1]$ if and only if*

$$P(x) = \begin{cases} P_m^2(x) + x(1 - x)Q_{m-1}^2(x), & n = 2m, \\ xP_m^2(x) + (1 - x)Q_m^2(x), & n = 2m + 1, \end{cases}$$

*where $P_k, Q_k$ are polynomials of degree $\leq k$.*

*Proof* There are at least three different proofs of this important theorem. Karlin (1963) has a proof, using fixed point arguments, of a more general result which includes Theorem 1.4.2 as a special case. The present theorem also follows from analogous results for the trigonometric functions. These will be given below in Theorem 9.2.6 and Remark 9.2.9. Krein and Nudelman (1977) prove Karlin's results using some basic arguments of approximation theory. The proof presented here follows the reasoning of the last named authors.

Assume first that $n = 2m$ and that $P(x)$ is strictly positive. Then let $u_i(x) = x^i/\sqrt{P(x)}$, $i = 0, 1, \ldots, m$, and consider the problem of minimizing

$$\sup_{0 \leq x \leq 1} \left| \sum_{i=0}^{m} a_i u_i(x) \right| \qquad (1.4.2)$$

over the set of real coefficients $a_0, a_1, \ldots, a_m$ where $a_m = 1$. The system of functions $u_0, u_1, \ldots, u_{m-1}$ is a *Haar system* or *Chebyshev-system* on the interval $[0, 1]$, which means that every linear combination of the functions $u_0, \ldots, u_{m-1}$ has at most $m - 1$ zeros in $[0, 1]$. Results of approximation theory (see Karlin and Studden, 1966, p. 280) imply that a minimal solution

$$u^*(x) = \frac{\displaystyle\sum_{i=0}^{m} a_i^* x^i}{\sqrt{P(x)}}$$

of (1.4.2) exists. Moreover, if $b$ denotes the minimal value of (1.4.2), then

$$|u^*(x)| \leq b \qquad \text{for all } x \in [0, 1],$$

and there exist $m + 1$ points $0 \leq t_0 < \cdots < t_m \leq 1$ for which $u^*$ has absolute value $b$ and is alternating in sign, i.e.

$$u^*(t_i) = \varepsilon(-1)^i b,$$

where $\varepsilon = \pm 1$. For the polynomial $P_m(x) := (\sum_{i=0}^{m} a_i^* x^i)/b$, we have

$$P_m^2(x) = \left(\frac{u^*(x)}{b}\right)^2 P(x) \leq P(x) \qquad \text{for all } x \in [0, 1],$$

and equality holds at $x = t_i$, $i = 0, 1, \ldots, m$. This shows that the nonnegative polynomial $P(x) - P_m^2(x)$ of degree $\leq 2m$ has zeros of multiplicity two at each interior point $t_i$. Counting multiplicities, this implies that $t_0 = 0$ and $t_m = 1$ and

$$P(x) = P_m^2(x) + Ax(1 - x) \prod_{i=1}^{m-1} (x - t_i)^2.$$

The required result then holds on letting

$$Q_{m-1}^2(x) = A \prod_{i=1}^{m-1} (x - t_i)^2.$$

If $P(x)$ is not strictly positive, we can use $P_\varepsilon(x) = P(x) + \varepsilon$ and then let $\varepsilon \to 0$. Note that for each $\varepsilon$ the corresponding polynomials $P_m(x)$ and $Q_{m-1}(x)$ are bounded, so their coefficients are bounded.

The odd case $n = 2m + 1$ follows from the even case by considering the new polynomial $xP(x)$ of degree $2m + 2$. Then

$$xP(x) = P_{m+1}^2(x) + x(1-x)Q_m^2(x).$$

Clearly $P_{m+1}$ must vanish at $x = 0$, and the result follows on cancelling $x$ from both sides. ∎

The *Hausdorff moment problem* is to decide if a given sequence $c = (c_1, c_2, \ldots)$ is in fact the moment sequence of a probability measure $\mu \in \mathcal{P}$. It is well known (see Shohat and Tamarkin, 1944) that the Hausdorff moment problem is determined, that is, there exists at most one probability measure on the interval $[0, 1]$ representing $c$. A sequence $c$ is called a *Hausdorff moment sequence* if $c \in M$. The representation of an arbitrary nonnegative polynomial in Theorem 1.4.2 is crucial for the characterization of Hausdorff moment sequences. It permits one to check, using Theorem 1.4.1, whether a moment point $c_n \in M_n$ by checking the various polynomials in Theorem 1.4.2.

For $c_k \in M_k$ and for the polynomial $P(x) = \sum_{i=0}^k a_i x^i$, define

$$L(P) = a_0 + \sum_{i=1}^k a_i c_i.$$

Then, if $n = 2m$, Theorem 1.4.1 and 1.4.2 imply that $c_n \in M_n$ if and only if

$$L(P_m^2) \quad \text{and} \quad L(x(1-x)Q_{m-1}^2)$$

are nonnegative for all polynomials $P_m$ and $Q_{m-1}$. Now, if $a_0, \ldots, a_m$ denote the coefficients of the polynomial $P_m$, it follows that

$$L(P_m^2) = \sum_{i=0}^m \sum_{j=0}^m a_i a_j c_{i+j} \geq 0$$

for all $(a_0, \ldots, a_m) \in \mathbb{R}^{m+1}$. Thus the *moment matrix* $(c_{i+j})_{i,j=0}^m$ is nonnegative definite. Similarly (by a consideration of the other polynomial) the matrix

$$(c_{i+j+1} - c_{i+j+2})_{i,j=0}^{m-1}$$

is nonnegative definite. Thus, if $n = 2m$, then $c_n \in M_n$ if and only if the matrices

$$(c_{i+j})_{i,j=0}^m \quad \text{and} \quad (c_{i+j+1} - c_{i+j+2})_{i,j=0}^{m-1}$$

are nonnegative definite. Similarly, if $n = 2m + 1$, then $c_n \in M_n$ if and only if the matrices

$$(c_{i+j+1})_{i,j=0}^m \quad \text{and} \quad (c_{i+j} - c_{i+j+1})_{i,j=0}^m$$

are nonnegative definite. We therefore introduce the *Hankel determinants*

$$\underline{H}_{2m} = \begin{vmatrix} c_0 & \cdots & c_m \\ \vdots & & \vdots \\ c_m & \cdots & c_{2m} \end{vmatrix}, \quad \overline{H}_{2m+1} = \begin{vmatrix} c_0 - c_1 & \cdots & c_m - c_{m+1} \\ \vdots & & \vdots \\ c_m - c_{m+1} & \cdots & c_{2m} - c_{2m+1} \end{vmatrix},$$

$$(1.4.3)$$

$$\underline{H}_{2m+1} = \begin{vmatrix} c_1 & \cdots & c_{m+1} \\ \vdots & & \vdots \\ c_{m+1} & \cdots & c_{2m+1} \end{vmatrix}, \quad \overline{H}_{2m} = \begin{vmatrix} c_1 - c_2 & \cdots & c_m - c_{m+1} \\ \vdots & & \vdots \\ c_m - c_{m+1} & \cdots & c_{2m-1} - c_{2m} \end{vmatrix},$$

and obtain the following characterization for Hausdorff moment sequences:

**Theorem 1.4.3**

1. $c_n \in M_n$ if and only if $\underline{H}_i$ and $\overline{H}_i$ are nonnegative for $i = 1, \ldots, n$.
2. $c_n \in Int\ M_n$ if and only if $\underline{H}_i$ and $\overline{H}_i$ are positive for $i = 1, \ldots, n$.

*Proof*  Recall that a symmetric matrix is nonnegative definite if and only if the leading principal minors are nonnegative. From the considerations preceding the theorem, it follows that $c_n \in M_n$ if and only if $\underline{H}_{n-2k}$ and $\overline{H}_{n-2k}$ are nonnegative provided $n - 2k \geq 1$. The proof can be completed by noting that $c_n \in M_n$ implies also that $c_{n-1} \in M_{n-1}$. The second assertion of the theorem is proved by similar arguments.  ∎

Note that a Hausdorff moment sequence is characterized by the property that $\underline{H}_i$ and $\overline{H}_i$ are nonnegative for all $i \in \mathbb{N}$. It is also worthwhile to mention that $\underline{H}_n = 0$ or $\overline{H}_n = 0$ implies that $\overline{H}_k = \underline{H}_k = 0$ for all $k \geq n + 1$. This follows because a vanishing Hankel determinant of index $n$ implies that $c_n \in \partial M_n$ or equivalently $I(c_n) \leq n/2$. Therefore the (unique) representing measure of $c_n$ has at most $\lfloor n/2 + 1 \rfloor$ support points. On the other hand, the Hankel determinants vanish whenever the number of support points of the representing measure is less than the size of the matrix, and a careful inspection of the different cases yields $\overline{H}_k = \underline{H}_k = 0$ for all $k \geq n + 1$.

An examination of the Hankel determinants in Theorem 1.4.3 reveals that if $c_{n-1} \in Int\ M_{n-1}$, then $\underline{H}_n$ is an increasing function of $c_n$. Therefore, for fixed $c_{n-1} \in Int\ M_{n-1}$, the $n$th moment $c_n$ has a lower bound $c_n^-$ determined by varying $c_n$ in $\underline{H}_n$ so that $\underline{H}_n = 0$. Similarly the upper bound $c_n^+$ is determined by varying $c_n$ in $\overline{H}_n$ so that $\overline{H}_n = 0$. This explains the upper and lower bar notation. The following result provides an explicit representation of the canonical moments in terms of Hankel determinants:

**Theorem 1.4.4**  *If $c_{n-1} \in Int\, M_{n-1}$, then*

$$c_n - c_n^- = \frac{\underline{H}_n}{\underline{H}_{n-2}}, \quad c_n^+ - c_n = \frac{\overline{H}_n}{\overline{H}_{n-2}}, \tag{1.4.4}$$

*and*

$$p_n = \frac{\underline{H}_n \overline{H}_{n-2}}{\underline{H}_n \overline{H}_{n-2} + \overline{H}_n \underline{H}_{n-2}}, \tag{1.4.5}$$

*where $\underline{H}_{-1} = \overline{H}_{-1} = \underline{H}_0 = \overline{H}_0 = 1$.*

*Proof*  The value of $p_n$ follows from the expressions for $c_n - c_n^-$ and $c_n^+ - c_n$ and the definition of $p_n$ in (1.3.2). To obtain the expression for $c_n - c_n^-$, we note again that $\underline{H}_n$ would be zero if we replace $c_n$ by $c_n^-$. Then writing $c_n = c_n^- + (c_n - c_n^-)$ for the last element in $\underline{H}_n$ gives $\underline{H}_n = (c_n - c_n^-)\underline{H}_{n-2}$. The value of $c_n^+ - c_n$ is verified in a similar manner.  ∎

The expression in the denominator of $p_n$ in (1.4.5) can be written in a different form by noting the following basic relationship between the Hankel determinants. The proof is given in Corollary 2.3.2.

**Theorem 1.4.5**  *For all $n \geq 0$,*

$$\underline{H}_n \overline{H}_n = \underline{H}_{n-1} \overline{H}_{n+1} + \overline{H}_{n-1} \underline{H}_{n+1}$$

The following Corollary provides an explicit representation of the canonical moments in terms of the moments and also shows that the the canonical moments are continuous functions of the ordinary moments:

**Corollary 1.4.6**  *For all $1 \leq n \leq N(c)$,*

$$c_n^+ - c_n^- = \frac{\underline{H}_{n-1} \overline{H}_{n-1}}{\overline{H}_{n-2} \underline{H}_{n-2}}, \quad p_n = \frac{\underline{H}_n \overline{H}_{n-2}}{\underline{H}_{n-1} \overline{H}_{n-1}}, \quad q_n = 1 - p_n = \frac{\underline{H}_{n-2} \overline{H}_n}{\underline{H}_{n-1} \overline{H}_{n-1}}.$$

Another useful set of quantities is defined by

$$\zeta_0 = 1, \zeta_1 = p_1, \quad \zeta_k = q_{k-1} p_k, \qquad k \geq 2,$$
$$\gamma_0 = 1, \gamma_1 = q_1, \quad \gamma_k = p_{k-1} q_k, \qquad k \geq 2. \tag{1.4.6}$$

It will gradually be apparent that these quantities are in some sense more basic than the canonical moments $p_k$. Their representation in terms of Hankel determinants is an immediate consequence of Corollary 1.4.6.

**Corollary 1.4.7**   *For all* $1 \leq n \leq N(c)$,

$$\zeta_n = \frac{\underline{H}_n \underline{H}_{n-3}}{\underline{H}_{n-1} \underline{H}_{n-2}}, \quad \gamma_n = \frac{\overline{H}_n \overline{H}_{n-3}}{\overline{H}_{n-1} \overline{H}_{n-2}},$$

*where* $\underline{H}_{-2} = \overline{H}_{-2} = \underline{H}_{-1} = \overline{H}_{-1} = \underline{H}_0 = \overline{H}_0 = 1$.

The next corollary is interesting and in fact rather crucial for later developments. It expresses a sort of duality between the lower and upper boundaries and the expressions $\zeta_i$ and $\gamma_{i+1}$. In order to describe the result, we write the expression for $\zeta_i$ and $\gamma_{i+1}$ given in Corollary 1.4.7 as functions of the moments: Let

$$\zeta_i = \zeta_i(c_0, c_1, \ldots, c_i),$$

$$\gamma_i = \gamma_i(c_0, c_1, \ldots, c_i).$$

We have generally assumed that $c_0 = \int_0^1 d\mu(x) = 1$; however, for the following result it is convenient to let $\mu$ be an arbitrary finite measure. Note that by (1.3.2), the canonical moments do not depend on the standardization of the measure which implies that $\zeta_i(\lambda c_0, \lambda c_1, \ldots, \lambda c_n) = \zeta_i(c_0, c_1, \ldots, c_n)$.

**Corollary 1.4.8**   *If* $(c_1, \ldots, c_n) \in Int\ M_n$, *then*

$$\zeta_n(c_0 - c_1, c_1 - c_2, \ldots, c_n - c_{n+1}) = \gamma_{n+1}(c_0, c_1, c_2, \ldots, c_{n+1})$$

*Proof*   In the expression for $\zeta_n$ in Corollary 1.4.7, we replace the moments $c_i$ in the Hankel determinants $\underline{H}_n$ by the differences $c_i - c_{i+1}$. The resulting expression is $\gamma_{n+1}$.                                                                    ∎

The material in this section so far has concentrated on expressing the new quantities $p_k$, $\zeta_k$ and $\gamma_k$ in terms of the original moments using the Hankel determinants. In Section 2.4 of Chapter 2 a complete discussion will be given of the inverse map expressing the moments $c_1, \ldots, c_n$ in terms of the canonical moments $p_1, \ldots, p_n$. Some elementary results in this direction are presented here. Theorems 1.4.9 and 1.4.10 are from Skibinsky (1968).

**Theorem 1.4.9**   *For all* $1 \leq n \leq N(c)$,

$$c_n^+ - c_n^- = \prod_{i=1}^{n-1} p_i q_i,$$

$$c_n - c_n^- = \prod_{i=1}^{n} \zeta_i,$$

$$c_n^+ - c_n = \prod_{i=1}^{n} \gamma_i.$$

*Proof* From Example 1.3.1 we obtain $c_2^+ - c_2^- = c_1(1 - c_1) = p_1 q_1$, and by Corollary 1.4.6,

$$c_n^+ - c_n^- = p_{n-1} q_{n-1}(c_{n-1}^+ - c_{n-1}^-) = \prod_{i=1}^{n-1} p_i q_i$$

whenever $2 \leq n \leq N(\mathbf{c})$. The other two expressions follow in a similar manner using Theorem 1.4.4 and Corollary 1.4.7. ∎

The first expression in Theorem 1.4.9 is noteworthy as it gives the "$n$-width" of the moment space $M_n$ for given $\mathbf{c}_{n-1} \in M_{n-1}$. The following result expresses the Hankel determinants in terms of the canonical moments:

**Theorem 1.4.10**
*If $2m \leq N(\mathbf{c})$, then*

$$\underline{H}_{2m} = \prod_{i=1}^{m} (\zeta_{2i-1} \zeta_{2i})^{m-i+1} \quad and \quad \overline{H}_{2m} = \prod_{i=1}^{m} (\gamma_{2i-1} \gamma_{2i})^{m-i+1}.$$

*If $2m + 1 \leq N(\mathbf{c})$, then*

$$\overline{H}_{2m+1} = \prod_{i=0}^{m} (\gamma_{2i} \gamma_{2i+1})^{m-i+1} \quad and \quad \underline{H}_{2m+1} = \prod_{i=0}^{m} (\zeta_{2i} \zeta_{2i+1})^{m-i+1}.$$

*Proof* The results are all similar, and only the proof for $\underline{H}_{2m}$ is given. By Corollary 1.4.7,

$$\frac{\underline{H}_{2m}}{\underline{H}_{2m-2}} = \zeta_{2m} \zeta_{2m-1} \frac{\underline{H}_{2m-2}}{\underline{H}_{2m-4}} = \prod_{i=1}^{m} (\zeta_{2i-1} \zeta_{2i}),$$

which implies that

$$\underline{H}_{2m} = \prod_{i=1}^{m} (\zeta_{2i-1} \zeta_{2i})^{m-i+1}.$$ ∎

**EXAMPLE 1.4.11** The following is a simple example which shows, using the uniform distribution, that one can calculate the quantities $\zeta_n$ and hence the canonical moments by knowing or evaluating the determinants $\underline{H}_n$. For the uniform measure on $[0, 1]$, $c_k = 1/(k + 1)$, $k \geq 0$. We will indicate below that for $m \geq 1$,

$$\underline{H}_{2m} = \frac{(m!)^4}{(2m)!(2m + 1)!} \underline{H}_{2m-2}.$$

and

$$\underline{H}_{2m+1} = \frac{(m!(m+1)!)^2}{(2m+1)!(2m+2)!} \underline{H}_{2m-1}$$

It is then easy to show from Corollary 1.4.7 that

$$\zeta_{2m} = \frac{m}{2(2m+1)}$$

and

$$\zeta_{2m+1} = \frac{m+1}{2(2m+1)}.$$

One can then prove inductively that

$$p_{2k-1} = \frac{1}{2}, \quad p_{2k} = \frac{k}{2k+1}, \quad k \geq 1.$$

Finally we will indicate only the proof of the recursion formula for $\underline{H}_{2m}$; the odd case is very similar. This is obtained by replacing the $i$th column $C_i$ in $\underline{H}_{2m}$ by $C_i - C_{m+1}, i = 1, \ldots, m$ and then factoring out, in an obvious way, the quantity $(m!)^2/(2m+1)!$. The result will follow by doing the same operations on the rows of $\underline{H}_{2m}$ and by an expansion of the resulting determinant with respect to the last row.

**EXAMPLE 1.4.12**  *The volume of the nth moment space $M_n$.* Using the expressions in Theorem 1.4.9 and some of the previous discussion regarding the relationship between the canonical moments $(p_1, \ldots, p_n)$ and $(c_1, \ldots, c_n)$, it is easy to calculate the volume of the $n$th moment space $M_n$ expressed in equation (1.1.2) as

$$\int_{M_n} \cdots \int dc_1 dc_2 \ldots dc_n. \tag{1.4.7}$$

In this form it is difficult to evaluate. However, the integral can be quickly transformed to the canonical moments $p_1, \ldots, p_n$ and easily evaluated. Note that Int $M_n$ maps onto the set $0 < p_i < 1, i = 1, 2, \ldots, n$ and that $c_k$ depends only on $p_1, \ldots, p_k$. Consequently the Jacobian matrix is triangular, and its determinant is given by

$$\left| \frac{\partial(c_1, \ldots, c_n)}{\partial(p_1, \ldots, p_n)} \right| = \prod_{k=1}^{n} \frac{\partial c_k}{\partial p_k}.$$

From the second expression in Theorem 1.4.9, it follows that $\partial c_1/\partial p_1 = 1$,

$$\frac{\partial c_k}{\partial p_k} = \prod_{i=1}^{k-1}(p_i q_i), \qquad k \geq 2,$$

in which case

$$\left|\frac{\partial(c_1, \ldots, c_n)}{\partial(p_1, \ldots, p_n)}\right| = \prod_{i=1}^{n-1}(p_i q_i)^{n-i}.$$

By integrating this expression over $0 < p_i < 1$, $i = 1, \ldots, n$ the volume of the $n$th moment space $M_n$ is then given by

$$\text{Vol } M_n = \prod_{k=1}^{n} B(k, k), \tag{1.4.8}$$

where $B(\alpha, \beta)$ is the Beta function defined by (1.3.10). For example, we obtain Vol $M_2 = 1/6$, Vol $M_3 = 1/180, \ldots$ and for $n \to \infty$,

$$\text{Vol } M_n \approx \text{const} \cdot 2^{-n^2},$$

which demonstrates that the $n$th moment space forms a very small part of the $n$th unit cube $[0, 1]^n$.

An alternative calculation of the volume of $M_n$ that involves *Selberg's integral*

$$I_m = \int_0^1 \cdots \int_0^1 \prod_{1 \leq k < j \leq m}(t_k - t_j)^4 dt_1 \ldots dt_m$$

$$= \prod_{k=1}^{m} \frac{\Gamma(2k+1)[\Gamma(2k-1)]^2}{2\Gamma(2m+2k-2)} \tag{1.4.9}$$

(Selberg, 1944) is given in Karlin and Shapley (1953). These authors showed that

$$\text{Vol } M_{2m-1} = \frac{I_m}{(2m-1)!m!},$$

and consequently our derivation of (1.4.8) provides an alternative and surprisingly simple proof of Selberg's identity (1.4.9).

## 1.5 Q-D ALGORITHM

The *Q-D algorithm* will be used in this section for a number of purposes. It will provide a recursive algorithm to construct a table allowing us to calculate the quantities $\zeta_1, \zeta_2, \ldots$ (and hence the canonical moments) from the moments $c_1, c_2, \ldots$ of some measure $\mu$. The calculations done in constructing the table will also permit one to see the corresponding values of $\zeta_1, \zeta_2, \ldots$ for the measure $x^n d\mu(x)$. The algorithm is described more completely in Henrici (1977, vol. 1, p. 608).

In order to present a transparent discussion, we consider a measure with infinite support that corresponds to a moment point $c \in \text{Int } M$ or to an infinite sequence of canonical moments. For measures corresponding to a terminating sequence of canonical moments, the following discussion is completely analogous provided that the corresponding quantities are well defined. Recall that $\zeta_k = q_{k-1} p_k$, $k \geq 1$, where $q_0 := 1$, and from Corollary 1.4.7

$$\zeta_k = \frac{\underline{H}_k \underline{H}_{k-3}}{\underline{H}_{k-1} \underline{H}_{k-2}}, \tag{1.5.1}$$

where we use $\underline{H}_k := 1$ if $k \leq 0$. To describe the algorithm, further Hankel determinants are needed, namely $H_0^{(n)} := 1$, and for $k \geq 1$,

$$H_k^{(n)} := \begin{vmatrix} c_n & c_{n+1} & \cdots & c_{n+k-1} \\ c_{n+1} & c_{n+2} & \cdots & c_{n+k} \\ \vdots & \vdots & & \vdots \\ c_{n+k-1} & c_{n+k} & \cdots & c_{n+2k-2} \end{vmatrix}. \tag{1.5.2}$$

Note that $H_k^{(n)}$ starts with $c_n$ in the upper left corner and is of size or order $k$. The determinants $H_k^{(0)}$ and $H_k^{(1)}$ have already been defined and used in Section 1.4, that is,

$$H_k^{(0)} = \underline{H}_{2k-2} \quad \text{and} \quad H_k^{(1)} = \underline{H}_{2k-1}. \tag{1.5.3}$$

Now define $e_0^{(n)} := 0$ and for $n \geq 0$ and $m \geq 1$,

$$q_m^{(n)} = \frac{H_m^{(n+1)} H_{m-1}^{(n)}}{H_m^{(n)} H_{m-1}^{(n+1)}},$$

$$\tag{1.5.4}$$

$$e_m^{(n)} = \frac{H_{m+1}^{(n)} H_{m-1}^{(n+1)}}{H_m^{(n)} H_m^{(n+1)}}.$$

The Q-D algorithm is usually written in a triangular array

$$
\begin{array}{cccccccc}
 & & q_1^{(0)} & & & & & \\
0 & & & e_1^{(0)} & & & & \\
 & q_1^{(1)} & & & q_2^{(0)} & & & \\
0 & & e_1^{(1)} & & & e_2^{(0)} & & \\
 & q_1^{(2)} & & q_2^{(1)} & & & \ddots & \\
0 & & e_1^{(2)} & & e_2^{(1)} & & & \\
 & q_1^{(3)} & & q_2^{(2)} & & \ddots & & \\
 & \vdots & e_1^{(3)} & & e_2^{(2)} & & & \\
 & & \vdots & q_2^{(3)} & \vdots & \ddots & & \\
 & & & \vdots & & & &
\end{array}
\qquad (1.5.5)
$$

Observing (1.5.2) and (1.5.4), it follows that the values of $q_1^{(n)}$, $n \geq 0$ in the first nonzero column are

$$
q_1^{(n)} = \frac{c_{n+1}}{c_n}, \qquad (1.5.6)
$$

and by (1.5.1) and (1.5.3) the values of $q_m^{(0)}$ and $e_m^{(0)}$ on the first diagonal are

$$
q_m^{(0)} = \zeta_{2m-1}, \qquad e_m^{(0)} = \zeta_{2m}. \qquad (1.5.7)
$$

The algorithm will recursively define the table allowing us to go from the moments $c_1, c_2, \ldots,$ or $q_1^{(n)}$ in (1.5.6) to the values of $\zeta_1, \zeta_2, \ldots$ in (1.5.7) or to the canonical moments. The reverse direction is also available but will not be used as it is somewhat cumbersome. Other algorithms for this purpose are given in Chapter 2. The reverse direction will also allow a discussion of the canonical moments for $x^n d\mu$ in terms of those for $\mu$.

**Theorem 1.5.1 (Q-D algorithm)**  *If $q_m^{(n)}$ and $e_m^{(n)}$ are defined as in (1.5.4), then for $m \geq 1$ and $n \geq 0$,*

$$
e_m^{(n)} = (q_m^{(n+1)} - q_m^{(n)}) + e_{m-1}^{(n+1)},
$$

$$
q_{m+1}^{(n)} = \frac{e_m^{(n+1)}}{e_m^{(n)}} q_m^{(n+1)}. \qquad (1.5.8)
$$

Two different proofs of the algorithm will be given. One here and another later in Remark 1.5.7 at the end of this section. The "quotient" or second equation in (1.5.8) can be checked directly from the definitions in (1.5.4). To verify the difference or first equation, use is made of a particular form of *Sylvester's identity*. This identity will also be used in Section 4.4.

Let $A = (a_{ij})_{i,j=1}^{n}$ denote an $n \times n$ matrix with determinant

$$A\begin{pmatrix} 1, & 2, & \cdots, & n \\ 1, & 2, & \cdots, & n \end{pmatrix} = \begin{vmatrix} a_{11} & a_{12} & \cdots & a_{1n} \\ a_{21} & a_{22} & \cdots & a_{2n} \\ \vdots & \vdots & & \vdots \\ a_{n1} & a_{n2} & \cdots & a_{nn} \end{vmatrix},$$

and let

$$A\begin{pmatrix} i_1, & \cdots, & i_p \\ k_1, & \cdots, & k_p \end{pmatrix} = \begin{vmatrix} a_{i_1 k_1} & a_{i_1 k_2} & \cdots & a_{i_1 k_p} \\ \vdots & \vdots & & \vdots \\ a_{i_p k_1} & a_{i_p k_2} & \cdots & a_{i_p k_p} \end{vmatrix}$$

denote a corresponding minor. The particular form of Sylvester's identity that we need is the following:

**Lemma 1.5.2 (Sylvester)**  *If $1 \leq i < j \leq n$ and $1 \leq k < \ell \leq n$, then*

$$A\begin{pmatrix} 1, 2, \ldots, n \\ 1, 2, \ldots, n \end{pmatrix} A\begin{pmatrix} 1, \ldots, i-1, i+1, \ldots, j-1, j+1, \ldots, n \\ 1, \ldots, k-1, k+1, \ldots, \ell-1, \ell+1, \ldots, n \end{pmatrix}$$

$$= \begin{vmatrix} A\begin{pmatrix} 1, \ldots, j-1, j+1, \ldots, n \\ 1, \ldots, \ell-1, \ell+1, \ldots, n \end{pmatrix} & A\begin{pmatrix} 1, \ldots, j-1, j+1, \ldots, n \\ 1, \ldots, k-1, k+1, \ldots, n \end{pmatrix} \\ A\begin{pmatrix} 1, \ldots, i-1, i+1, \ldots, n \\ 1, \ldots, \ell-1, \ell+1, \ldots, n \end{pmatrix} & A\begin{pmatrix} 1, \ldots, i-1, i+1, \ldots, n \\ 1, \ldots, k-1, k+1, \ldots, n \end{pmatrix} \end{vmatrix}$$

*The second determinant on the left side is referred to as the pivot block.*

*Proof*  See Gantmacher (1959, p. 31).

*Proof of Theorem 1.5.1*  It has already been indicated that the "quotient" or second equation of (1.5.8) follows from (1.5.4). To verify the "difference" or first equation in (1.5.8), Lemma 1.5.2 is applied to give

$$H_{k-1}^{(n+1)} H_{k+1}^{(n-1)} = H_k^{(n-1)} H_k^{(n+1)} - (H_k^{(n)})^2. \tag{1.5.9}$$

This is accomplished by starting with $H_{k+1}^{(n-1)}$ and omitting the first and last columns and rows in order to obtain the pivot block.

To verify the first part of (1.5.8), we calculate

$$H_{m-1}^{(n+2)} H_m^{(n+1)} H_m^{(n)} H_{m-1}^{(n+1)} \left[ q_m^{(n+1)} - e_m^{(n)} - (q_m^{(n)} - e_{m-1}^{(n+1)}) \right]$$

$$= (H_{m-1}^{(n+1)})^2 \left[ H_m^{(n+2)} H_m^{(n)} - H_{m-1}^{(n+2)} H_{m+1}^{(n)} \right]$$

$$+ (H_m^{(n+1)})^2 \left[ H_m^{(n)} H_{m-2}^{(n+2)} - H_{m-1}^{(n)} H_{m-1}^{(n+2)} \right] = 0,$$

where we used the definitions of the determinants in (1.5.4) and two applications of (1.5.9). ∎

**EXAMPLE 1.5.3**   This example is concerned with verifying the canonical moments of the Beta distribution or *Jacobi measure* on the interval $[0, 1]$ given in Example 1.3.6, namely

$$w_{(\alpha,\beta)}(x) = \frac{1}{B(\alpha + 1, \beta + 1)} x^\beta (1 - x)^\alpha, \qquad 0 < x < 1,$$

where $\beta > -1$ and $\alpha > -1$. By an application of the Q-D algorithm, we will show that the corresponding canonical moments are given, for $m \geq 1$, by

$$p_{2m} = \frac{m}{2m + 1 + \alpha + \beta}, \qquad p_{2m-1} = \frac{\beta + m}{2m + \alpha + \beta}.$$

These values will be verified by showing that the associated sequence $\{\zeta_m\}_{m \geq 1} = \{q_{m-1} p_m\}_{m \geq 1}$ ($q_0 = 1$) is in fact

$$\zeta_{2m} = \frac{(\alpha + m)m}{(2m + \alpha + \beta)(2m + 1 + \alpha + \beta)},$$

$$\zeta_{2m-1} = \frac{(m + \alpha + \beta)(\beta + m)}{(2m - 1 + \alpha + \beta)(2m + \alpha + \beta)}. \tag{1.5.10}$$

An alternative derivation for the canonical moments of the Beta distribution is given in Example 3.3.4.

The ordinary moments of the Beta distribution are given in Example 1.3.6. The quantities $c_{n+1}/c_n$, $n \geq 0$, needed in the first column of the array (1.5.5) are obtained from Example 1.3.6 as

$$q_1^{(n)} = \frac{c_{n+1}}{c_n} = \frac{\beta + n + 1}{\alpha + \beta + n + 2}.$$

An induction argument using (1.5.8) then shows that

$$e_m^{(n)} = \frac{(\alpha + m)m}{(2m + n + \alpha + \beta)(2m + 1 + n + \alpha + \beta)},$$

$$q_m^{(n)} = \frac{(m + n + \alpha + \beta)(\beta + m + n)}{(2m - 1 + n + \alpha + \beta)(2m + n + \alpha + \beta)}.$$

The values of $\zeta_{2m}$ and $\zeta_{2m-1}$ in (1.5.10) now follow from (1.5.7).

In the above discussion the array (1.5.5) was used to establish the values for $\zeta_1, \zeta_2, \ldots$ in the upper diagonal from the moment ratios in the first nonzero column. It is possible to proceed in the opposite direction using the rules (1.5.8) and successively produce the diagonals using $e_0^{(1)} = 0$ and knowledge of the first diagonal.

The elements in the $n$th diagonal are easily seen to be the corresponding values for the measure $x^n d\mu$. For example, by Corollary 1.4.7 it follows that

$$q_m^{(1)} = \frac{H_m^{(2)} H_{m-1}^{(1)}}{H_m^{(1)} H_{m-1}^{(2)}} = \zeta_{2m-1}', \quad e_m^{(1)} = \frac{H_{m+1}^{(1)} H_{m-1}^{(2)}}{H_m^{(1)} H_m^{(2)}} = \zeta_{2m}', \tag{1.5.11}$$

where $\zeta_j' = q_{j-1}' p_j'$ ($j \geq 1, q_0' = 1$) and $p_1', p_2', \ldots$ denote the canonical moments of the measure $x d\mu$. In later chapters we will have interest in the measures $x d\mu$, $(1-x) d\mu$, and $x(1-x) d\mu$. The ordinary and canonical moments for these three measures will be denoted by $c_i', c_i'', c_i^*$ and $p_i', p_i'', p_i^*$, respectively. The corresponding values for $\zeta_i = q_{i-1} p_i$ will be denoted by $\zeta_i'$ $\zeta_i''$ and $\zeta_i^*$, where $\zeta_i' = q_{i-1}' p_i'$, $\zeta_i'' = q_{i-1}'' p_i''$, and so on. This correspondence is indicated in the following display:

$$
\begin{array}{lll}
d\mu & p_1, p_2, \ldots & \zeta_1, \zeta_2, \ldots \\
x d\mu & p_1', p_2', \ldots & \zeta_1', \zeta_2', \ldots \\
(1-x) d\mu & p_1'', p_2'', \ldots & \zeta_1'', \zeta_2'', \ldots \\
x(1-x) d\mu & p_1^*, p_2^*, \ldots & \zeta_1^*, \zeta_2^*, \ldots
\end{array}
$$

It should be remarked once more that all these quantities and the entries in the array (1.5.5) are independent of the "normalizing" constant on the measure or the value of $c_0 = \int_0^1 d\mu(x)$.

For the measure $x d\mu$ we have from (1.3.3),

$$c_0' = \int_0^1 x d\mu(x) = p_1 \quad \text{and} \quad p_1' = \frac{H_1^{(2)}}{H_1^{(1)}} = p_1^{-1} \int_0^1 x^2 d\mu(x) = p_1 + q_1 p_2.$$

Similarly we obtain for the measure $(1 - x)d\mu(x)$,

$$c_0'' = q_1 \quad \text{and} \quad p_1'' = q_1^{-1} \int_0^1 x(1-x)d\mu(x) = p_1q_2.$$

It is fairly easy to express the value of $\zeta_k', \zeta_k''$, and $\zeta_k^*$ in terms of the original canonical moments. These quantities will be used extensively in Chapters 2 and 10.

**Theorem 1.5.4** *Suppose that $\mu$ is supported on the interval $[0, 1]$ and has canonical moments $p_1, p_2, \ldots$. Then the measure $(1 - x)d\mu$ has values $\zeta_k''$ given for $k \geq 1$ by*

$$\zeta_k'' = q_{k-1}'' p_k'' = \gamma_{k+1} = p_k q_{k+1}. \tag{1.5.12}$$

*Proof* This result is precisely Corollary 1.4.8. ∎

**Theorem 1.5.5** *The values of $\zeta_k'$ associated with the measure $xd\mu$ are given from the relations*

$$\zeta_1' = \zeta_1 + \zeta_2,$$

*and for $k \geq 1$,*

$$\zeta_{2k-1}'\zeta_{2k}' = \zeta_{2k}\zeta_{2k+1},$$

$$\zeta_{2k}' + \zeta_{2k+1}' = \zeta_{2k+1} + \zeta_{2k+2}.$$

*Proof* The first identity follows in a relatively straightforward manner by the relations (1.5.8) in Theorem 1.5.1 and the identifications (1.5.11) and (1.5.7). Using the same arguments, we obtain by two applications of the identity (1.5.9)

$$H_{m+1}^{(1)} H_m^{(2)} H_m^{(1)} H_{m+1}^{(0)} \left( \zeta_{2m}' + \zeta_{2m+1}' - (\zeta_{2m+1} + \zeta_{2m+2}) \right)$$

$$= (H_m^{(1)})^2 \left[ H_{m+1}^{(0)} H_{m+1}^{(2)} - H_{m+2}^{(0)} H_m^{(2)} \right]$$

$$+ (H_{m+1}^{(1)})^2 \left[ H_{m+1}^{(0)} H_{m-1}^{(2)} - H_m^{(0)} H_m^{(2)} \right] = 0,$$

which completes the proof. ∎

**Corollary 1.5.6** *The values for $\zeta_k^*$ associated with the measure $x(1 - x)d\mu$ are given by*

$$\zeta_1^* = \gamma_2 + \gamma_3,$$

*and for $k \geq 1$*

$$\zeta_{2k-1}^*\zeta_{2k}^* = \gamma_{2k+1}\gamma_{2k+2},$$

$$\zeta_{2k}^* + \zeta_{2k+1}^* = \gamma_{2k+2} + \gamma_{2k+3}.$$

*Proof*  The representations for the expressions $\zeta_i^* = q_{i-1}^* p_i^*$ $(q_0^* = 1)$ follow by combining Theorems 1.5.4 and 1.5.5.                                                    ∎

**REMARK 1.5.7**  A second interesting proof of the quotient-difference relations in (1.5.8) will be given here. Note first that Theorem 1.5.5 is a special case of Theorem 1.5.1 for $n = 0$. The two theorems are actually equivalent in that a repeated application of Theorem 1.5.5 would produce the relations in (1.5.8).

Our second proof will now show that Theorem 1.5.5 also follows from Theorem 1.5.4 and some symmetry relations in Theorem 1.3.3. Note that the equations defining $\zeta_k'$, $k \geq 0$, in Theorem 1.5.5 can be (inductively) rewritten as

$$q_1' = q_1 q_2,$$

and for $k \geq 1$,

$$p_{2k-1}' p_{2k}' = p_{2k} p_{2k+1}, \qquad (1.5.13)$$

$$q_{2k}' q_{2k+1}' = q_{2k+1} q_{2k+2}.$$

These equations for the canonical moments $p_j'$ of the measure $x d\mu$ match more closely the equations in (1.5.12) for $(1 - x) d\mu$.

Now Theorem 1.3.3 says that if $\mu^{(r)}$ corresponds on $[0, 1]$ to the measure induced from $\mu$ via the transformation $y = 1 - x$, then $\mu^{(r)}$ has canonical moments given by $p_{2k}^{(r)} = p_{2k}$ and $p_{2k-1}^{(r)} = q_{2k-1}$. The proof of (1.5.13) uses this reversal twice with an application of Theorem 1.5.4 in between. Thus we start with $\mu$, reverse to $\mu^{(r)}$, then use Theorem 1.5.4 to get $(1 - x) d\mu^{(r)} = d\nu$, and then reverse again to get $d\nu^{(r)} = x d\mu$. In the following diagram the values of $\zeta_1, \zeta_2, \ldots$ are given for the measure $d\mu$, $d\mu^{(r)}$, and $d\nu = (1 - x) d\mu^{(r)}$:

| | | | | | |
|---|---|---|---|---|---|
| $d\mu$ | | $p_1$ | $q_1 p_2$ | $q_2 p_3$ | $q_3 p_4$ $\cdots$ |
| $d\mu^{(r)}$ | | $q_1$ | $p_1 p_2$ | $q_2 q_3$ | $p_3 p_4$ $\cdots$ |
| $d\nu = (1 - x) d\mu^{(r)}$ | | $q_1 q_2$ | $p_2 p_3$ | $q_3 q_4$ | $p_4 p_5$ $\cdots$ |

The final step is to reverse the odd canonical moments of $d\nu$ to give the expressions for $d\nu^{(r)} = x d\mu$ on the right-hand side of (1.5.13).

## 1.6  STATISTICAL DESIGN OF EXPERIMENTS AND MAXIMIZATION OF THE HANKEL DETERMINANTS

In this section a brief introduction to some applications of canonical moments to optimal design theory is given. A more complete discussion will be presented in Chapters 5 and 6.

One of the simplest problems arising in design theory is essentially solved by maximizing the Hankel determinant $\underline{H}_{2m}$ over the moment space $M_{2m}$. In terms of the canonical moments this is a fairly trivial matter. From Theorem 1.4.10 the value of $\underline{H}_{2m}$ is given by

$$\underline{H}_{2m} = \prod_{i=1}^{m} (\zeta_{2i-1}\zeta_{2i})^{m-i+1}.$$

Using $\zeta_i = q_{i-1}p_i$, this is rewritten as

$$\underline{H}_{2m} = \prod_{j=1}^{m} (q_{2j-1}p_{2j-1})^{m-j+1} \prod_{j=1}^{m-1} (q_{2j}^{m-j} p_{2j}^{m-j+1}) p_{2m}.$$

The maximum is readily seen to occur when

$$p_{2j-1} = \frac{1}{2}, \qquad p_{2j} = \frac{m-j+1}{2(m-j)+1}, \qquad j = 1, 2, \ldots, m. \tag{1.6.1}$$

Note that $p_{2m} = 1$ and $0 < p_i < 1$, $i = 1, \ldots, 2m - 1$, so that the corresponding $\mu^*$ is an upper principal representation with support on $m + 1$ points $x_0 = 0 < x_1 < \ldots < x_m = 1$. General methods for retrieving the corresponding measure $\mu$ on $[0, 1]$ will be described later in Section 3.6. The solution will be shown in Section 4.3 to be supported at the zeros of the polynomial

$$x(x-1)P'_m(x) \tag{1.6.2}$$

where $P_k(x)$ is the $k$th *Legendre polynomial* orthogonal with respect to Lebesgue measure on the interval $[0, 1]$, that is,

$$\int_0^1 P_n(x)P_m(x)dx = 0$$

whenever $n \neq m$. For example, orthogonalizing the functions $1, x, x^2, \ldots$ with respect to the Lebesgue measure yields for the Legendre polynomials up to degree 3,

$$P_1(x) = 2x - 1,$$

$$P_2(x) = 6x^2 - 6x + 1,$$

$$P_3(x) = 20x^3 - 30x^2 + 12x - 1.$$

Knowing that $p_{2m} = 1$ and the unique representation $\mu^*$ is on $m + 1$ points, say,

$x_0, x_1, \ldots x_m$, which is the size of determinant $\underline{H}_{2m}$, allows us to write

$$
\underline{H}_{2m} = 
\begin{vmatrix}
1 & 1 & \cdots & 1 \\
x_0 & x_1 & \cdots & x_m \\
\vdots & \vdots & \ddots & \vdots \\
x_0^m & x_1^m & \cdots & x_m^m
\end{vmatrix}
\begin{vmatrix}
\mu_0 & 0 & \cdots & 0 \\
0 & \mu_1 & \cdots & 0 \\
\vdots & \vdots & \ddots & \vdots \\
0 & 0 & \cdots & \mu_m
\end{vmatrix}
\begin{vmatrix}
1 & 1 & \cdots & 1 \\
x_0 & x_1 & \cdots & x_m \\
\vdots & \vdots & \ddots & \vdots \\
x_0^m & x_1^m & \cdots & x_m^m
\end{vmatrix}
$$

$$
= (\prod_{i=0}^{m} \mu_i) V^2(x_0, x_1, \ldots, x_m),
$$

where $\mu_i$ is the weight of the upper principal representation $\mu^*$ at the point $x_i$ $(i = 0, \ldots, m)$ and

$$
V(x_0, x_1, \ldots, x_m) = \prod_{0 \le i < j \le m} (x_i - x_j)
$$

is the usual *Vandermonde determinant* (note that $c_k = \sum_{i=0}^{m} \mu_i x_i^k$). The maximizing weights $\mu_i$ must then be all equal to $\mu_i = 1/(m+1)$. Thus the measure $\mu^* \in \mathcal{P}$ maximizing $\underline{H}_{2m}$ has equal masses at the zeros of the polynomial in (1.6.2).

It might be appropriate to also point out here that the sequence of canonical moments in (1.6.1) in reverse order (omitting $p_{2m} = 1$) is given by

$$
\frac{1}{2}, \frac{2}{3}, \frac{1}{2}, \frac{3}{5}, \frac{1}{2}, \frac{4}{7}, \frac{1}{2}, \cdots
$$

The canonical moments for the uniform measure in the usual order are

$$
\frac{1}{2}, \frac{1}{3}, \frac{1}{2}, \frac{2}{5}, \frac{1}{2}, \cdots
$$

(see Example 1.4.11). It will be shown later that the measures corresponding to the sequences $(p_1, p_2, \ldots, p_k, 1)$ and $(q_k, q_{k-1}, \ldots, q_1, 1)$ have the same support, and in the above situation this common support is given by the roots of the polynomial in (1.6.2).

As a corollary the above maximization of $\underline{H}_{2m}$ gives the solution of maximizing $T = |V(x_0, x_1, \ldots, x_m)|$ over $x_i$ as the roots of (1.6.2). This problem is described in Szegö (1959, p. 139) in an electrostatic setting. If $m+1$ unit *charges* are placed on $[0, 1]$, then $\log T^{-1}$ can be interpreted as the *energy* of the system. The position minimizing $\log T^{-1}$ or maximizing $T$ corresponds to an *electrostatic equilibrium*.

The statistical design viewpoint of the determinant $\underline{H}_{2m}$ is entirely different and arises in the following context: An experimenter wants to estimate an unknown function $g(x)$ for $0 < x < 1$ which is assumed for simplicity to be a polynomial

$$g(x) = \sum_{i=0}^{m} \beta_i x^i.$$

The unknowns are then the coefficients $\beta_0, \beta_1, \ldots, \beta_m$. At selected points $x_i, n_i$ random observations of $g(x_i)$ are observed, $i = 1, \ldots, r$, $\sum_{i=1}^{r} n_i = n$. The observations $Y_1, Y_2, \ldots, Y_n$ are written as $Y_i = g(x_i) + \epsilon_i$, where $\epsilon_i$ is the error in the $i$th measurement ($i = 1, \ldots, n, n > m$). Thus the values $g(x_i)$ are not given exactly. If they were, then any $m + 1$ values of $g(x_i)$ would completely determine $\beta_0, \beta_1, \ldots, \beta_m$. The "errors" in the measurements give rise to the problem of choosing the $\{x_i, n_i\}_{i=1}^{n}$ in some optimal manner. The coefficients in many cases are estimated by *least squares*; that is, they are chosen to minimize

$$\sum_{j=1}^{n} \left( Y_j - \sum_{i=0}^{m} \beta_i x_j^i \right)^2.$$

One of the quantities in the theory of statistical design of experiments that is used to measure the accuracy of the resulting estimates $\hat{\beta}_0, \ldots, \hat{\beta}_m$ is the determinant of the matrix

$$\left| \sum_{i=1}^{r} \frac{n_i}{n} f(x_i) f^T(x_i) \right|,$$

where $f^T(x) = (1, x, \ldots, x^m)$ is the vector of monomials up to the order $m$. This can be written as $\underline{H}_{2m}(\mu)$, where $\mu$ is the measure with masses $n_i/n$ at the points $x_i, i = 1, \ldots, r$. Instead of maximizing $\underline{H}_{2m}(\mu)$ over $\{x_i, n_i\}_1^r$ and $r$, the quantity $\underline{H}_{2m}(\mu)$ is maximized over the set $\mathcal{P}$ of all probability measures on the interval $[0, 1]$ giving rise to an approximate solution $\mu^*$ with equal masses at the $m + 1$ zeros of the polynomial in (1.6.2). Thus for $m = 2$ we have $P_2'(x) = 12x - 6$, and the solution is to take $\frac{1}{3}$ of the $n$ observations at $x_0 = 0$, $x_1 = \frac{1}{2}$ and $x_2 = 1$. For $m = 3$ the experimenter has to take $\frac{1}{4}$ of the observations at $x = 0$, $x_1 = \frac{1}{2} - \frac{1}{2}\sqrt{5}$, $x_2 = \frac{1}{2} + \frac{1}{2}\sqrt{5}$, and $x_3 = 1$, respectively (note that $P_3'(x) = 60x^2 - 60x + 12$). In practice this can only be done exactly if $n$ is a multiple of 3 or 4, respectively. In the other cases one has to use a procedure for rounding.

## 1.7 CANONICAL MOMENTS, CONTINUED FRACTIONS, AND ORTHOGONAL POLYNOMIALS

In this section some connections between canonical moments, continued fractions, and orthogonal polynomials will be briefly described. Further details and a rigorous treatment of the theory are given in Chapters 2 and 3.

The literature on continued fractions and orthogonal polynomials is extensive. The recent book by Brezinski (1991) gives an excellent review of their history and has a very long list of references. The use of continued fractions goes back to antiquity, while orthogonal polynomials date from the late eighteenth century. In recent years there has been a renewed interest in the subject of continued fractions due to their algorithmic nature and the advent of computers. The major classic works on continued fractions are Perron (1954) and Wall (1948). More recent works include Khovanski (1963), Jones and Thron (1980), Henrici (1977), and Lorentzen and Waadeland (1992). Numerous applications in statistics are given in the recent book by Bowman and Shenton (1989).

A finite or terminating *continued fraction* is an expression of the form

$$\frac{a_1}{\lvert b_1} + \frac{a_2}{\lvert b_2} + \cdots + \frac{a_n}{\lvert b_n} = \frac{A_n}{B_n}. \tag{1.7.1}$$

The notation is best explained by example; if $n = 3$, then by the left side of (1.7.1) is meant

$$\cfrac{a_1}{b_1 + \cfrac{a_2}{b_2 + \cfrac{a_3}{b_3}}}.$$

The general case is understood in the same way. An infinite expression of the same type is defined as the limit of $A_n/B_n$ if it exists. A more complete discussion is given in Chapter 3.

If $c_1, c_2, \ldots$ denotes the sequence of moments of a probability measure $\mu$ on the interval $[0, 1]$, then the measure, the ordinary moments, and the canonical moments are related by a Taylor series and a continued fraction expansion of the *Stieltjes transform*

$$\int_0^1 \frac{d\mu(x)}{z - x} = \frac{1}{z} \sum_{k=0}^{\infty} \frac{c_k}{z^k}$$

$$= \frac{1}{\lvert z} - \frac{\zeta_1}{\lvert 1} - \frac{\zeta_2}{\lvert z} - \frac{\zeta_3}{\lvert 1} - \cdots \tag{1.7.2}$$

$$= \frac{1}{\lvert z - \zeta_1} - \frac{\zeta_1 \zeta_2}{\lvert z - \zeta_2 - \zeta_3} - \frac{\zeta_3 \zeta_4}{\lvert z - \zeta_4 - \zeta_5} - \cdots$$

Expansions of this type will be derived in Chapter 3 and can be used for identifying the canonical moments of a measure $\mu$ from its corresponding Stieltjes transform.

**EXAMPLE 1.7.1**    Euler in the eighteenth century was presumably aware of the continued fraction expansion

$$\int_0^1 \frac{dx}{z-x} = \log\left(\frac{z}{z-1}\right) = \frac{1}{z} + \frac{1}{2z^2} + \frac{1}{3z^3} + \cdots$$

$$= \frac{1}{\mid z} - \frac{\frac{1}{2}}{\mid 1} - \frac{\frac{1}{2}\frac{1}{3}}{\mid z} - \frac{\frac{2}{3}\frac{1}{2}}{\mid 1} - \frac{\frac{1}{2}\frac{2}{5}}{\mid z} - \frac{\frac{3}{5}\frac{1}{2}}{\mid 1} - \cdots$$

Observing (1.7.2), the canonical moments of the uniform measure can be read from this expression as

$$p_{2k-1} = \frac{1}{2}, \quad p_{2k} = \frac{k}{2k+1}, \qquad k = 1, 2, 3, \ldots \ldots$$

There are standard algorithms for expanding the power series in (1.7.2) into a continued fraction by a repeated division or inversion of the power series. Stieltjes showed that the problem of expanding the power series in (1.7.2) into a continued fraction was intimately related to the problem of providing an *LU decomposition* of the moment matrix $C = (c_{i+j})_{i,j=0}^\infty$. That is, the matrix $C$ is written as the product of a lower triangular matrix times an upper triangular matrix where the triangular matrices involve the canonical moments. This will provide a number of formulas expressing $c_{i+j}$ in terms of the canonical moments. These formulas will be explained more fully in Section 2.4 of Chapter 2 and will be related to standard formula for simple random walks in Section 8.5 of Chapter 8.

If the last expression in (1.7.2) is truncated after $n$ terms, it can be expressed as a single ratio of two polynomials $P_{n-1}^{(1)}$ and $P_n$ of degree $n-1$ and $n$, i.e.

$$\frac{1}{\mid z - \zeta_1} - \frac{\zeta_1\zeta_2}{\mid z - \zeta_2 - \zeta_3} - \cdots - \frac{\zeta_{2n-3}\zeta_{2n-2}}{\mid z - \zeta_{2n-2} - \zeta_{2n-1}} = \frac{P_{n-1}^{(1)}(z)}{P_n(z)}.$$

The quantities $P_n(z)$ in the denominator are the polynomials, with leading coefficient 1, which are *orthogonal with respect to the measure* $\mu$, that is

$$\int_0^1 P_n(x)P_m(x)d\mu(x) = 0$$

whenever $n \neq m$. A more complete discussion of the relation between orthogonal polynomials and canonical moments will be given in Chapter 2. It will be shown in Corollary 2.3.4 that these polynomials satisfy the recursive relations

$$P_0(z) = 1, \quad P_1(z) = z - \zeta_1 \tag{1.7.3}$$

$$P_{k+1}(z) = (z - \zeta_{2k} - \zeta_{2k+1})P_k(z) - \zeta_{2k-1}\zeta_{2k}P_{k-1}(z).$$

Thus, if the recursion formulas (1.7.3) are available for the orthogonal polynomials with respect to a given measure $\mu$, then the canonical moments can be obtained from the values

$$\zeta_1, \zeta_{2k-1}\zeta_{2k} \quad \text{and} \quad \zeta_{2k} + \zeta_{2k+1}, \qquad k \geq 1.$$

**EXAMPLE 1.7.2**  *Chebyshev Polynomials and the Arc-Sine Law.* The simplest set of orthogonal polynomials is defined by

$$T_n(y) = \cos(n\theta), \quad y = \cos\theta.$$

These polynomials are called *Chebyshev polynomials of the first kind*, and they satisfy the simple recursion formulas

$$T_0(y) = 1, \quad T_1(y) = y, \tag{1.7.4}$$

$$T_{k+1}(y) = 2yT_k(y) - T_{k-1}(y), \qquad k \geq 1.$$

They are orthogonal on the interval $[-1, 1]$ with respect to the arc-sine distribution, with density

$$\frac{1}{\pi\sqrt{1-y^2}}, \quad y \in (-1, 1),$$

that is,

$$\int_{-1}^{1} \frac{T_n(y)T_m(y)}{\pi\sqrt{1-y^2}} dy = \begin{cases} 0 & \text{if } n \neq m, \\ 1 & \text{if } n = m = 0, \\ \frac{1}{2} & \text{if } n = m \geq 1. \end{cases}$$

It follows from Example 1.5.3 and Theorem 1.3.2 that the canonical moments of the arc-sine distribution are given by $p_k \equiv \frac{1}{2}$ ($k \geq 1$). We would now like to quickly identify these canonical moments from the recurrence relationship (1.7.4). Since the measure is symmetric it follows from Corollary 1.3.4 that $p_{2i-1} = \frac{1}{2}$ for $i \geq 1$. The version of (1.7.3) on the interval $[-1, 1]$ becomes

$$R_0(y) = 1, \quad R_1(y) = y + 1 - 2\zeta_1, \tag{1.7.5}$$

$$R_{k+1}(y) = (y + 1 - 2\zeta_{2k} - 2\zeta_{2k+1})R_k(y) - 4\zeta_{2k-1}\zeta_{2k}R_{k-1}(y).$$

For the symmetric case where $p_{2i-1} = \frac{1}{2}$, this simplifies to

$$R_0(y) = 1, \quad R_1(y) = y, \tag{1.7.6}$$

$$R_{k+1}(y) = yR_k(y) - q_{2k-2}p_{2k}R_{k-1}(y), \qquad k \geq 1 \ (q_0 = 1).$$

The polynomials $T_k$ in (1.7.4) have *leading coefficient* equal to $2^{k-1}$ except for $k = 0$. Rewriting (1.7.4) in the form with leading coefficient 1 gives

$$\tilde{T}_0(x) = 1, \quad \tilde{T}_1(x) = x, \quad \tilde{T}_2(x) = x^2 - \tfrac{1}{2},$$

$$\tilde{T}_{k+1}(x) = x\tilde{T}_k(x) - \tfrac{1}{4}\tilde{T}_{k-1}(x), \quad k \geq 2,$$

where $\tilde{T}_k(x) = 2^{1-k}T_k(x)$, $k \geq 1$. Comparing this recursion with (1.7.6) yields $p_2 = \tfrac{1}{2}$ and $q_{2k-2}p_{2k} = \tfrac{1}{4}$ for $k \geq 2$. In this case $p_{2k} \equiv \tfrac{1}{2}$ for all $k \geq 1$.

## 1.8 RANDOM WALKS OR BIRTH AND DEATH CHAINS

The study of birth and death processes is intimately connected to orthogonal polynomials and hence to continued fractions and canonical moments. Some discussion of birth and death chains or simple random walks in this context will be given in Chapter 8.

A *simple random walk*, indexed by $\mathbb{N}_0 = \{0, 1, \ldots\}$, is a stochastic process $\{X_n\}_{n\in\mathbb{N}_0}$ whose transition probabilities $P_{k\ell} = Pr(X_{n+1} = \ell | X_n = k)$ are given by

$$P_{k\ell} = \begin{cases} d_\ell, & \ell = k - 1, \\ h_\ell, & \ell = k, \\ u_\ell, & \ell = k + 1, \\ 0, & \text{otherwise}, \end{cases}$$

where $d_\ell + h_\ell + u_\ell \leq 1$. Here the case $d_\ell + h_\ell + u_\ell < 1$ is interpreted as a permanent absorbing state $\ell^*$ that can be reached with probability $1 - (d_\ell + h_\ell + u_\ell)$ from state $\ell$. It is convenient to define a system of polynomials $Q_k$ associated with $\{X_n\}_{n\in\mathbb{N}_0}$ by

$$Q_{-1}(y) = 0, \quad Q_0(y) = 1,$$

$$(1.8.1)$$

$$yQ_k(y) = d_kQ_{k-1}(y) + h_kQ_k(y) + u_kQ_{k+1}(y).$$

If these polynomials are rewritten with leading coefficient 1, we obtain the recursion

$$P_0(y) = 1, \quad P_1(y) = y - h_0,$$

$$P_{k+1}(y) = (y - h_k)P_k(y) - u_{k-1}d_kP_{k-1}(y).$$

Comparing this recursion with (1.7.5), one can make the identifications

$$h_0 = -1 + 2\zeta_1,$$
$$h_k = -1 + 2\zeta_{2k} + 2\zeta_{2k+1}, \tag{1.8.2}$$
$$u_{k-1}d_k = 4\zeta_{2k-1}\zeta_{2k}, \qquad k \geq 1.$$

It will be shown that these equations can be solved for the canonical moments $p_i \in (0, 1)$, which implies that the system of polynomials is orthogonal with respect to some measure $\psi$ on the interval $[-1, 1]$. This measure has canonical moments $p_i$ and is called the *spectral measure* of the random walk $\{X_n\}_{n \in \mathbb{N}_0}$.

It will also be proved in Chapter 8 that the probabilities for going from state $i$ to state $j$ in $n$-steps are given by

$$P_{ij}^n = P(X_n = j | X_0 = i) = \frac{\int_{-1}^1 y^n Q_i(y) Q_j(y) d\psi(y)}{\int_{-1}^1 Q_j^2(y) d\psi(y)} \tag{1.8.3}$$

$(i, j = 0, 1, \ldots)$. Consequently the return probabilities to the state zero are

$$P_{00}^n = \int_{-1}^1 y^n d\psi(y),$$

which of course are the moments of the spectral measure $\psi$. If $h_k = 0$ for $k \geq 0$, then it follows from (1.8.2) that $p_{2k-1} = \frac{1}{2}$ for $k \geq 1$, the spectral measure $\psi$ is symmetric, $P_{00}^{2k+1} = 0$ for $k \geq 1$, and

$$P_{00}^{2m} = \int_{-1}^1 y^{2m} d\psi(y) = \int_0^1 x^m d\mu(x),$$

where the measures $\mu$ and $\psi$ are related by (1.3.7), i.e.

$$\psi([-x, x]) = \mu([0, x^2]).$$

If there is reflection at the origin and no absorption, we have $d_0 = 0$, $u_0 = 1$, and the equations (1.8.2) can be solved for the canonical moments of $\psi$ which are given by $p_{2k-1} = \frac{1}{2}, p_{2k} = d_k$. Using Theorem 1.3.5, the canonical moments of the corresponding measure $\mu$ on the interval $[0, 1]$ are given directly by the sequence of downward probabilities $\{d_k\}_{k \geq 1}$.

Therefore there is a one-to-one correspondence between measures $\mu$ on the interval $[0, 1]$ with canonical moments $\{d_k\}_{k \geq 1}$ and simple random walks on $\mathbb{N}_0$ with no absorption, vanishing holding probabilities $h_k$, and reflecting barrier at zero. These relations will be discussed in more detail in Chapter 8.

**EXAMPLE 1.8.1** *The Ehrenfest Urn.* Consider two urns, say, A and B containing $N$ balls. The chain is in state $i$ if there are $i$ balls in urn A. At each time $n$ one ball is chosen with equal probabilities. If it is from urn A, it is placed in urn B. If it is from urn B, it is put into urn A. This model is called the *classical Ehrenfest urn* (e.g., see Kemperman 1961); it was used as a model of heat exchange between two isolated bodies (Ehrenfest 1907). The state space of the corresponding random walk is $E = \{0, \ldots, N\}$ and the transition probabilities are

$$u_i = 1 - \frac{i}{N}, \quad d_i = \frac{i}{N}, \quad h_i = 0 \qquad (i = 0, \ldots, N).$$

From the preceding discussion we obtain for the canonical moments of the spectral measure $\psi$:

$$p_{2i-1} = \frac{1}{2}, \quad p_{2i} = \frac{i}{N} \qquad (i = 1, \ldots, N).$$

The corresponding measure $\mu$ on the interval $[0, 1]$ then has canonical moments $1/N, 2/N, \ldots, (N-1)/N, 1$. From Example 1.3.7 we see that $\psi$ is the Binomial distribution with parameters $N$ and $\frac{1}{2}$ transferred to the interval $[-1, 1]$, that is,

$$\psi(x_j) = b\left(j; N, \frac{1}{2}\right) = \binom{N}{j}\left(\frac{1}{2}\right)^N, \qquad j = 0, \ldots, N,$$

where the support points are given by $x_j = 1 - (2j)/N, \ j = 0, \ldots, N$. The orthogonal polynomials with respect to the Binomial distribution will be discussed in Example 2.3.8, and (1.8.3) provides an explicit representation of the $n$-step transition probabilities. For example, the return probabilities to state 0 are given by

$$P_{00}^{2n} = \int_{-1}^{1} x^{2n} d\psi(x) = \left(\frac{1}{2}\right)^N \sum_{j=0}^{N} \binom{N}{j}\left(1 - \frac{2}{N}j\right)^{2n}.$$

CHAPTER 2

# Orthogonal Polynomials

## 2.1 INTRODUCTION

In the previous chapter canonical moments $p_1, p_2, \ldots$ were defined for a given sequence of ordinary moments $c_1, c_2, \ldots$, and the ordinary moments were associated with some probability measure $\mu$ on the interval $[0, 1]$. The canonical moments were defined geometrically and expressed algebraically in terms of certain Hankel determinants. The two kinds of sequences were related in a one-to-one manner. For the most part Chapter 1 considered expressing the canonical moments in terms of the ordinary moments. The present chapter will discuss the reverse problem of expressing the ordinary moments and the measure $\mu$ in terms of the canonical moments. This will be accomplished with the aid of four sequences of orthogonal polynomials associated with the measure $\mu$.

There are basically three main objectives in the present chapter. The first is to show that the support of the measures associated with the sequences of canonical moments $(p_1, \ldots, p_n, 0)$ and $(p_1, \ldots, p_n, 1)$ are the roots of certain orthogonal polynomials. The second is to provide a recursive formula defining the moment $c_n$ in terms of $p_1, p_2, \ldots, p_n$ using the recursive formula for the orthogonal polynomials corresponding to the underlying measure of orthogonality. Finally it is shown that the support of certain reversed sequences are the same. More specifically, the support of the measures corresponding to the sequence $(p_1, \ldots, p_n, 0)$ and to the reversed sequence $(p_n, \ldots, p_1, 0)$ are the same, and the support of the measures corresponding to the sequences $(p_1, \ldots, p_n, 1)$ and to $(q_n, \ldots, q_1, 1)$ are the same. Throughout this chapter we will often talk about the *support of a sequence* $(p_1, \ldots, p_n, \tau_n)$ (where $\tau_n = 0$ or 1) with the understanding that this means the support of the measure associated with the given sequence of canonical moments.

We will start with a brief description of some basic properties about orthogonal polynomials. There are numerous introductory books considering various aspect of the theory of orthogonal polynomials. Among others we refer to the classical work of Szegö (1975), Freud (1971), and Chihara (1978). Because the basic results given in this section do not rely on the support and the normalization of the underlying measures, we do not restrict ourselves to

42

measures supported on the interval $[0, 1]$. Let $\mu$ be a measure on the real line such that all moments

$$c_n = \int_{-\infty}^{+\infty} x^n d\mu(x), \qquad n = 0, 1, 2, \ldots,$$

are finite. The *support* of $\mu$ is defined as

$$\text{supp}(\mu) = \{x \in \mathbb{R} | \forall \varepsilon > 0 \ \mu((x - \varepsilon, x + \varepsilon)) > 0\}. \tag{2.1.1}$$

Consider a sequence of polynomials $\{P_k(x)\}_{k \geq 0}$, where $P_k$ is of exact degree $k$. The sequence is said to be *orthogonal with respect to the measure $\mu$* if and only if

$$\int_{-\infty}^{+\infty} P_m(x)P_k(x)d\mu(x) = 0, \qquad m \neq k. \tag{2.1.2}$$

If the leading coefficient of $P_m(x)$ is 1 (for all $m \geq 0$), then the polynomials are called the *monic orthogonal polynomials* with respect to the measure $\mu$. If the sequence satisfies (2.1.2) and additionally

$$\int_{-\infty}^{+\infty} (P_m(x))^2 d\mu(x) = 1, \qquad m \geq 0,$$

then the sequence of polynomials is called *orthonormal* with respect to the measure $\mu$.

By the equations (2.1.2) the polynomials are uniquely defined, up to a multiplicative constant, by orthogonalizing the sequence of powers 1, $x$, $x^2, \ldots$ in the space of functions $L_2(\mu)$ using the familiar *Gram-Schmidt procedure*. The $m$th polynomial is clearly determined by the $m$ equations

$$\int_{-\infty}^{+\infty} x^i P_m(x)d\mu(x) = 0, \qquad i = 0, 1, \ldots, m - 1. \tag{2.1.3}$$

The solution is given, up to a constant factor, by the polynomial

$$\underline{H}_{2m-1}(x) := \begin{vmatrix} c_0 & \cdots & c_{m-1} & 1 \\ \vdots & & \vdots & \vdots \\ c_m & \cdots & c_{2m-1} & x^m \end{vmatrix}. \tag{2.1.4}$$

The polynomial $\underline{H}_{2m-1}(x)$ is readily seen to satisfy (2.1.3), since

$$\int_{-\infty}^{+\infty} x^i \underline{H}_{2m-1}(x)d\mu(x) = \begin{vmatrix} c_0 & \cdots & c_{m-1} & c_i \\ \vdots & & \vdots & \vdots \\ c_m & \cdots & c_{2m-1} & c_{i+m} \end{vmatrix},$$

and this is equal to zero for $i = 0, 1, \ldots, m - 1$, since two columns in the determinant are equal.

The coefficient of $x^m$ in the polynomial $\underline{H}_{2m-1}(x)$ is the Hankel determinant $\underline{H}_{2m-2}$ defined in (1.4.3). The solution will be of exact degree $m$ and will be unique, up to a constant, provided that $\underline{H}_{2m-2} > 0$. This follows since, if the polynomial is given by $P_m(x) = a_0 + a_1 x + \ldots + a_m x^m$, then (2.1.3) is equivalent to the system of equations

$$\sum_{k=0}^{m} c_{i+k} a_k = 0, \qquad i = 0, 1, \ldots, m - 1.$$

This is a system of $m$ equations in $m + 1$ unknowns $a_0, \ldots, a_m$, and the solution is proportional to the determinants obtained by deleting successive columns from the $m$ by $m + 1$ matrix

$$\begin{pmatrix} c_0 & \cdots & c_m \\ \vdots & & \vdots \\ c_{m-1} & \cdots & c_{2m-1} \end{pmatrix},$$

where the coefficient $a_m$ is proportional to $\underline{H}_{2m-2} \geq 0$. Finally $\underline{H}_{2m-2} > 0$ if and only if the support of the underlying measure $\mu$ contains at least $m$ points. This means that in the case of a measure with an infinite support the orthogonal polynomials are well defined for arbitrary degree $k \in \mathbb{N}_0$. On the other hand, if the support of the measure consists only of a finite number of support points, say, $N$, then the orthogonal polynomials are only well defined up to degree $N$.

**EXAMPLE 2.1.1**   For $x \in [-1, 1]$ let $\theta = \text{arc-cos } x$, and define, for $n \geq 0$, the functions

$$T_n(x) = \cos(n\theta), \tag{2.1.5}$$

$$U_n(x) = \frac{\sin[(n + 1)\theta]}{\sin \theta}, \tag{2.1.6}$$

$$V_n(x) = \frac{\sin[(2n + 1)\theta/2]}{\sin(\theta/2)},$$

$$W_n(x) = \frac{\cos[(2n + 1)\theta/2]}{\cos(\theta/2)}.$$

A simple induction shows that $T_n, U_n, V_n$, and $W_n$ can be calculated recursively.

For example, we have

$$T_0(x) = 1, \quad T_1(x) = x,$$

$$\begin{aligned}
T_{n+1}(x) &= \cos(n\theta)\cos\theta - \sin(n\theta)\sin\theta \\
&= 2\cos(n\theta)\cos\theta - \cos((n-1)\theta) \\
&= 2xT_n(x) - T_{n-1}(x) \quad (n \geq 1).
\end{aligned} \tag{2.1.7}$$

The functions $U_n, V_n, W_n$ satisfy the same recurrence relation with different initial conditions, i.e.

$$U_0(x) = 1, \quad U_1(x) = 2x,$$
$$U_{n+1}(x) = 2xU_n(x) - U_{n-1}(x) \quad (n \geq 1), \tag{2.1.8}$$

$$V_0(x) = 1, \quad V_1(x) = 2x + 1,$$
$$V_{n+1}(x) = 2xV_n(x) - V_{n-1}(x) \quad (n \geq 1), \tag{2.1.9}$$

$$W_0(x) = 1, \quad W_1(x) = 2x - 1,$$
$$W_{n+1}(x) = 2xW_n(x) - W_{n-1}(x) \quad (n \geq 1). \tag{2.1.10}$$

Consequently $T_n, U_n, V_n,$ and $W_n$ are polynomials of precise degree $n$. $T_n(x)$ and $U_n(x)$ are called the *nth Chebyshev polynomial* of *the first* and *second kind*, respectively, and will play an important role in many places of this book. The following table gives these polynomials up to degree 4:

| $n$ | $T_n(x)$ | $U_n(x)$ |
|---|---|---|
| 1 | $x$ | $2x$ |
| 2 | $2x^2 - 1$ | $4x^2 - 1$ |
| 3 | $4x^3 - 3x$ | $8x^3 - 4x$ |
| 4 | $8x^4 - 8x^2 + 1$ | $16x^4 - 12x^2 + 1$ |

and an induction shows that for the Chebyshev polynomials of the first kind,

$$T_m(x) = \frac{m}{2}\sum_{k=0}^{\lfloor m/2 \rfloor}(-1)^k \frac{\Gamma(m-k)}{\Gamma(k+1)\Gamma(m-2k+1)}(2x)^{m-2k}. \tag{2.1.11}$$

If

$$w(x) = \frac{1}{\pi}(1-x^2)^{-1/2}, \quad x \in (-1,1),$$

denotes the density of the arc-sine distribution on the interval $[-1, 1]$, then

$$\int_{-1}^{1} T_n(x) T_m(x) w(x) dx = \frac{1}{\pi} \int_0^{\pi} \cos(n\theta) \cos(m\theta) d\theta$$

$$= \begin{cases} 0 & \text{if } n \neq m, \\ 1 & \text{if } n = m = 0, \\ \frac{1}{2} & \text{if } n = m \geq 1, \end{cases}$$

(2.1.12)

which means that the polynomials $\{T_n(x)\}_{n\geq 0}$ are orthogonal with respect to the arc-sine measure. Similarly it can be shown that the polynomials $\{U_n(x)\}_{n\geq 0}$, $\{V_n(x)\}_{n\geq 0}$, $\{W_n(x)\}_{n\geq 0}$ are orthogonal with respect to measures with densities

$$\frac{2}{\pi}\sqrt{1 - x^2}, \quad \frac{1}{\pi}\sqrt{\frac{1 - x}{1 + x}}, \quad \frac{1}{\pi}\sqrt{\frac{1 + x}{1 - x}}, \qquad x \in (-1, 1), \qquad (2.1.13)$$

respectively.

**EXAMPLE 2.1.2**   The polynomials in the previous example are special cases of a more general system of orthogonal polynomials. These were introduced by Jacobi in (1859) and are therefore called *Jacobi polynomials*. They are orthogonal with respect to the Beta distribution on the interval $(-1, 1)$ with density

$$w_{(\alpha,\beta)}(x) = \frac{2^{-1-\alpha-\beta}}{B(\alpha + 1, \beta + 1)} (1 - x)^{\alpha}(1 + x)^{\beta} \qquad x \in (-1, 1), \qquad (2.1.14)$$

$(\alpha, \beta > -1)$. The Jacobi polynomials are explicitly given by

$$P_n^{(\alpha,\beta)}(x) = \frac{(-1)^n}{2^n n!} (1 - x)^{-\alpha}(1 + x)^{-\beta} \left(\frac{d^n}{dx^n}\right) \left[(1 - x)^{n+\alpha}(1 + x)^{n+\beta}\right]$$

$$= 2^{-n} \sum_{j=0}^{n} \binom{n + \alpha}{j} \binom{n + \beta}{n - j} (x - 1)^{n-j}(x + 1)^j \qquad (2.1.15)$$

$$= \frac{\Gamma(n + \alpha + 1)}{n! \Gamma(n + \alpha + \beta + 1)} \sum_{j=0}^{n} \binom{n}{j} \frac{\Gamma(\alpha + \beta + n + j + 1)}{\Gamma(\alpha + j + 1)} \left(\frac{x - 1}{2}\right)^j.$$

The identity between the first and the second line follows from Leibniz's rule, while the third line is obtained from the second by comparing coefficients of $(x - 1)^j$. Integrating the right-hand side of the first line $n$ times by parts yields that the Jacobi polynomials satisfy the orthogonality relation (2.1.3) for the Beta distribution on the interval $[-1, 1]$. More precisely, we have

$$\int_{-1}^{1} P_n^{(\alpha,\beta)}(x) P_m^{(\alpha,\beta)}(x) w_{(\alpha,\beta)}(x) dx$$

$$= \begin{cases} 0 & \text{if } n \neq m, \\ \dfrac{(\alpha+1)_n(\beta+1)_n(n+\alpha+\beta+1)}{n!(\alpha+\beta+2)_n(2n+\alpha+\beta+1)} & \text{if } n = m, \end{cases} \tag{2.1.16}$$

where $(a)_0 = 1$, $(a)_n = a(a+1)\ldots(a+n-1)$ denotes the *Pochhammer symbol*. The first line in (2.1.15) is called a *Rodrigues type formula*, and it was first observed by Rodrigues in (1816) for $\alpha = \beta = 0$. Note that for $\alpha, \beta \in \{-\frac{1}{2}, \frac{1}{2}\}$ (2.1.14) gives the arc-sine density and the densities in (2.1.13), which means that the corresponding orthogonal polynomials differ only by a constant of proportionality. Comparing the leading coefficients yields

$$T_n(x) = \kappa_n P_n^{(-1/2,-1/2)}(x), \quad U_n(x) = \frac{\kappa_{n+1}}{2} P_n^{(1/2,1/2)}(x),$$

$$\tag{2.1.17}$$

$$V_n(x) = \kappa_n P_n^{(1/2,-1/2)}(x), \quad W_n(x) = \kappa_n P_n^{(-1/2,1/2)}(x),$$

where $\kappa_0 = 1$,

$$\kappa_n = \frac{2 \cdot 4 \cdots 2n}{1 \cdot 3 \cdots (2n-1)}, \quad n = 1, 2, 3, \ldots.$$

For historical reasons the Jacobi polynomials for $\alpha = \beta$ are normalized differently:

$$C_n^{(\lambda)}(x) = \begin{cases} \dfrac{\Gamma(\lambda+\frac{1}{2})\Gamma(2\lambda+n)}{\Gamma(2\lambda)\Gamma(\lambda+n+\frac{1}{2})} P_n^{(\lambda-1/2,\lambda-1/2)}(x), & \lambda \neq 0, \lambda > -\dfrac{1}{2}, \\ \dfrac{2}{n} \kappa_n P_n^{(-1/2,-1/2)}(x), & \lambda = 0, \end{cases} \tag{2.1.18}$$

and are called *ultraspherical* or *Gegenbauer polynomials*. Note that by (2.1.17) and (2.1.18),

$$U_n(x) = C_n^{(1)}(x),$$

$$\tag{2.1.19}$$

$$T_n(x) = \frac{n}{2} C_n^{(0)}(x).$$

Finally, in the case $\alpha = \beta = 0$ $(\lambda = \frac{1}{2})$ $w_{(0,0)}(x) = \frac{1}{2} I_{(-1,1)}(x)$ is the density of the uniform distribution, and the corresponding orthogonal polynomials are called *Legendre polynomials* because they were first investigated by Legendre in 1785:

$$P_n(x) = P_n^{(0,0)}(x) = C_n^{(1/2)}(x).$$
(2.1.20)

We will now mention some general properties of orthogonal polynomials. Results that are relevant for the development of the theory of canonical moments will be proved in the subsequent sections for measures on the interval $[0,1]$. For the remaining properties we refer to the literature (see Szegö, 1975; Freud, 1971; Chihara, 1978).

**Lemma 2.1.3** *The zeros of orthogonal polynomials are real and simple and belong to the interval $(a,b)$, where $a$ and $b$ are, respectively, the infimum and supremum of* supp$(\mu)$. *If*

$$a < x_1^{(n)} < x_2^{(n)} < \ldots < x_n^{(n)} < b$$

*denote the zeros of $P_n(x)$, then $x_j^{(n+1)} < x_j^{(n)} < x_{j+1}^{(n+1)}$ $(j = 1, \ldots, n)$.*

**Lemma 2.1.4** *A sequence of polynomials $\{P_k(x)\}_{k \geq 0}$ is orthogonal with respect to a measure $\mu$ on the real line if and only if it satisfies a three-term recurrence relation with real coefficients*

$$P_{n+1}(x) = (A_n x + B_n)P_n(x) - C_n P_{n-1}(x), \qquad n = 0, 1, 2, \ldots,$$
(2.1.21)

*where $P_0(x)$ is a nonzero constant, $P_{-1}(x) = 0$, and the positivity condition $A_{n-1}A_n C_n > 0$ holds for $n = 1, 2, \ldots$.*

**EXAMPLE 2.1.5** It will be shown in Example 2.3.10 that for $\lambda \neq 0$ the three-term recurrence relation for the ultraspherical polynomials is given by

$$C_{-1}^{(\lambda)}(x) = 0, \quad C_0^{(\lambda)}(x) = 1,$$
(2.1.22)
$$(n+1)C_{n+1}^{(\lambda)}(x) = 2(n+\lambda)xC_n^{(\lambda)}(x) - (n+2\lambda-1)C_{n-1}^{(\lambda)}(x).$$

Thus the ultraspherical polynomials up to degree 3 are

$$C_1^{(\lambda)}(x) = 2\lambda x,$$

$$C_2^{(\lambda)}(x) = \lambda[2(\lambda+1)x^2 - 1],$$

$$C_3^{(\lambda)}(x) = \lambda(\lambda+1)\left[\frac{4}{3}(\lambda+2)x^3 - 2x\right],$$

while the polynomial of degree $n$ can be expressed as

$$C_n^{(\lambda)}(x) = \sum_{m=0}^{\lfloor n/2 \rfloor} (-1)^m \frac{\Gamma(n-m+\lambda)}{\Gamma(\lambda)\Gamma(m+1)\Gamma(n-2m+1)} (2x)^{n-2m}. \qquad (2.1.23)$$

This representation is derived in Szegö (1975, p. 84) and can easily be seen to be equivalent to (2.1.22). A straightforward application of (2.1.23) shows that the derivatives of ultraspherical polynomials are again ultraspherical polynomials with a different parameter, i.e. for $\lambda \neq 0$,

$$\frac{d}{dx} C_n^{(\lambda)}(x) = 2\lambda C_{n-1}^{(\lambda+1)}(x), \quad n = 1, 2, 3, \dots. \qquad (2.1.24)$$

An alternative proof of this relation can be obtained from Rodrigues's formula in (2.1.15) observing (2.1.18), which also shows that (2.1.24) can be generalized to the Jacobi polynomials.

Lemma 2.1.4 is usually called *Favard's theorem*. It was first announced by Favard (1935) and discovered about the same time independently by Shohat and I. Natanson. A special case of this result is proved in Theorem 8.2.4 of Chapter 8 using the theory of canonical moments. Lemma 2.1.4 shows that the recurrence relation (2.1.21) implies the existence of a finite measure $\mu$ with respect to which the polynomials are orthogonal. This measure is not necessarily unique. It will be unique if and only if the *Hamburger moment problem* associated with this measure is determined. This means that there exists no other measure with the same moments as $\mu$. A sufficient condition for uniqueness is *Carleman's condition*

$$\sum_{n=0}^{\infty} c_{2n}^{-1/2n} = \infty. \qquad (2.1.25)$$

Note that (2.1.25) implies that the Hausdorff moment problem on a compact interval is determined, i.e. on a compact interval, say, $[a, b]$, a probability measure is always determined by its moments.


## 2.2  SUPPORT POLYNOMIALS

The remaining part of this chapter considers probability measures on the interval $[0, 1]$. In places where $[-1, 1]$ or more general intervals are of interest, this will be mentioned explicitly. In Chapter 1 four sequences of Hankel determinants, $\underline{H}_{2m}$, $\underline{H}_{2m+1}$, $\overline{H}_{2m}$ and $\overline{H}_{2m+1}$ were defined [see (1.4.3)]. The subscripts in each case determined the highest moment $c_k$ that was involved. These determinants were used to characterize the moment spaces and provide bounds on the $(n+1)$th moment $c_{n+1}$ for a given sequence $c_1, \ldots, c_n$ of moments up to the order $n$. For example, if $n = 2m - 1$, then the inequality $\underline{H}_{2m} \geq 0$ provides a lower bound on $c_{2m}$.

There are correspondingly four sets of polynomials that play an important role in the ensuing discussion. The polynomial $\underline{H}_{2m-1}(x)$ was defined in Equation (2.1.4). The other three polynomials are given by $\underline{H}_0(x) = \overline{H}_0(x) = 1$ and for $m \geq 1$ by

$$
\underline{H}_{2m}(x) = \begin{vmatrix} c_1 & \cdots & c_m & 1 \\ \vdots & & \vdots & \vdots \\ c_{m+1} & \cdots & c_{2m} & x^m \end{vmatrix}, \tag{2.2.1}
$$

$$
\overline{H}_{2m}(x) = \begin{vmatrix} c_0 - c_1 & \cdots & c_{m-1} - c_m & 1 \\ \vdots & & \vdots & \vdots \\ c_m - c_{m+1} & \cdots & c_{2m-1} - c_{2m} & x^m \end{vmatrix},
$$

$$
\overline{H}_{2m+1}(x) = \begin{vmatrix} c_1 - c_2 & \cdots & c_m - c_{m+1} & 1 \\ \vdots & & \vdots & \vdots \\ c_{m+1} - c_{m+2} & \cdots & c_{2m} - c_{2m+1} & x^m \end{vmatrix}.
$$

Throughout the text it will usually be more convenient to normalize the polynomials so that their leading coefficient is equal to one. We therefore define four systems of monic (orthogonal) polynomials, $\underline{P}_m(x)$, $\underline{Q}_m(x)$, $\overline{P}_m(x)$, and $\overline{Q}_m(x)$ for $m \geq 0$. These are given by

$$
\underline{P}_0(x) = \underline{Q}_0(x) = \overline{P}_0(x) = \overline{Q}_0(x) = 1,
$$

and for $m \geq 1$,

$$
\underline{P}_m(x) = \frac{\underline{H}_{2m-1}(x)}{\underline{H}_{2m-2}},
$$

$$
\underline{Q}_m(x) = \frac{\underline{H}_{2m}(x)}{\underline{H}_{2m-1}},
$$

$$
\overline{P}_m(x) = \frac{\overline{H}_{2m}(x)}{\overline{H}_{2m-1}}, \tag{2.2.2}
$$

$$
\overline{Q}_m(x) = \frac{\overline{H}_{2m+1}(x)}{\overline{H}_{2m}}.
$$

We will always assume, when using any of these expressions, that the determinant in the denominator is nonzero and that the polynomial is of exact degree $m$.

Note also that if, for example, $\underline{H}_{2m-2} > 0$, then $\underline{H}_{2k} > 0$ for $k = 0, \ldots, m-1$. Consequently $\underline{P}_k(x)$ is well defined and of exact degree $k$ for $k = 0, 1, \ldots, m$, and by the discussion in Section 2.1, the polynomials $\underline{P}_k(x)$, $k = 0, \ldots, m$ are orthogonal with respect to the measure $\mu$.

If the moment sequence is associated with respect to a given probability measure $\mu$ on the interval $[0, 1]$, then the polynomials in (2.2.2) are given the following association:

$$
\begin{array}{cc}
\underline{P}_m & d\mu \\
\underline{Q}_m & x d\mu \\
\overline{P}_m & (1-x)d\mu \\
\overline{Q}_m & x(1-x)d\mu
\end{array}
\qquad (2.2.3)
$$

**Theorem 2.2.1** *The sequences of polynomials defined by (2.2.2) are orthogonal with respect to the associated measure in (2.2.3) whenever the polynomials are defined.*

*Proof* The proof for the sequence $\{\underline{P}_m(x)\}_{m\geq0}$ was given preceding the statement of the theorem (see also the discussion on page 43). The other cases follow in exactly the same manner. ∎

The following theorem gives the $L_2$-norm of the polynomials with respect to the corresponding measures:

**Theorem 2.2.2** *The $L_2$-norms of the monic orthogonal polynomials in (2.2.2) with respect to the corresponding measure in (2.2.3) are given by*

$$
\int_0^1 \underline{P}_m^2(x)d\mu(x) = \frac{\underline{H}_{2m}}{\underline{H}_{2m-2}} = \zeta_1 \cdots \zeta_{2m},
$$

$$
\int_0^1 \underline{Q}_m^2(x)x d\mu(x) = \frac{\underline{H}_{2m+1}}{\underline{H}_{2m-1}} = \zeta_1 \cdots \zeta_{2m+1},
$$

$$
\int_0^1 \overline{P}_m^2(x)(1-x)d\mu(x) = \frac{\overline{H}_{2m+1}}{\overline{H}_{2m-1}} = \gamma_1 \cdots \gamma_{2m+1},
$$

$$
\int_0^1 \overline{Q}_m^2(x)x(1-x)d\mu(x) = \frac{\overline{H}_{2m+2}}{\overline{H}_{2m}} = \gamma_1 \cdots \gamma_{2m+2},
$$

*whenever these polynomials are well defined.*

*Proof* The value of the ratio of the Hankel determinants in each case is expressed with the aid of Theorem 1.4.10. The first equality in each case, follows from (2.2.1) and (2.2.2), for example, by writing

$$\int_0^1 \underline{P}_m^2(x)d\mu(x) = \int_0^1 x^m \underline{P}_m(x)d\mu(x) = \underline{H}_{2m}/\underline{H}_{2m-2}. \qquad \blacksquare$$

With the aid of these expressions, we can immediately show that the roots or support points of sequences of the form $(p_1, \ldots, p_n, 0)$ and $(p_1, \ldots, p_n, 1)$ are, except for end points, the zeros of these polynomials. When an expression such as $(p_1, \ldots, p_n, 0)$ or $(p_1, \ldots, p_n, 1)$ is used, it will always be assumed that the corresponding moment sequence $(c_1, \ldots, c_n)$ is an interior point of the $n$th moment space $M_n$. By the discussion in Section 1.3 this is equivalent to the assumption $0 < p_j < 1$ for all $j = 1, \ldots, n$.

**Theorem 2.2.3**

1. *The measure corresponding to the sequence $(p_1, \ldots, p_{2m-1}, 0)$ is supported by the m distinct zeros of the polynomial $\underline{P}_m(x)$.*
2. *The measure corresponding to the sequence $(p_1, \ldots, p_{2m}, 0)$ is supported by the $m + 1$ distinct zeros of the polynomial $x\underline{Q}_m(x)$.*
3. *The measure corresponding to the sequence $(p_1, \ldots, p_{2m}, 1)$ is supported by the $m + 1$ distinct zeros of the polynomial $(1 - x)\overline{P}_m(x)$.*
4. *The measure corresponding to the sequence $(p_1, \ldots, p_{2m-1}, 1)$ is supported by the $m + 1$ distinct zeros of the polynomial $x(1 - x)\overline{Q}_{m-1}(x)$.*

*Proof*   The proofs are all similar, and we consider only case 3 as a representative example. Since $p_{2m+1} = 1$ or equivalently $q_{2m+1} = 0$, we have $\gamma_{2m+1} = p_{2m}q_{2m+1} = 0$, and the extreme right expression in the third equation of Theorem 2.2.2 is zero. In this case

$$\int_0^1 (1 - x)\overline{P}_m^2(x)d\mu(x) = 0,$$

and the support of the measure $\mu$ corresponding to $(p_1, \ldots, p_{2m}, 1)$ must be contained in the set of zeros of the polynomial $(1 - x)\overline{P}_m(x)$. We have assumed that $(c_1, \ldots, c_{2m})$ is in Int $M_{2m}$, and from $p_{2m+1} = 1$ it follows that the next moment is $c_{2m+1}^+$, which gives $\overline{H}_{2m+1} = 0$. The measure $\mu$ is correspondingly the upper principal representation of $(c_1, \ldots, c_{2m})$ and is supported on $x = 1$ and $m$ distinct interior points.                                                                  $\blacksquare$

The discussion in the proof of Theorem 2.2.3 also leads to the fact that the integrands in Theorem 2.2.2 must be the *supporting polynomials* to the $n$th moment space $M_n$ at the various boundary points. We will say that a polynomial $P(x) = \sum_{i=0}^n a_i x^i$ determines or corresponds to a supporting hyperplane at the point $c_n = (c_1, \ldots, c_n) \in \partial M_n$ if it is nonnegative on the interval $[0, 1]$, i.e.

$P \in \mathcal{P}_{n+1}$ and

$$a_0 + \sum_{i=1}^{n} a_i c_i = 0.$$

In this case the coefficients in the polynomial determine the supporting hyperplane, and we call $P$ a supporting polynomial to $M_n$ at the boundary point $c_n \in \partial M_n$. For example, continuing with case 3 in the proof of Theorem 2.2.3, we determine $(a_0, a_1, \ldots, a_{2m+1})$ by

$$P(x) = \sum_{i=0}^{2m+1} a_i x^i = (1-x)\overline{P}_m^2(x) \geq 0.$$

If $\mu$ has moments $c_1, \ldots, c_{2m+1}$ such that $c_{2m} \in \operatorname{Int} M_{2m}$ and $c_{2m+1} \in \partial M_{2m+1}$, corresponding to $(p_1, \ldots, p_{2m}, 1)$, then

$$\int_0^1 P(x)d\mu(x) = a_0 + \sum_{i=1}^{2m+1} a_i c_i = 0,$$

and $(1-x)\overline{P}_{2m}^2(x)$ determines the hyperplane supporting $M_{2m+1}$ at the boundary point $(c_1, \ldots, c_{2m+1})$. Note also that the supporting polynomial of degree $2m+1$ is uniquely defined up to a multiplicative constant. For example, in case 3, $(p_1, \ldots, p_{2m})$ was tacitly assumed to be in $(0,1)^{2m}$, and hence $(p_1, \ldots, p_{2m}, 1)$ is supported on $x = 1$ and $m$ distinct interior points. Consequently every nonnegative polynomial of degree $2m+1$ with these roots is a supporting polynomial to $M_{2m+1}$ at $(c_1, \ldots, c_{2m+1})$. There are of course numerous polynomials of higher degree vanishing on these points.

**Corollary 2.2.4** *The boundary points and corresponding supporting hyperplanes or polynomials are given by*

| | |
|---|---|
| $(p_1, \ldots, p_{2m-1}, 0)$ | $\underline{P}_m^2(x)$ |
| $(p_1, \ldots, p_{2m}, 0)$ | $x\underline{Q}_m^2(x)$ |
| $(p_1, \ldots, p_{2m}, 1)$ | $(1-x)\overline{P}_m^2(x)$ |
| $(p_1, \ldots, p_{2m+1}, 1)$ | $x(1-x)\overline{Q}_m^2(x)$ |

*The polynomials of minimal degree are uniquely defined up to a multiplicative constant.*

Theorem 2.2.3 contains also some useful information regarding the strict interlacing of the roots of successive polynomials in (2.2.2). By this we mean, for example, that if $t_1^{(m)} < \cdots < t_m^{(m)}$ and $t_1^{(m+1)} < \cdots < t_{m+1}^{(m+1)}$ are the roots of

$\underline{P}_m(x)$ and $\underline{P}_{m+1}(x)$, respectively, then

$$t_j^{(m+1)} < t_j^{(m)} < t_{j+1}^{(m+1)}, \qquad j = 1, \ldots, m. \qquad (2.2.4)$$

Note that Theorem 2.2.3 implies that $\underline{P}_m$, $\underline{Q}_m$, $\overline{P}_m$, $\overline{Q}_m$ all have $m$ distinct simple roots in $(0, 1)$. Results of this type are a consequence of the following lemma:

**Lemma 2.2.5** *Let $\sigma$ denote a principal representation of a moment point $(c_1, \ldots, c_n) \in \text{Int } M_n$. Let $t_j < t_{j+1}$ denote either two adjacent points in the support of $\sigma$ or an end point of $[0, 1]$ and the nearest point in the support of $\sigma$. If $\mu$ is any other measure of finite support with the same moments $(c_1, \ldots, c_n)$, then $\mu$ has a support point in $(t_j, t_{j+1})$.*

*Proof*   A careful check of various cases reveals that there exists a polynomial $P$ of degree at most $n$ which vanishes precisely on the support of $\sigma$ and is negative on $(t_j, t_{j+1})$ and nonnegative on $[0, 1] \setminus (t_j, t_{j+1})$. Since the measures $\sigma$ and $\mu$ both have moments $c_0, c_1, \ldots, c_n$, it follows that

$$0 = \int_0^1 P(t)d\mu(t) - \int_0^1 P(t)d\sigma(t) = \int_0^1 P(t)d\mu(t). \qquad (2.2.5)$$

However, if $\mu$ has no support points in $(t_j, t_{j+1})$, then

$$\int_0^1 P(t)d\mu(t) \geq 0,$$

and (2.2.5) can hold only if the support of $\mu$ is contained in the support of $\sigma$. Because $\sigma$ is a principal representation of $(c_1, \ldots, c_n) \in \text{Int } M_n$, Theorem 1.2.9 implies that $\mu \equiv \sigma$, contradicting our assumption that $\mu$ and $\sigma$ are different measures. ∎

**Theorem 2.2.6**   *The roots of the polynomials $\underline{P}_m(x)$ and $\underline{P}_{m+1}(x)$ strictly interlace. Similarly the roots of successive polynomials from $\underline{Q}_m(x)$, $\overline{P}_m(x)$, and $\overline{Q}_m(x)$ strictly interlace.*

*Proof*   If the polynomials $\underline{P}_m(x)$ and $\underline{P}_{m+1}(x)$ are orthogonal with respect to some measure $\sigma$ with moments $c_1, c_2, \ldots$, then by Theorem 2.2.3 the roots of $\underline{P}_m(x)$ are the support points of the lower principal representation $\sigma_{2m-1}^-$ of $(c_1, c_2, \ldots, c_{2m-1}) \in \text{Int } M_{2m-1}$. Similarly the zeros of $\underline{P}_{m+1}(x)$ are the support points of $\sigma_{2m+1}^-$ with the same moments $c_i$, $i = 1, \ldots, 2m - 1$. By Lemma 2.2.5, $\sigma_{2m+1}^-$ has a support point between every pair of support points of $\sigma_{2m-1}^-$ and between each end point and the nearest support point of $\sigma_{2m-1}^-$. Thus the roots of $\underline{P}_m(x)$ and $\underline{P}_{m+1}(x)$ strictly interlace. The other cases are similar. ∎

**REMARK 2.2.7**   Note that Lemma 2.2.5 can also be used to show that any pair
of the polynomials

$$\underline{P}_m(x), \quad x\underline{Q}_m(x), \quad (1-x)\overline{P}_m(x), \quad x(1-x)\overline{Q}_m,$$

whose degrees are the same or differ by one will strictly interlace. By this we
mean, for example, $\underline{P}_m(x)$ and $x\underline{Q}_m(x)$ or $x\underline{Q}_m(x)$ and $(1-x)\overline{P}_m(x)$. The proof
is the same as that illustrated in Theorem 2.2.6.


## 2.3   RECURSIVE FORMULA

The polynomials $\underline{H}_n(x), \overline{H}_n(x)$ defined in (2.1.4) and (2.2.1) and the corresponding normalized polynomials defined in (2.2.2) are expressed in terms of the
original ordinary moments $c_1, c_2, \ldots$ . In this section a number of recursive
formulas are derived expressing these polynomials in terms of the canonical
moments.

In general, any system of orthogonal polynomials satisfies a three-term
recursive formula. Suppose that $\{P_k(x)\}_{k\geq 0}$ is a system of polynomials orthogonal with respect to some measure $\mu$ and that $P_k(x)$ is of exact degree $k$ for each
$k$. Since $xP_m(x)$ is a polynomial of degree $m+1$, we may write

$$xP_m(x) = \sum_{i=0}^{m+1} a_{m,i} P_i(x). \tag{2.3.1}$$

If both sides are multiplied by $P_k$ and integrated with respect to the measure $\mu$,
we find that

$$\int_0^1 xP_k(x)P_m(x)d\mu = a_{m,k} \int_0^1 P_k^2(x)d\mu(x). \tag{2.3.2}$$

The left side is zero if $xP_k(x)$ has degree at most $m-1$, and hence $a_{m,k} = 0$ for
$k \leq m-2$, provided that $\int_0^1 P_k^2(x)d\mu(x) > 0$. In this case (2.3.1) reduces to

$$xP_m(x) = a_{m,m+1}P_{m+1}(x) + a_{m,m}P_m(x) + a_{m,m-1}P_{m-1}(x), \tag{2.3.3}$$

and $P_{m+1}$ can be recursively defined in terms of $P_m$ and $P_{m-1}$. The coefficients
$a_{m,k}$ in (2.3.3) can be given in terms of integrals involving $P_k$ and then expressed
in terms of canonical moments.

These recursive formula will be derived below in a more basic manner using
Sylvester's identity in Lemma 1.5.2. With the aid of Sylvester's identity one can
prove the recursive formula for orthogonal polynomials without resorting to any
orthogonality properties. These results will be given in Theorem 2.3.1 and its
corollaries. Note that Corollary 2.3.2 coincides with Theorem 1.4.5 which was
stated in Section 1.4 without a proof.

**Theorem 2.3.1** *The polynomials defined in (2.1.4) and (2.2.1) satisfy $\underline{H}_0(x) = 1$, $\underline{H}_1(x) = x - \zeta_1$ and the following relations:*

$$\underline{H}_{2m-1}\underline{H}_{2m+1}(x) = x\underline{H}_{2m}\underline{H}_{2m}(x) - \underline{H}_{2m+1}\underline{H}_{2m-1}(x), \qquad m \geq 1,$$

$$\underline{H}_{2m}\underline{H}_{2m+2}(x) = \underline{H}_{2m+1}\underline{H}_{2m+1}(x) - \underline{H}_{2m+2}\underline{H}_{2m}(x), \qquad m \geq 0.$$

*Proof* To obtain the first expression, Sylvester's identity is applied to the polynomial $\underline{H}_{2m+1}(x)$ defined in (2.1.4) using the pivot block obtained by omitting the first and last row and the last two columns. For the second expression the case $m = 0$ can be verified directly. For $m \geq 1$ we first write

$$\underline{H}_{2m+2}(x) = \begin{vmatrix} 1 & 0 & \cdots & 0 & 0 \\ c_0 & c_1 & \cdots & c_{m+1} & 1 \\ \vdots & \vdots & & \vdots & \vdots \\ c_{m+1} & c_{m+2} & \cdots & c_{2m+2} & x^{m+1} \end{vmatrix}.$$

Then we proceed as before, obtaining the pivot block by omitting the first and last rows and the last two columns. ∎

**Corollary 2.3.2**

$$\underline{H}_n\overline{H}_n = \underline{H}_{n-1}\overline{H}_{n+1} + \overline{H}_{n-1}\underline{H}_{n+1}, \qquad n \geq 1.$$

*Proof* The result follows from Theorem 2.3.1 by inserting $x = 1$. The expressions for $\underline{H}_n(1)$ are evaluated by replacing the first row by the first row minus the second, then the second row by the second minus the third, and so on. The determinant is then expanded by the last column to give $\underline{H}_n(1) = \overline{H}_n$. ∎

The expressions in Theorem 2.3.1 can now be normalized using equations (2.2.2) and the expressions for $\zeta_i$ in Corollary 1.4.7. These give rise to a pair of recursive equations for the monic orthogonal polynomials $\underline{P}_k(x)$ and $\underline{Q}_k(x)$.

**Corollary 2.3.3** *The polynomials defined in (2.2.2) satisfy $\underline{P}_0(x) = \underline{Q}_0(x) = \overline{P}_0(x) = \overline{Q}_0(x) = 1$ and for $m \geq 0$,*

1.         $$\underline{P}_{m+1}(x) = x\underline{Q}_m(x) - \zeta_{2m+1}\underline{P}_m(x),$$

            $$\underline{Q}_{m+1}(x) = \underline{P}_{m+1}(x) - \zeta_{2m+2}\underline{Q}_m(x);$$

2.         $$\overline{P}_{m+1}(x) = x\overline{Q}_m(x) - \gamma_{2m+2}\overline{P}_m(x),$$

            $$\overline{Q}_{m+1}(x) = \overline{P}_{m+1}(x) - \gamma_{2m+3}\overline{Q}_m(x).$$

*Proof* The first two expressions follow from the definition (2.2.2) by normal-izing Theorem 2.3.1 and an application of Corollary 1.4.7. The second pair follows from the first two by starting with the sequence $c_0 - c_1, c_1 - c_2, \ldots$ in place of $c_0, c_1, c_2, \ldots$. By Corollary 1.4.8 the definition of $\zeta_i$ changes over to $\gamma_{i+1}$, and by (2.2.1), (2.2.2), the polynomials $\underline{P}_i$ and $\underline{Q}_i$ change to $\overline{P}_i$ and $\overline{Q}_i$. ∎

From Corollary 2.3.3 recursive relations involving only one kind of polynomial can be derived.

**Corollary 2.3.4** *The individual sequences of monic orthogonal polynomials* $\{\underline{P}_i(x)\}_{i \geq 0}$, $\{\underline{Q}_i(x)\}_{i \geq 0}$, $\{\overline{P}_i(x)\}_{i \geq 0}$, $\{\overline{Q}_i(x)\}_{i \geq 0}$ *orthogonal with respect to the respective measures* $d\mu$, $xd\mu$, $(1 - x)d\mu$, *and* $x(1 - x)d\mu$ *given in* (2.2.2) *satisfy* $(\underline{Q}_{-1}(x) = \overline{Q}_{-1}(x) = 0, \underline{P}_0(x) = \underline{Q}_0(x) = \overline{P}_0(x) = \overline{Q}_0(x) = 1)$

1.  $\underline{P}_1(x) = x - \zeta_1$, *and for* $m \geq 1$,

    $$\underline{P}_{m+1}(x) = (x - \zeta_{2m} - \zeta_{2m+1})\underline{P}_m(x) - \zeta_{2m-1}\zeta_{2m}\underline{P}_{m-1}(x);$$

2.  *for* $m \geq 0$,

    $$\underline{Q}_{m+1}(x) = (x - \zeta_{2m+1} - \zeta_{2m+2})\underline{Q}_m(x) - \zeta_{2m}\zeta_{2m+1}\underline{Q}_{m-1}(x);$$

3.  $\overline{P}_1(x) = x - \gamma_2$, *and for* $m \geq 1$,

    $$\overline{P}_{m+1}(x) = (x - \gamma_{2m+1} - \gamma_{2m+2})\overline{P}_m(x) - \gamma_{2m}\gamma_{2m+1}\overline{P}_{m-1}(x);$$

4.  *for* $m \geq 0$,

    $$\overline{Q}_{m+1}(x) = (x - \gamma_{2m+2} - \gamma_{2m+3})\overline{Q}_m(x) - \gamma_{2m+1}\gamma_{2m+2}\overline{Q}_{m-1}(x).$$

*Proof* These equations are obtained by separating out the two sets of poly-nomials in parts 1 and 2 of Corollary 2.3.3. The proof is indicated only for the first pair. The representation of $\underline{P}_1(x)$ is obvious. From the first expression in Corollary 2.3.3 we have

$$\underline{Q}_m(x) = x^{-1}(\underline{P}_{m+1}(x) + \zeta_{2m+1}\underline{P}_m(x)).$$

This is substituted for $m$ and $m - 1$ into the second equation written as

$$\underline{P}_m(x) = \underline{Q}_m(x) + \zeta_{2m}\underline{Q}_{m-1}(x).$$

A slight rearrangement of the terms gives part 1 of the present Corollary 2.3.4. ∎

In later chapters it will also be convenient to have the recursion formula for the corresponding polynomials transferred to the interval $[-1, 1]$. The (monic) polynomials corresponding to $\underline{P}_m$, $\underline{Q}_m$, $\overline{P}_m$, and $\overline{Q}_m$ on the interval $[-1, 1]$ will be denoted by $\underline{R}_m$, $\underline{S}_m$, $\overline{R}_m$, and $\overline{S}_m$. If $\mu$ denotes the measure associated with the systems on $[0, 1]$, this is transformed linearly to $[-1, 1]$ and denoted by $\nu$. The systems of polynomials on $[-1, 1]$ are orthogonal with respect to the following measures

$$
\begin{array}{ll}
\underline{R}_m & d\nu \\
\underline{S}_m & (1 + y)d\nu \\
\overline{R}_m & (1 - y)d\nu \\
\overline{S}_m & (1 - y^2)d\nu
\end{array}
\tag{2.3.4}
$$

The recursion formula can be obtained from Corollary 2.3.4. Only the first case is considered. If $0 < x < 1$ is transformed to $[-1, 1]$ by $y = 2x - 1$, we substitute $x = (y + 1)/2$ in the equations for $\underline{P}_m(x)$ and use the fact that

$$
2^{-m}\underline{R}_m(y) = \underline{P}_m\left(\frac{y + 1}{2}\right)
\tag{2.3.5}
$$

(note that $\underline{R}_m(x)$ is assumed to have leading coefficient 1). The resulting expression gives the recursion formula for the monic orthogonal polynomials $\{\underline{R}_m(x)\}_{m \geq 0}$ with respect to the measure $d\nu$.

**Corollary 2.3.5**  *The monic orthogonal polynomials on the interval $[-1, 1]$ in display (2.3.4) satisfy the following recursion formula ($\underline{S}_{-1}(y) = \overline{S}_{-1}(y) = 0$, $\underline{S}_0(y) = \overline{S}_0(y) = \underline{R}_0(y) = \overline{R}_0(y) = 1$):*

1.
$$
\underline{R}_1(y) = y + 1 - 2\zeta_1,
$$

$$
\underline{R}_{m+1}(y) = (y + 1 - 2\zeta_{2m} - 2\zeta_{2m+1})\underline{R}_m(y)
$$
$$
- 4\zeta_{2m-1}\zeta_{2m}\underline{R}_{m-1}(y), \qquad m \geq 1;
$$

2.
$$
\underline{S}_{m+1}(y) = (y + 1 - 2\zeta_{2m+1} - 2\zeta_{2m+2})\underline{S}_m(y)
$$
$$
- 4\zeta_{2m}\zeta_{2m+1}\underline{S}_{m-1}(y), \qquad m \geq 0;
$$

3.
$$
\overline{R}_1(y) = y + 1 - 2\gamma_2,
$$

$$
\overline{R}_{m+1}(y) = (y + 1 - 2\gamma_{2m+1} - 2\gamma_{2m+2})\overline{R}_m(y)
$$
$$
- 4\gamma_{2m}\gamma_{2m+1}\overline{R}_{m-1}(y), \qquad m \geq 1;
$$

4.
$$
\overline{S}_{m+1}(y) = (y + 1 - 2\gamma_{2m+2} - 2\gamma_{2m+3})\overline{S}_m(y)
$$
$$
- 4\gamma_{2m+1}\gamma_{2m+2}\overline{S}_{m-1}(y), \qquad m \geq 0.
$$

If the measure $\nu$ is symmetric about zero, then by Corollary 1.3.4, $p_{2i+1} = \frac{1}{2}$, $i \geq 0$. The polynomials $\underline{R}_m$ and $\overline{S}_m$ orthogonal with respect to $d\nu$ and $(1 - y^2)d\nu$ are even or odd functions according as $m$ is even or odd. The corresponding recursion equations are particularly simple.

**Corollary 2.3.6** *If $\nu$ is symmetric then the monic orthogonal polynomials $\underline{R}_i$ and $\overline{S}_i$ on the interval $[-1, 1]$ with respect to the measures $\nu$ and $(1 - y^2)d\nu$ satisfy the recursive relations*

1.
$$\underline{R}_0(y) = 1, \underline{R}_1(y) = y,$$
$$\underline{R}_{m+1}(y) = y\underline{R}_m(y) - q_{2m-2}p_{2m}\underline{R}_{m-1}(y), \qquad m \geq 1;$$

2.
$$\overline{S}_0(y) = 1, \overline{S}_1(y) = y,$$
$$\overline{S}_{m+1}(y) = y\overline{S}_m(y) - p_{2m}q_{2m+2}\overline{S}_{m-1}(y), \qquad m \geq 1.$$

**REMARK 2.3.7** In later chapters it will be useful to know the $L^2$-norm of the polynomials $\overline{R}_j(x)$, $\underline{R}_j(x)$, $\underline{S}_j(x)$, $\overline{S}_j(x)$, which can easily be obtained from Theorem 2.2.2. For example, if $\mu$ and $\nu$ are related by the linear transformation $y = 2x - 1$, then it follows from (2.3.5) that

$$\frac{H_{2m}}{H_{2m-2}} = \int_{-1}^{1} \underline{R}_m^2(y)d\nu(y) = 2^{2m}\int_0^1 \underline{P}_m^2(x)d\mu(x) = 2^{2m}\prod_{j=1}^{m}\zeta_{2j-1}\zeta_{2j}.$$

Similarly we have ($\zeta_0 = \gamma_0 = 1$),

$$\int_{-1}^{1} \underline{S}_m^2(y)(1 + y)d\nu(y) = 2^{2m+1}\prod_{j=0}^{m}\zeta_{2j}\zeta_{2j+1},$$

$$\int_{-1}^{1} \overline{R}_m^2(y)(1 - y)d\nu(y) = 2^{2m+1}\prod_{j=0}^{m}\gamma_{2j}\gamma_{2j+1},$$

$$\int_{-1}^{1} \overline{S}_m^2(y)(1 - y^2)d\nu(y) = 2^{2m+2}\prod_{j=1}^{m+1}\gamma_{2j-1}\gamma_{2j},$$

and the $L^2$-norm of the monic orthogonal polynomials with respect to a measure on a general interval $[a, b]$ is obtained in the same way.

**EXAMPLE 2.3.8** *Krawtchouk Polynomials.* If the orthogonal polynomials with respect to a given distribution are known, the corresponding recursive relations for the monic orthogonal polynomials can be used to identify the canonical moments of the corresponding measure of orthogonality. In order to illustrate this method, consider as a first example the Binomial distribution

$$b(x; N, p) = \binom{N}{x} p^x (1-p)^{N-x}, \qquad x = 0, \ldots, N \qquad (2.3.6)$$

where $0 < p < 1$, $N \in \mathbb{N}_0$. The corresponding canonical moments were first determined by Skibinsky (1969) using the inverse map described in Section 2.4. The orthogonal polynomials with respect to the Binomial distribution are the *Krawtchouk polynomials* (see Krawtchouk, 1929)

$$k_n(x) = k_n(x, p, N) = \sum_{k=0}^{n} \frac{(-n)_k}{k!} \frac{(-x)_k}{(-N)_k} \left(\frac{1}{p}\right)^k.$$

These polynomials can be defined up to degree $N$ and satisfy the recursive relation (see Karlin and McGregor, 1961) $k_0(x) = 1$, $k_1(x) = 1 - x/(Np)$:

$$(N-n)pk_{n+1}(x) = [-x + nq + (N-n)p]k_n(x) - nqk_{n-1}(x) \qquad (2.3.7)$$

$(n = 1, \ldots, N-1)$. Now consider the corresponding monic orthogonal polynomials:

$$\tilde{k}_n(x) = \tilde{k}_n(x, p, N) = p^n(-N)_n k_n(x, p, N), \qquad n = 0, 1, \ldots, N.$$

Equation (2.3.7) then yields $\tilde{k}_0(x) = 1$, $\tilde{k}_1(x) = x - Np$, and

$$\tilde{k}_{n+1}(x) = [x - nq - (N-n)p]\tilde{k}_n(x) - (N-n+1)npq\tilde{k}_{n-1}(x) \qquad (2.3.8)$$

$(n = 1, \ldots, N-1)$. Finally we define

$$\underline{P}_n(x) = \left(\frac{1}{N}\right)^n \tilde{k}_n(Nx, p, N), \qquad n = 0, \ldots, N,$$

which gives a system of monic polynomials orthogonal with respect to the Binomial distribution on the interval $[0, 1]$ [i.e., with respect to the distribution with masses (2.3.6) at the points $x = 0, 1/N, \ldots, (N-1)/N, 1$]. From (2.3.8) we obtain the recursive relation $\underline{P}_0(x) = 1$, $\underline{P}_1(x) = x - p$:

$$\underline{P}_{n+1}(x) = \left[x - \frac{n}{N}q - \frac{N-n}{N}p\right]\underline{P}_n(x) - \frac{n}{N}\frac{N-n+1}{N}pq\underline{P}_{n-1}(x)$$

$(n = 1, \ldots, N-1)$. Comparing this recursion with the general recurrence relation in Corollary 2.3.4, it follows that the canonical moments of the Binomial distribution can be obtained by solving the system of equations

$$\zeta_1 = p_1 = p,$$

$$q_{2n-1}p_{2n} + q_{2n}p_{2n+1} = \frac{n}{N}q + \frac{N-n}{N}p,$$

$$q_{2n-2}p_{2n-1}q_{2n-1}p_{2n} = \frac{n}{N}\frac{(N-n+1)}{N}pq, \qquad n = 1, \ldots, N.$$

An induction argument shows that the solution of this system is unique and given by

$$p_{2j} = \frac{j}{N}, \quad p_{2j-1} = p, \qquad j = 1, \ldots, N,$$

which provides the canonical moments of the Binomial distribution.

**EXAMPLE 2.3.9** *Hahn Polynomials and the Hypergeometric Distribution.* In this example we consider a discrete probability measure that generalizes some of the commonly used discrete distributions. For $x = 0, \ldots, N$ define

$$p(x, \alpha, \beta, N) = \frac{\binom{x+\alpha}{x}\binom{N-x+\beta}{N-x}}{\binom{N+\alpha+\beta+1}{N}}, \tag{2.3.9}$$

where $\alpha, \beta$ are given numbers such that $\alpha, \beta > -1$ or $\alpha, \beta < -N$. These conditions are needed in order to guarantee that the masses $p(x, \alpha, \beta, N)$ are positive. It can be shown (see Karlin and McGregor, 1961) that

$$\sum_{x=0}^{N} p(x, \alpha, \beta, N) = 1$$

and that $p$ defines a discrete probability measure on the set $\{0, \ldots, N\}$. For $\alpha = \beta = 0$ this measure gives the *discrete uniform distribution* on the set $\{0, \ldots, N\}$. A straightforward calculation yields

$$\lim_{t \to \infty} p(x, pt, (1-p)t, N) = \binom{N}{x}p^x(1-p)^{N-x}$$

and shows that the Binomial distribution appears as a limiting case of the measure in (2.3.9). Note also that (2.3.9) is a generalization of the *Hypergeometric distribution*. In this case we consider an urn with $r + w$ balls of which $w$ are white and $r$ are red. If a sample of $n$ balls is drawn at random without replacing any balls in the urn at any stage, then the probability that the number of red balls in the sample of $n$ balls is equal to $x \in \{\max\{0, n - w\}, \ldots, \min\{r, n\}\}$ is given by

$$h(x, r, w, n) = \frac{\binom{r}{x}\binom{w}{n-x}}{\binom{r+w}{n}}. \tag{2.3.10}$$

In order to identify the Hypergeometric distribution as a special case of (2.3.9), we distinguish four cases and identify the parameters $\alpha, \beta,$ and $N$ in (2.3.9) according to the following display:

|           |           | $N$         | $\alpha$        | $\beta$         |
|-----------|-----------|-------------|-----------------|-----------------|
| $n \le r$ | $n \le w$ | $n$         | $-r - 1$        | $-w - 1$        |
| $n \le r$ | $n > w$   | $w$         | $-r - w - 1 + n$| $-n - 1$        |
| $n > r$   | $n \le w$ | $r$         | $-n - 1$        | $-w - r - 1 + n$|
| $n > r$   | $n > w$   | $r + w - n$ | $-w - 1$        | $-r - 1$        |

$$\tag{2.3.11}$$

For example, if $n \le r, n \le w$, the Hypergeometric distribution is supported on $\{0, \ldots, n\}$, and we obtain form (2.3.11) ($N = n, \alpha = -r - 1, \beta = -w - 1$):

$$p(x, -r - 1, -w - 1, n)$$

$$= \frac{(-1)^x r(r-1) \ldots (r - x + 1)(-1)^{n-x} w(w-1) \ldots (w + 1 + x - n)n!}{x!(n-x)!(-1)^n(r+w) \ldots (r + w - n + 1)}$$

$$= \frac{\binom{r}{x}\binom{w}{n-x}}{\binom{r+w}{n}} = h(x, r, w, n), \qquad x = 0, \ldots, n.$$

The other cases are very similar and left to the reader.

The orthogonal polynomials $Q_n(x, \alpha, \beta, N)$ with respect to the discrete probability measure (2.3.9) are called *Hahn polynomials* (Hahn, 1949) if $\alpha, \beta > -1$ and *Hahn-Eberlein polynomials* if $\alpha, \beta < -N$. These functions can be represented as a *Hypergeometric series*, that is,

$$Q_n(x) = Q_n(x, \alpha, \beta, N) = \sum_{k=0}^{n} \frac{(-n)_k(n + \alpha + \beta + 1)_k(-x)_k}{k!(\alpha + 1)_k(-N)_k}$$

($n = 0, \ldots, N$). The Krawtchouk polynomials can be obtained as limits from the Hahn polynomials ($p = 1 - q$)

$$k_n(x, p, N) = \lim_{t \to \infty} Q_n(x, pt, qt, N), \qquad n = 0, \ldots, N. \tag{2.3.12}$$

The three-term recurrence relation is given in Weber and Erdélyi (1952) (see also Karlin and McGregor, 1961) as

$$Q_0(x) = 1, \quad Q_1(x) = 1 - \frac{(\alpha + \beta + 2)}{N(\alpha + 1)} x$$

$$-xQ_n(x) = \frac{A_n Q_{n+1}(x) + B_n Q_{n-1}(x) - (A_n + B_n)Q_n(x)}{(2n + \alpha + \beta)_3} \tag{2.3.13}$$

($n = 1, \ldots, N - 1$), where the coefficients $A_n$ and $B_n$ are given by

$$A_n = (N - n)(\alpha + \beta + 2n)(\alpha + n + 1)(\alpha + \beta + n + 1),$$

$$B_n = n(\beta + n)(\alpha + \beta + N + n + 1)(\alpha + \beta + 2n + 2). \tag{2.3.14}$$

Define

$$\tilde{Q}_n(x) = \tilde{Q}_n(x, \alpha, \beta, N) = \frac{(-N)_n (\alpha + 1)_n}{(n + \alpha + \beta + 1)_n} Q_n(x, \alpha, \beta, N)$$

as the $n$th monic orthogonal polynomial with respect to the measure (2.3.9); then the recurrence relation for these polynomials is obtained from (2.3.13) as

$$\tilde{Q}_0(x) = 1, \quad \tilde{Q}_1(x) = x - \frac{N(\alpha + 1)}{\alpha + \beta + 2},$$

$$\tilde{Q}_{n+1}(x) = \left[ x - \frac{A_n + B_n}{(2n + \alpha + \beta)_3} \right] \tilde{Q}_n(x) - \frac{A_{n-1} B_n}{(2n + \alpha + \beta - 2)_3 (2n + \alpha + \beta)_3} \tilde{Q}_{n-1}(x)$$

($n = 1, \ldots, N - 1$). For the corresponding monic orthogonal polynomials on the interval $[0, 1]$,

$$\underline{P}_n(x) = \left( \frac{1}{N} \right)^n \tilde{Q}_n(Nx, \alpha, \beta, N), \qquad n = 0, \ldots, N,$$

it follows that $\underline{P}_{-1}(x) = 0$, $\underline{P}_0(x) = 1$ and that

$$\underline{P}_{n+1}(x) = (x - \alpha_n)\underline{P}_n(x) - \beta_n \underline{P}_{n-1}(x), \qquad n = 0, \ldots, N - 1,$$

where

$$\alpha_n = \frac{(N-n)(\alpha+n+1)(\alpha+\beta+n+1)}{N(\alpha+\beta+2n+1)_2} + \frac{n(\beta+n)(\alpha+\beta+N+n+1)}{N(\alpha+\beta+2n)_2},$$

$$\beta_n = \frac{(N-n+1)(\alpha+n)(\alpha+\beta+n)n(\beta+n)(\alpha+\beta+N+n+1)}{N^2(2n+\alpha+\beta-1)(2n+\alpha+\beta)^2(2n+\alpha+\beta+1)}$$

$(n = 0, \ldots, N-1)$. The canonical moments can now be identified by a comparison of this recurrence relation with the general statement in part 1 of Corollary 2.3.4. This yields $(q_0 = 1), p_{2N} = 1$,

$$p_1 = \alpha_0 = \frac{\alpha+1}{\alpha+\beta+2},$$

and for $k = 1, \ldots, N-1$,

$$q_{2k-2}p_{2k-1}q_{2k-1}p_{2k} = \beta_k,$$

$$q_{2k-1}p_{2k} + q_{2k}p_{2k+1} = \alpha_k.$$

An induction argument shows that the solution of this system is unique and given by

$$p_{2k-1} = \frac{k+\alpha}{2k+\alpha+\beta},$$

$$p_{2k} = \frac{k}{N}\frac{N+k+\alpha+\beta+1}{\alpha+\beta+2k+1} = 1 - \frac{N-k}{N}\frac{\alpha+\beta+k+1}{\alpha+\beta+2k+1}$$

(2.3.15)

$(k = 1, \ldots, N)$ which provides the canonical moments of the discrete distribution in (2.3.9).

It is interesting to note that for $N \to \infty$ the canonical moments in (2.3.15) converge to the canonical moments of the Beta distribution in (1.3.11). Note that the canonical moments are in one-to-one correspondence with the ordinary moments and that this relation is continuous (see Corollary 1.4.6). This means that the moments of the distribution with masses (2.3.9) at the points $0, 1/N, \ldots, (N-1)/N, 1$ converge to the moments of the Beta distribution on the interval $[0, 1]$. Because this distribution is obviously determined by its moments, it follows that the measure with masses (2.3.9) at the points $\{x/N\}_{x=0}^{N}$ converges (weakly) to the Beta distribution.

In order to obtain the canonical moments of the hypergeometric distribution, we choose in (2.3.15) the parameters $\alpha$, $\beta$, and $N$ according to the display (2.3.11). For example, if $n \leq r, n \leq w$, we obtain for the canonical moments of the Hypergeometric distribution

$$p_{2k-1} = \frac{r+1-k}{r+w+2-2k}, \quad p_{2k} = \frac{k}{n}\frac{r+w+1-n-k}{r+w+1-2k} \quad (2.3.16)$$

$(k = 1, \ldots, n)$. Finally the canonical moments of the discrete uniform distribution on $N+1$ points are obtained from (2.3.15) by putting $\alpha = \beta = 0$, i.e.

$$p_{2k-1} = \frac{1}{2}, \quad p_{2k} = \frac{k}{N}\frac{N+k+1}{2k+1}, \quad k = 1, \ldots, N. \quad (2.3.17)$$

**EXAMPLE 2.3.10**   In Example 2.3.8 and 2.3.9 we used the recurrence relation of the orthogonal polynomials for identifying the canonical moments corresponding to the measure of orthogonality. In the present example we demonstrate how the recurrence relation can be obtained from the canonical moments. Consider the Beta distribution in (2.1.14) with parameters $\alpha = \beta = \lambda - \frac{1}{2}$, $\lambda > -\frac{1}{2}$. It is shown in Example 1.5.3 that the canonical moments are

$$p_{2j-1} = \frac{1}{2}, \quad p_{2j} = \frac{j}{2(j+\lambda)}, \quad j = 1, 2, 3, \ldots.$$

From Corollary 2.3.6 we obtain for the corresponding monic orthogonal polynomials $\hat{C}_0^{(\lambda)}(x) = 1$, $\hat{C}_1^{(\lambda)}(x) = x$, and for $n \geq 1$,

$$\hat{C}_{n+1}^{(\lambda)}(x) = x\hat{C}_n^{(\lambda)}(x) - \frac{(n-1+2\lambda)n}{4(n+\lambda-1)(n+\lambda)}\hat{C}_{n-1}^{(\lambda)}(x).$$

The ultraspherical polynomials in (2.1.18) are orthogonal with respect to the measure $(1-x^2)^{\lambda-1/2}dx$ and, if $\lambda \neq 0$, have the leading coefficient

$$\frac{\Gamma(\lambda+\frac{1}{2})\Gamma(2n+2\lambda)}{2^n\Gamma(n+1)\Gamma(2\lambda)\Gamma(\lambda+n+1/2)} = \frac{2^n\Gamma(n+\lambda)}{\Gamma(\lambda)\Gamma(n+1)},$$

where the equality follows from the *duplication formula for the Gamma function*

$$\Gamma(2z) = \frac{2^{(2z-1)}}{\sqrt{\pi}}\Gamma(z)\Gamma\left(z+\frac{1}{2}\right). \quad (2.3.18)$$

This implies that

$$C_n^{(\lambda)}(x) = \frac{2^n\Gamma(n+\lambda)}{\Gamma(\lambda)\Gamma(n+1)}\hat{C}_n^{(\lambda)}(x), \quad \lambda \neq 0, \quad (2.3.19)$$

and a straightforward induction gives the recurrence relation for the ultrespherical polynomials, i.e. $C_0^{(\lambda)}(x) = 1$, $C_1^{(\lambda)}(x) = 2\lambda x$, and for $n \geq 1$,

$$(n+1)C_{n+1}^{(\lambda)}(x) = 2(n+\lambda)xC_n^{(\lambda)}(x) - (n+2\lambda-1)C_{n-1}^{(\lambda)}(x) \qquad (2.3.20)$$

Finally we remark that the remaining case $\lambda = 0$ corresponds to the arc-sine measure and the Chebyshev polynomials considered in Example 2.1.1. Therefore (2.1.19) and (2.1.7) yield $C_0^{(0)}(x) = 1$, $C_1^{(0)}(x) = x$, and for $n \geq 1$,

$$(n+1)C_{n+1}^{(0)}(x) = 2nxC_n^{(0)}(x) - (n-1)C_{n-1}^{(0)}(x).$$

## 2.4 INVERSE MAP

In Chapter 1 the canonical moments were defined explicitly in terms of the ordinary moments using Hankel determinants. A recursive method for calculating $\zeta_k = q_{k-1}p_k$ in terms of $c_1, \ldots, c_k$ was described using the Q-D algorithm. We now indicate how the recursion formula for the orthogonal polynomials can be used to produce a recursion formula for the ordinary moments in terms of the $\zeta_k$. One of the main results along these lines is Theorem 2.4.4.

Let $M = (c_{i+j})_{i,j=0}^\infty$ denote an infinite moment matrix where $c_k$, $k \geq 0$, are moments of some measure $\rho$. The corresponding monic orthogonal polynomials satisfy a recursion formula given by $P_0(x) \equiv 1$, $P_1(x) = x - \alpha_1$, and

$$P_{m+1}(x) = (x - \alpha_{m+1})P_m(x) - \beta_{m+1}P_{m-1}(x), \qquad m \geq 1. \qquad (2.4.1)$$

Let $J$ denote the *Jacobi* or *tri-diagonal matrix* of coefficients

$$J = \begin{pmatrix} \alpha_1 & 1 & 0 & 0 & \cdots \\ \beta_2 & \alpha_2 & 1 & 0 & \cdots \\ 0 & \beta_3 & \alpha_3 & 1 & \cdots \\ \vdots & \vdots & \vdots & & \end{pmatrix} \qquad (2.4.2)$$

and $S$ denote the matrix

$$S = \begin{pmatrix} 0 & 1 & 0 & 0 & \cdots \\ 0 & 0 & 1 & 0 & \cdots \\ 0 & 0 & 0 & 1 & \cdots \\ \vdots & \vdots & \vdots & & \end{pmatrix}.$$

Note that for an infinite matrix $A$ the operation $SA$ moves the rows of $A$ up by one or replaces the $i$th row by the $(i+1)$th row and $AS$ has all zeros in the first column and the $(j+1)$th column replaced by the $j$th for $j \geq 1$.

**Theorem 2.4.1**   *If $d_i = \int P_i^2(x)d\rho(x) > 0$, $i \geq 0$, then*

$$M = KDK^T, \qquad (2.4.3)$$

*where $D$ is a diagonal matrix with diagonal elements $d_i$ and $K$ is a lower triangular matrix, with ones on the main diagonal, which is defined recursively by*

$$SK = KJ. \qquad (2.4.4)$$

*Proof* All of the operations with the infinite matrices are actually finite operations in the sense that the summation in a product of matrices is performed over a finite number of terms. Let $f^T(x) = (1, x, x^2, \ldots)$ be the vector of powers, $p^T(x) = (P_0(x), P_1(x), P_2(x), \ldots)$, the corresponding vector of monic orthogonal polynomials, and let $L$ denote the lower triangular matrix of coefficients of the orthogonal polynomials, i.e.

$$p(x) = Lf(x).$$

The recursion formula for the orthogonal polynomials in (2.4.1) can be rewritten in matrix form as

$$LS = JL. \qquad (2.4.5)$$

Because $d_i > 0$ for all $i \geq 0$, the inverse matrix $K := L^{-1}$ exists and satisfies (2.4.4) and $f(x) = Kp(x)$. Since

$$\int p(x)p^T(x)d\rho(x) = D,$$

we find for the moment matrix of $\rho$ that

$$M = \int f(x)f^T(x)d\rho(x) = KDK^T. \qquad \blacksquare$$

**Corollary 2.4.2**   *If $d_i = \int P_i^2(x)d\rho(x) > 0$, $i \geq 0$, and $K = (k_{ij})_{i,j \geq 0}$ denotes the lower triangular matrix in Theorem 2.4.1, then for all $i, j \geq 0$,*

$$c_{i+\ell} = \sum_{j=0}^{\min\{i,\ell\}} d_j k_{ij} k_{\ell j}$$

$$= d_0 k_{i0} k_{\ell 0} + d_1 k_{i1} k_{\ell 1} + d_2 k_{i2} k_{\ell 2} + \ldots.$$

*In particular, if $d_0 = 1$, then*

$$c_n = k_{n0}.$$

**EXAMPLE 2.4.3** Suppose that $\rho = \mu$ is a probability measure on the interval $[0, 1]$. The elements $\alpha_m$ and $\beta_m$ in $J$ are given, from Corollary 2.3.4, by $\alpha_1 = \zeta_1$, $\alpha_{m+1} = \zeta_{2m} + \zeta_{2m+1}$, $\beta_{m+1} = \zeta_{2m-1}\zeta_{2m}$, $m \geq 1$. Moreover by Theorem 2.2.2, $d_0 = 1$ and $d_i = \zeta_1\zeta_2 \ldots \zeta_{2i}$ for $i \geq 1$.

Then $k_{ii} = 1$, $i \geq 0$, $k_{10} = \alpha_1$, and

$$k_{20} = \alpha_1 k_{10} + \beta_2 k_{11} = \alpha_1^2 + \beta_2,$$

$$k_{21} = k_{10} + \alpha_2 k_{11} = \alpha_1 + \alpha_2,$$

$$k_{30} = \alpha_1 k_{20} + \beta_2 k_{21} = \alpha_1(\alpha_1^2 + \beta_2) + \beta_2(\alpha_1 + \alpha_2), \qquad (2.4.6)$$

$$k_{31} = k_{20} + \alpha_2 k_{21} + \beta_3 k_{22} = \alpha_1^2 + \beta_2 + \alpha_2(\alpha_1 + \alpha_2) + \beta_3,$$

$$k_{32} = k_{21} + \alpha_3 k_{22} = \alpha_1 + \alpha_2 + \alpha_3.$$

The first three ordinary moments are then given by

$$c_1 = k_{10} = \zeta_1,$$

$$c_2 = k_{20} = d_0 k_{10} k_{10} + d_1 k_{11}^2 = \zeta_1^2 + \zeta_1\zeta_2 = \zeta_1(\zeta_1 + \zeta_2),$$

$$c_3 = k_{30} = d_0 k_{20} k_{10} + d_1 k_{21} k_{11} = \alpha_1(\alpha_1^2 + \beta_2) + \beta_2(\alpha_1 + \alpha_2),$$

$$= \zeta_1[\zeta_1(\zeta_1 + \zeta_2) + \zeta_2(\zeta_1 + \zeta_2 + \zeta_3)].$$

Note also that in Corollary 2.4.2 we have $(d_0 = 1)$

$$c_{i+\ell} = c_i c_\ell + d_1 k_{i1} k_{\ell 1} + \ldots$$

expressing $c_{i+\ell}$ as the product of $c_i$ and $c_\ell$ plus a positive remainder.

A number of more basic formula arise if $\rho$ is a symmetric measure on the interval $[-1, 1]$ and Theorem 1.3.5 is used. To describe these, let $S_{i,j}$ be given by $S_{i,j} = 0$, $0 \leq j < i$, $S_{0,j} \equiv 1$, $j \geq 0$, and

$$S_{i,j} = S_{i,j-1} + \zeta_{j-i+1} S_{i-1,j}, \qquad 1 \leq i \leq j. \qquad (2.4.7)$$

The following theorem is from Skibinsky (1968) who proved the result by a different method:

**Theorem 2.4.4** *If $c_n = (c_1, \ldots, c_n) \in Int\ M_n$, then*

$$c_n = S_{n,n} \qquad (2.4.8)$$

$$= \sum_{i=0}^{[n/2]} S_{i,n-i}^2 \prod_{j=1}^{n-2i} \zeta_j.$$

*Proof* The proof makes use of the fact that the measure $\mu$ on $[0, 1]$ can be associated with a symmetric measure $\mu^{(s)}$ on $[-1, 1]$ by (1.3.7). The canonical moments of $\mu^{(s)}$ are given by $p_{2i-1}^{(s)} = \frac{1}{2}$, $p_{2i}^{(s)} = p_i$ for $1 \le i \le N(c)$. The moments $b_k$ of $\mu^{(s)}$ and $c_k$ of $\mu$ are related by $c_k = b_{2k}$. Theorem 2.4.1 and Corollary 2.4.2 are now applied to $\mu^{(s)}$, and the results expressing $b_{2k}$ in terms of $p_i^{(s)}$ are expressed back in terms of $c_k$ and $p_i$.

For the symmetric measure $\mu^{(s)}$ the recursive formula in Corollary 2.3.6 gives the elements of the Jacobi matrix $J$ as $\alpha_k = 0$ and $\beta_{k+1} = q_{2k-2}^{(s)} p_{2k}^{(s)} = q_{k-1} p_k = \zeta_k$ for $k \ge 1$. Since the orthogonal polynomials $P_m(x)$ are odd or even functions according as the degree $m$ is odd or even, it follows that $k_{ij} = 0$ if $i + j$ is odd. Going down the diagonal in (2.4.4) from the upper left to the lower right yields that (2.4.4) is equivalent to

$$k_{i+2j,i} = k_{i+2j-1,i-1} + \beta_{i+2}k_{i+2j-1,i+1}, \qquad i,j \ge 0 \qquad (2.4.9)$$

($k_{0,-1} = 0$). The recursive formulas (2.4.7) and (2.4.9) are the same if we identify

$$k_{i+2j,i} = S_{j,i+j}, \qquad i \ge 0, j \ge 0. \qquad (2.4.10)$$

The first part of Theorem 2.4.4 follows from Corollary 2.4.2, since

$$c_n = b_{2n} = d_0 k_{2n,0} = S_{n,n}.$$

The second part is a result of

$$c_n = b_{2n} = b_{n+n}$$

$$= d_0 k_{n0}^2 + d_1 k_{n1}^2 + \ldots + d_n k_{nn}^2$$

by re-indexing the summation and applying (2.4.10) and Remark 2.3.7 to the measure $\mu^{(s)}$, which yields

$$d_i = \int_{-1}^{1} P_i^2(x) d\mu^{(s)}(x) = \prod_{k=1}^{i} q_{2k-2}^{(s)} p_{2k}^{(s)} = \prod_{k=1}^{i} q_{k-1} p_k = \prod_{k=1}^{i} \zeta_k. \qquad \blacksquare$$

## 2.5   REVERSED CANONICAL MOMENT SEQUENCES

In Chapter 1 a number of results were given that were related to invariance properties of the canonical moments. It will now be shown that for finite sequences $(p_1, \ldots, p_n, 0)$ and $(p_1, \ldots, p_n, 1)$, the *reversed sequences* $(p_n, \ldots, p_1, 0)$ and $(q_n, \ldots, q_1, 1)$ have, correspondingly, the same support. In Chapter 4 these results are applied in an essential way in identifying the support for certain finite sequences using some basic facts about classical orthogonal polynomials. Viewed geometrically the results say, for example, that the moment points $(p_1, \ldots, p_n, 0)$ and $(p_n, \ldots, p_1, 0)$ lie in the same "face" of the boundary of the moment space $M_n$.

**Theorem 2.5.1**  *If $c_n \in Int\ M_n$, then*

1.  *The measures associated with the sequences $(p_1, \ldots, p_n, 0)$ and $(p_n, \ldots, p_1, 0)$ have the same support.*
2.  *The measures associated with the sequences $(p_1, \ldots, p_n, 1)$ and $(q_n, \ldots, q_1, 1)$ have the same support.*

The proof will make use of two auxiliary lemmas that involve determinants of tridiagonal $(n+1) \times (n+1)$ matrices of the form

$$
K\begin{pmatrix} -\zeta_1 & -\zeta_2 & \cdots & -\zeta_n \\ x & 1 & x & \cdots & \tau_n \end{pmatrix} = \begin{vmatrix} x & -1 & 0 & \cdots & 0 & 0 \\ -\zeta_1 & 1 & -1 & \cdots & 0 & 0 \\ 0 & -\zeta_2 & x & \cdots & 0 & 0 \\ \vdots & \vdots & \vdots & \ddots & \vdots & \vdots \\ 0 & 0 & 0 & \cdots & \tau_{n-1} & -1 \\ 0 & 0 & 0 & \cdots & -\zeta_n & \tau_n \end{vmatrix},
$$

$$(2.5.1)$$

where

$$
\tau_n = \begin{cases} 1 & n \text{ odd,} \\ x & n \text{ even.} \end{cases}
$$

By expanding the determinant by the last row or column, it is easily seen that

$$
K\begin{pmatrix} -\zeta_1 & \cdots & -\zeta_n \\ x & 1 & \cdots & \tau_n \end{pmatrix} = \tau_n K\begin{pmatrix} -\zeta_1 & \cdots & -\zeta_{n-1} \\ x & 1 & \cdots & \tau_{n-1} \end{pmatrix} \qquad (2.5.2)
$$

$$
- \zeta_n K\begin{pmatrix} -\zeta_1 & \cdots & -\zeta_{n-2} \\ x & 1 & \cdots & \tau_{n-2} \end{pmatrix}.
$$

The following lemmas will be used extensively here and in Chapter 3.

**Lemma 2.5.2**

1.  $\underline{P}_m(x) = K\begin{pmatrix} -\zeta_1 & \cdots & -\zeta_{2m-1} \\ x & 1 & \cdots & x & 1 \end{pmatrix}$,

    $x\underline{Q}_m(x) = K\begin{pmatrix} -\zeta_1 & \cdots & -\zeta_{2m} \\ x & 1 & \cdots & 1 & x \end{pmatrix}$.

2.  $(x-1)\overline{P}_m(x) = K\begin{pmatrix} -\zeta_1 & \cdots & -\zeta_{2m}-q_{2m} \\ x & 1 & \cdots & x & 1 \end{pmatrix}$,

    $x(x-1)\overline{Q}_{m-1}(x) = K\begin{pmatrix} -\zeta_1 & \cdots & -\zeta_{2m-1}-q_{2m-1} \\ x & 1 & \cdots & 1 & x \end{pmatrix}$.

*Proof*   The pair of equations in part 1 can be shown as follows. Recall that the sequences $\underline{P}_m$ and $\underline{Q}_m$ satisfy the first recurrence formula pair in Corollary 2.3.3. Using the recursion (2.5.2), it can be shown that the corresponding pair of determinants on the right satisfy the same recurrence formulas. Since the initial case $m = 1$ can be verified directly, the result follows.

To verify the second pair in part 2, we insert $p_n = 1$ appropriately in the first pair. Thus, if $m$ is replaced by $m + 1$ in the first equation of part 1 and we set $p_{2m+1} = 1$, then the result for $(x - 1)\overline{P}_m(x)$ follows provided that the roots of certain representations are continuous functions of the moment points. To be more explicit, we consider a sequence of moment points $(c_1, \ldots, c_{2m}, c_{2m+1}^{(k)}) \in$ Int $M_{2m+1}$ where $c_{2m+1}^{(k)} \longrightarrow c_{2m+1}^{+}$. We denote the corresponding canonical moment sequence by $p^{(k)} = (p_1, \ldots, p_{2m}, p_{2m+1}^{(k)})$, and the unique lower principal representation of $p^{(k)}$ by $\mu_{2m+1}^{(k)}$. For each $k$ we have from the first part of the proof

$$\underline{P}_{m+1}^{(k)}(x) = K\begin{pmatrix} -\zeta_1 & & \cdots & & -\zeta_{2m} & & -\zeta_{2m+1}^{(k)} & \\ x & 1 & \cdots & & & x & & 1 \end{pmatrix},$$

and the right-hand side clearly converges to

$$K\begin{pmatrix} -\zeta_1 & & \cdots & & -\zeta_{2m} & -q_{2m} & \\ x & 1 & \cdots & & x & & 1 \end{pmatrix}.$$

In order to establish the first identity in part 2, we need to show that $\underline{P}_{m+1}^{(k)}(x)$ converges to $(x - 1)\overline{P}_m(x)$. The sequence $\mu_{2m+1}^{(k)}$ has convergent subsequences, and any limit measure will have moments $(c_1, \ldots, c_{2m}, c_{2m+1}^{+})$. Since the measure corresponding to this boundary point is unique, it follows that $\mu_{2m+1}^{(k)}$ must converge to the upper principal representation $\mu_{2m}^{+}$ of $(c_1, \ldots, c_{2m})$. By Theorem 2.2.3 the support points of $\mu_{2m}^{+}$ are the zeros of the polynomial $(x - 1)\overline{P}_m(x)$, and it follows that $\underline{P}_{m+1}^{(k)}(x)$ must converge to $(x - 1)\overline{P}_m(x)$.

The second equation in part 2 follows in a similar manner; in this case we insert $p_{2m} = 1$ in the second equation of part 1.  ■

**Lemma 2.5.3**   *For all* $1 \leq n \leq N(c)$,

$$K\begin{pmatrix} -\zeta_1 & & \cdots & & -\zeta_n & \\ x & 1 & \cdots & & & \tau_n \end{pmatrix} = \tau_n K\begin{pmatrix} -\zeta_1 & & \cdots & & -\zeta_n & \\ 1 & x & \cdots & & & \tau_{n-1} \end{pmatrix}$$

$$= K\begin{pmatrix} -\zeta_n & & \cdots & & -\zeta_1 & \\ x & 1 & \cdots & & & \tau_n \end{pmatrix}.$$

*Proof* The lemma follows by a fairly straightforward induction argument. ■

**Lemma 2.5.4** *For all* $1 \leq n \leq N(c)$,

$$K\begin{pmatrix} -\zeta_1 & \cdots & -\zeta_n \\ x & 1 & \cdots & \tau_n \end{pmatrix} = K\begin{pmatrix} -\gamma_2 & \cdots & -\gamma_n - p_n \\ x & 1 & \cdots & \tau_n \end{pmatrix}$$

$$= K\begin{pmatrix} -p_n & -\gamma_n & \cdots & -\gamma_2 \\ x & 1 & \cdots & \tau_n \end{pmatrix}.$$

*Proof* The second equality follows from Lemma 2.5.3. The first equality can be shown by induction as in Lau (1983), Lemma 2.7.3. We give here a slightly different proof based on Theorem 1.5.4.

Let $\mu_n^-$ and $(c_1, \ldots, c_n, c_{n+1}^-)$ be the measure and moment section corresponding to $(p_1, \ldots, p_n, 0)$, and consider the measure $d\nu_n^- = q_1^{-1}(1 - x)d\mu_n^-$. Since $\mu_n^-$ is a lower principal measure, it does not have $x = 1$ in its support and $\mu_n^-$ and $\nu_n^-$ have the same support, and both are lower principal representations. The moments of $\nu_n^-$ are given by $c_i' = (c_i - c_{i+1})/q_1$, $i = 1, \ldots, n-1$ and $c_n' = (c_n - c_{n+1}^-)/q_1$. By Corollary 1.4.8 or Theorem 1.5.4 we obtain for the quantities $\zeta_i'$ corresponding to $\nu_n^-$, $\zeta_i' = \gamma_{i+1}$, $i = 1, \ldots, n$. Note that $\gamma_{n+1} = p_n$ because $\mu_n^-$ is the lower principal representation of $(c_1, \ldots, c_n)$, i.e. $p_{n+1} = 0$. By Theorem 2.2.3 and Lemma 2.5.2 the support of $\mu_n^-$ is given by the zeros of

$$K\begin{pmatrix} -\zeta_1 & \cdots & -\zeta_n \\ x & 1 & \cdots & \tau_n \end{pmatrix},$$

and the support of $\nu_n^-$ is given by the zeros of

$$K\begin{pmatrix} -\zeta_1' & \cdots & -\zeta_n' \\ x & 1 & \cdots & \tau_n \end{pmatrix} = K\begin{pmatrix} -\gamma_2 & \cdots & -\gamma_n - p_n \\ x & 1 & \cdots & \tau_n \end{pmatrix}.$$

Because both measures have the same support, the result follows. ■

*Proof of Theorem 2.5.1* The first part follows from Theorem 2.2.3, Lemma 2.5.2, and Lemma 2.5.4. The second part is obtained from Theorem 2.2.3 and Lemma 2.5.2. The rows and columns in the determinant expressions are reversed. The odd case is then immediate, while the even case needs an application of Lemma 2.5.3.

To be more explicit, suppose that $n = 2m - 1$, then by Theorem 2.2.3, part 4, and the second equation in part 2 of Lemma 2.5.2, the support of

$(p_1, \ldots, p_{2m-1}, 1)$ consists of the roots of the polynomial

$$K\begin{pmatrix} -\zeta_1 & \cdots & -\zeta_{2m-1} & -q_{2m-1} & \\ x & 1 & \cdots & 1 & x \end{pmatrix}.$$

If we let $p_i' = q_{2m-i}$, $i = 1, 2, \ldots, 2m - 1$, then the support of $(p_1', \ldots, p_{2m-1}', 1) = (q_{2m-1}, \ldots, q_1, 1)$ consists of the roots of

$$K\begin{pmatrix} -\zeta_1' & \cdots & -\zeta_{2m-1}' & -q_{2m-1}' & \\ x & 1 & \cdots & 1 & x \end{pmatrix},$$

where $\zeta_1' = p_1' = q_{2m-1}$, $\zeta_i' = q_{i-1}'p_i' = q_{2m-i}p_{2m-i+1} = \zeta_{2m-i+1}$ $(i = 1, \ldots, 2m - 1)$. These two determinants are equal by interchanging rows and columns or using Lemma 2.5.3. The even case is exactly the same except that an application of Lemma 2.5.3 is needed. ∎

The polynomials $\underline{P}_m, \underline{Q}_m, \overline{P}_m$ and $\overline{Q}_{m-1}$ can be written separately in determinant form using Lemma 2.5.2. To accomplish this, one further auxiliary result is needed.

**Lemma 2.5.5**

$$K\begin{pmatrix} -\zeta_1 & \cdots & -\zeta_n & -q_n & \\ x & 1 & \cdots & \tau_n & \tau_{n+1} \end{pmatrix}$$

$$= (x - 1)K\begin{pmatrix} -\gamma_2 & \cdots & -\gamma_n & \\ x & 1 & \cdots & \tau_{n+1} \end{pmatrix}.$$

*Proof* By setting $p_{n+1} = 1$, using Lemma 2.5.4 and an expansion of the determinant, we obtain

$$K\begin{pmatrix} -\zeta_1 & \cdots & -\zeta_n & -q_n & \\ x & 1 & \cdots & \tau_n & \tau_{n+1} \end{pmatrix}$$

$$= K\begin{pmatrix} -1 & 0 & -\gamma_n & \cdots & -\gamma_2 & \\ x & 1 & x & \cdots & \tau_{n+1} \end{pmatrix}$$

$$= (x - 1)K\begin{pmatrix} -\gamma_n & \cdots & -\gamma_2 & \\ x & 1 & \cdots & \tau_{n+1} \end{pmatrix}.$$

The result then follows using Lemma 2.5.3 ∎

**Theorem 2.5.6**

1. $\underline{P}_m(x) = K\begin{pmatrix} & -\zeta_1 & & \cdots & & -\zeta_{2m-1} & \\ x & & 1 & \cdots & x & & 1 \end{pmatrix}$

$= K\begin{pmatrix} & -\gamma_2 & & \cdots & -\gamma_{2m-1} & & -p_{2m-1} & \\ x & & 1 & \cdots & & x & & 1 \end{pmatrix},$

$\underline{Q}_m(x) = K\begin{pmatrix} & -\zeta_1 & & \cdots & & -\zeta_{2m} & \\ 1 & & x & \cdots & x & & 1 \end{pmatrix}$

$= K\begin{pmatrix} & -\gamma_2 & & \cdots & -\gamma_{2m} & & -p_{2m} & \\ 1 & & x & \cdots & & x & & 1 \end{pmatrix}.$

2. $\overline{P}_m(x) = K\begin{pmatrix} & -\gamma_2 & & \cdots & & -\gamma_{2m} & \\ x & & 1 & \cdots & x & & 1 \end{pmatrix},$

$\overline{Q}_{m-1}(x) = K\begin{pmatrix} & -\gamma_2 & & \cdots & & -\gamma_{2m-1} & \\ 1 & & x & \cdots & x & & 1 \end{pmatrix}.$

*Proof* The first identity in part 1 follows from Lemmas 2.5.2 and 2.5.4. The second uses Lemmas 2.5.2, 2.5.3, and 2.5.4. The other cases in part 2 are treated similarly using Lemma 2.5.5. ∎

CHAPTER 3

# Continued Fractions and the Stieltjes Transform

## 3.1 INTRODUCTION

The *Stieltjes transform* of a probability measure $\mu$ on the real line $(-\infty, \infty)$ is given by

$$S(z) = S(z, \mu) = \int_{-\infty}^{\infty} \frac{d\mu(x)}{z - x} \qquad (3.1.1)$$

and determines the measure $\mu$ uniquely. The transform originally studied by Stieltjes in his famous long paper "Investigations in Continuous Fractions" in 1884 was concerned with measures with support on the nonnegative real line $[0, \infty)$. We will actually be using measures on finite intervals and more specifically on $[0, 1]$ or $[-1, 1]$.

The transform $S(z, \mu)$ is an analytic function of $z$ in $\mathbb{C} \setminus \text{supp}(\mu)$ where $\mathbb{C}$ is the complex plane and $\text{supp}(\mu)$ denotes the support of the probability measure $\mu$ defined in Section 2.1. This is a result of the fact that the integral in (3.1.1) is uniformly convergent in any compact region that has a positive distance from $\text{supp}(\mu)$.

The integral (3.1.1) is closely related to the Cauchy integral

$$\frac{1}{2\pi i} \int_C \frac{f(t)}{z - t} dt.$$

Here $C$ is some contour in the complex plane, and integrating over a contour consisting of a large semicircle with diameter on the real line relates this integral to (3.1.1). The value or definition of (3.1.1) for real $z$ will generally not concern us here. This would involve a careful discussion of the *Cauchy principal value* of the integrals in question. A very detailed discussion of this is presented in Henrici (1977).

The Stieltjes transform is extremely useful in that its different expansions provide a nice connection between the ordinary moments $c_1, c_2, \ldots$ of the measure $\mu$ and the canonical moments $p_1, p_2, \ldots$. On the one hand, the power series expansion of (3.1.1) for a measure on the interval $[a, b]$ gives

$$\int_a^b \frac{d\mu(x)}{z - x} = \frac{c_0}{z} + \frac{c_1}{z^2} + \frac{c_2}{z^3} + \cdots \tag{3.1.2}$$

The series on the right-hand side of (3.1.2) is sometimes referred to as the generating function or the *z-transform* of the sequence $c_0, c_1, c_2, \ldots$. This series converges for any $|z| > \max\{|a|, |b|\}$ and is equal to $S(z, \mu)$. The validity of (3.1.2) can be seen by expanding the integrand in $S(z, \mu)$ and integrating termwise.

On the other hand, the transform $S(z, \mu)$ can also be expanded into a continued fraction, which for $[a, b] = [0, 1]$ takes the form

$$\int_0^1 \frac{d\mu(x)}{z - x} = \frac{1|}{|z} - \frac{\zeta_1|}{|1} - \frac{\zeta_2|}{|z} - \frac{\zeta_3|}{|1} - \cdots \tag{3.1.3}$$

$$:= \cfrac{1}{z - \cfrac{\zeta_1}{1 - \cfrac{\zeta_2}{z - \cfrac{\zeta_3}{1 - \cdots}}}},$$

where $\zeta_k = q_{k-1} p_k$ for $k \geq 1$ and $q_0 = 1$. It will be shown in Theorem 3.3.1 that (3.1.3) is valid for any $z \in \mathbb{C} \backslash [0, 1]$. Thus the ordinary moments and the canonical moments are linked in an intricate manner through the different expansions (3.1.2) and (3.1.3) of the Stieltjes transform $S(z, \mu)$.

The orthogonal polynomials introduced in Chapter 2 enter the picture by truncating the continued fraction in (3.1.3). If, for example, we take

$$\frac{1|}{|z} - \frac{\zeta_1|}{|1} - \frac{\zeta_2|}{|z} - \cdots - \frac{\zeta_{2m-1}|}{|1}$$

and write it as the ratio of two polynomials in $z$, then the polynomial in the denominator is precisely the $m$th monic polynomial $\underline{P}_m(z)$ orthogonal with respect to the measure $d\mu$. More generally, if $\mu$ has finite support, the continued fraction will terminate, and when written as a ratio of polynomials, the denominator polynomial must vanish for any $z \in \mathrm{supp}(\mu)$.

In the next section some elementary properties of continued fractions will be discussed. Further sections are devoted to the expansion (3.1.3) and its various uses and implications.

## 3.2 CONTINUED FRACTIONS

In this section a brief introduction to continued fractions is given. A more extensive treatment of this material can be found in Wall (1948), Perron (1954), Henrici (1977), Jones and Thron (1980), or Lorentzen and Waadeland (1992).

A finite or *terminating continued fraction* is an expression of the form

$$b_0 + \cfrac{a_1\,|}{|\,b_1} + \cfrac{a_2\,|}{|\,b_2} + \ldots + \cfrac{a_n\,|}{|\,b_n} = \frac{A_n}{B_n}. \tag{3.2.1}$$

A few simple examples suffice to explain this notation. Thus we have

$$\frac{A_0}{B_0} = \frac{b_0}{1}, \quad \frac{A_1}{B_1} = b_0 + \frac{a_1}{b_1} = \frac{b_0 b_1 + a_1}{b_1},$$

$$\frac{A_2}{B_2} = b_0 + \cfrac{a_1}{b_1 + \cfrac{a_2}{b_2}} = \frac{b_0 b_1 b_2 + b_0 a_2 + a_1 b_2}{b_1 b_2 + a_2}$$

$$= \frac{b_2 A_1 + a_2 A_0}{b_2 B_1 + a_2 B_0}.$$

The quantities $A_n$ and $B_n$ are called the *nth partial numerator* and *denominator*, respectively, and $A_n/B_n$ is called an *nth approximant*, *nth convergent* or *nth partial quotient*. There are basic recursive relations for the quantities $A_n$ and $B_n$ given by

$$A_n = b_n A_{n-1} + a_n A_{n-2},$$
$$B_n = b_n B_{n-1} + a_n B_{n-2}, \tag{3.2.2}$$

for $n \geq 1$, with initial conditions

$$A_{-1} = 1, \quad A_0 = b_0,$$
$$B_{-1} = 0, \quad B_0 = 1. \tag{3.2.3}$$

Thus, if $A_n$ and $B_n$ are defined by (3.2.2) and (3.2.3), then the ratio $A_n/B_n$ satisfies (3.2.1). A simple induction argument can be used to verify this. The case $n = 0$ is trivial. To proceed from $n$ to $n + 1$, $b_n$ must be replaced by $b_n + a_{n+1}/b_{n+1}$ which gives

$$\frac{A_{n+1}}{B_{n+1}} = \frac{\left(b_n + \frac{a_{n+1}}{b_{n+1}}\right) A_{n-1} + a_n A_{n-2}}{\left(b_n + \frac{a_{n+1}}{b_{n+1}}\right) B_{n-1} + a_n B_{n-2}}$$

$$= \frac{b_{n+1} A_n + a_{n+1} A_{n-1}}{b_{n+1} B_n + a_{n+1} B_{n-1}}.$$

Note that the recursive relations for the partial numerators and denominators in
(3.2.2) are the same, but they start with different initial conditions (3.2.3). An
*infinite continued fraction* of the form

$$b_0 + \frac{a_1}{\mid b_1} + \frac{a_2}{\mid b_2} + \ \dots \tag{3.2.4}$$

is said to converge to a limit $L$ if $B_n \neq 0$ for sufficiently large $n$ and the sequence
$\{A_n/B_n\}_{n \geq 0}$ of approximants in (3.2.1) converges with limit $L$, i.e.

$$\lim_{n \to \infty} \frac{A_n}{B_n} = L. \tag{3.2.5}$$

Whenever we use the notation of an infinite continued fraction in this book, we
will actually assume that it converges. In places where convergence is of some
special interest, this will be mentioned explicitly.

**EXAMPLE 3.2.1**   The value of simple square roots can be easily represented
by continued fractions. For example if $a > 0$, $a^2 + b > 0$ we obtain formally

$$\sqrt{a^2 + b} = a + (\sqrt{a^2 + b} - a)$$

$$= a + \frac{b}{a + \sqrt{a^2 + b}}$$

$$= a + \frac{b}{\mid 2a} + \frac{b}{\mid 2a} + \dots \ .$$

It will be shown in Lemma 3.2.3 that expressions of this type actually converge.
The convergence is more rapid with a judicious choice of $a$ and $b$. For example,
we would write $\sqrt{8} = \sqrt{3^2 - 1}$ rather than $\sqrt{8} = \sqrt{2^2 + 4}$. Using the recursive
relations in (3.2.2), a small table of the approximants can be constructed

$$\begin{array}{ccc} A_{-1} & A_0 & A_1 & \dots \\ B_{-1} & B_0 & B_1 & \dots \end{array}$$

If it is desired to find $\sqrt{2}$, then we choose $a = b = 1$ and calculate successively the
quantities

| $n$ | $-1$ | 0 | 1 | 2 | 3 | 4 | 5 | 6 | 7 | $\dots$ |
|---|---|---|---|---|---|---|---|---|---|---|
| $A_n$ | 1 | 1 | 3 | 7 | 17 | 41 | 99 | 239 | 577 | $\dots$ |
| $B_n$ | 0 | 1 | 2 | 5 | 12 | 29 | 70 | 169 | 408 | $\dots$ |

The fourth approximant gives $41/29 \approx 1.413793$, while the seventh approximant
gives $577/408 \approx 1.414216$.

The recursive formula can be manipulated in a very simple way to give the continued fraction expansions for $B_n/B_{n-1}$ and $A_n/A_{n-1}$. These formula will be useful in Section 3.5 where we show a direct relationship between the canonical moment sequence $p_1, p_2, \ldots, p_n$ and the combined set of support points of the upper and lower principal representations $\mu_n^+$ and $\mu_n^-$.

**Lemma 3.2.2** *If $b_0 = 0$ and $n \geq 2$,*

$$\frac{B_n}{B_{n-1}} = b_n + \frac{a_n}{\left| b_{n-1} \right.} + \frac{a_{n-1}}{\left| b_{n-2} \right.} + \cdots + \frac{a_2}{\left| b_1 \right.},$$

(3.2.6)

$$\frac{A_n}{A_{n-1}} = b_n + \frac{a_n}{\left| b_{n-1} \right.} + \frac{a_{n-1}}{\left| b_{n-2} \right.} + \cdots + \frac{a_3}{\left| b_2 \right.}.$$

*Proof*  These equations are nearly immediate from (3.2.2) and (3.2.3). For example, the relationship for $B_n$ can be written as

$$B_n/B_{n-1} = b_n + \frac{a_n}{B_{n-1}/B_{n-2}},$$

and the first result in (3.2.6) follows from an induction observing that $B_2/B_1 = b_2 + a_2/b_1$. For the second identity we proceed in the same way and note that $A_2/A_1 = b_2$ (because $b_0 = 0$, by assumption).  ∎

In the texts on continued fractions, there are two general viewpoints in proving various results. In older texts one proceeded algebraically as in the arguments above. More modern texts use linear fractional transformations or Moebius transformations. A general *Moebius transformation* is a mapping of the form

$$m(w) = \frac{cw + d}{ew + f},$$

(3.2.7)

where $c, d, e,$ and $f$ are given complex numbers. The above transformation maps the complex plane into itself. The matrix

$$M = \begin{pmatrix} c & d \\ e & f \end{pmatrix}$$

is called the *matrix of the Moebius transformation*. The usefulness of this viewpoint stems partially from the fact that the composition $m_1(m_2(w))$ of two such transformations is a transformation of the same form and that the matrix of the composition is the ordinary matrix product of the two matrices $M_1$ and $M_2$ corresponding to $m_1$ and $m_2$.

If we let

$$m_0(w) = b_0 + w, \quad m_k(w) = \frac{a_k}{b_k + w}, \tag{3.2.8}$$

then the corresponding matrices are

$$M_0 = \begin{pmatrix} 1 & b_0 \\ 0 & 1 \end{pmatrix}, \quad M_k = \begin{pmatrix} 0 & a_k \\ 1 & b_k \end{pmatrix} \quad k = 1, 2, \ldots,$$

and the expression in (3.2.1) is readily seen to be

$$\frac{A_n}{B_n} = m_0 \circ m_1 \circ \ldots \circ m_n(0). \tag{3.2.9}$$

Since the composition $m_0 \circ m_1 \circ \ldots \circ m_n(w)$ is a Moebius transformation, we conclude, using (3.2.9), that

$$m_0 \circ m_1 \circ \ldots \circ m_n(w) = \frac{C_n w + A_n}{D_n w + B_n}$$

and

$$M_0 M_1 \cdots M_n = \begin{pmatrix} C_n & A_n \\ D_n & B_n \end{pmatrix}.$$

Because the left-hand side of the above expression is $(M_0 M_1 \ldots M_{n-1}) M_n$, it follows that

$$\begin{pmatrix} C_n & A_n \\ D_n & B_n \end{pmatrix} = \begin{pmatrix} C_{n-1} & A_{n-1} \\ D_{n-1} & B_{n-1} \end{pmatrix} \begin{pmatrix} 0 & a_n \\ 1 & b_n \end{pmatrix}.$$

Then $C_n = A_{n-1}$, $D_n = B_{n-1}$, and we obtain the basic recursive equation for $A_n$:

$$A_n = b_n A_{n-1} + a_n C_{n-1}$$
$$= b_n A_{n-1} + a_n A_{n-2}.$$

The recursion equation for $B_n$ is derived in the same way. Thus we have

$$m_0 \circ m_1 \circ \ldots \circ m_n(w) = \frac{A_n + A_{n-1} w}{B_n + B_{n-1} w}. \tag{3.2.10}$$

Several other elementary properties of continued fractions follow from the composition in (3.2.10). For example, if the basic Moebius transformations in

(3.2.8) are rewritten as

$$m_k(w) = \frac{a_k}{b_k + w} = \frac{c_k a_k}{c_k b_k + c_k w} \qquad \text{for } c_k \neq 0,$$

then $m_0 \circ m_1 \circ \ldots \circ m_k(w)$ remains the same and the $k$th approximants must all remain equal. The "new" continued fraction is

$$b_0 + \frac{c_1 a_1}{\left| c_1 b_1 \right.} + \frac{c_1 c_2 a_2}{\left| c_2 b_2 \right.} + \frac{c_2 c_3 a_3}{\left| c_3 b_3 \right.} + \ldots \tag{3.2.11}$$

This is of course simply multiplying the numerator and denominator at a given stage by $c_k$. The new partial numerator and denominator $A'_n$ and $B'_n$ are

$$A'_n = c_1 \ldots c_n A_n \quad \text{and} \quad B'_n = c_1 \ldots c_n B_n.$$

Two continued fractions with the same approximants $A_k/B_k$ are called *equivalent*. The modification just described shows that the continued fractions in (3.2.4) and (3.2.11) are equivalent. One can use (3.2.11) to obtain an equivalent continued fraction so that, for example, all of its numerators $a_k$ (or denominators $b_k$) are equal to 1.

The determinant of the matrix of the composition $m_0 \circ m_1 \circ \ldots \circ m_n(w)$ can be calculated from (3.2.10). Since the determinant of the product of matrices is the product of the determinants of the individual terms, we find that

$$\begin{vmatrix} A_{n-1} & A_n \\ B_{n-1} & B_n \end{vmatrix} = \begin{vmatrix} 1 & b_0 \\ 0 & 1 \end{vmatrix} \cdot \prod_{k=1}^{n} \begin{vmatrix} 0 & a_k \\ 1 & b_k \end{vmatrix}.$$

This can be expressed (with $a_0 = 1$) as

$$A_{n-1} B_n - A_n B_{n-1} = (-1)^n a_0 a_1 \ldots a_n \tag{3.2.12}$$

or equivalently (provided that $B_n \neq 0$, $B_{n-1} \neq 0$),

$$\frac{A_n}{B_n} - \frac{A_{n-1}}{B_{n-1}} = (-1)^{n+1} \frac{a_0 a_1 \ldots a_n}{B_{n-1} B_n}. \tag{3.2.13}$$

Expression (3.2.12) is called the *basic determinant formula*. Summing (3.2.13), it follows that

$$\frac{A_n}{B_n} = b_0 + \frac{a_1}{B_0 B_1} - \frac{a_1 a_2}{B_1 B_2} + \ldots + (-1)^{n+1} \frac{a_1 \ldots a_n}{B_{n-1} B_n}. \tag{3.2.14}$$

Provided that $B_k \neq 0, k = 0, 1, 2, \ldots$, the infinite continued fraction (3.2.4) will converge if and only if the series in (3.2.14) converges.

It is often desirable to "contract" (or expand) a continued fraction. The *even contraction* is a continued fraction whose $m$th approximant is equal to the $2m$th approximant of the original continued fraction. The corresponding *odd contraction* is such that its $m$th approximant is equal to the $(2m - 1)$th approximant of the original.

Our main application will be for the even contraction of the continued fraction of the form

$$\frac{1}{\left| z \right.} - \frac{\zeta_1}{\left| 1 \right.} - \frac{\zeta_2}{\left| z \right.} - \cdots \tag{3.2.15}$$

The contraction in this case can be deduced from the identity

$$z - \frac{\zeta_{2m-1}}{1 - \frac{\zeta_{2m}}{\lambda}} = z - \zeta_{2m-1} - \frac{\zeta_{2m-1}\zeta_{2m}}{\lambda - \zeta_{2m}},$$

and it is given by

$$\frac{1}{\left| z - \zeta_1 \right.} - \frac{\zeta_1\zeta_2}{\left| z - \zeta_2 - \zeta_3 \right.} - \frac{\zeta_3\zeta_4}{\left| z - \zeta_4 - \zeta_5 \right.} - \cdots \tag{3.2.16}$$

The general even contraction of (3.2.4) is given by

$$b_0 + \frac{a_1b_2}{\left| b_1b_2 + a_2 \right.} - \frac{a_2a_3b_4}{\left| (b_2b_3 + a_3)b_4 + b_2a_4 \right.} - \frac{a_4a_5b_2b_6}{\left| (b_4b_5 + a_5)b_6 + b_4a_6 \right.}$$

$$- \frac{a_6a_7b_4b_8}{\left| (b_6b_7 + a_7)b_8 + b_6a_8 \right.} - \cdots, \tag{3.2.17}$$

while the corresponding odd contraction is

$$\frac{a_1 + b_0b_1}{b_1} - \frac{a_1a_2b_3/b_1}{\left| (b_1b_2 + a_2)b_3 + b_1a_3 \right.} - \frac{a_3a_4b_1b_5}{\left| (b_3b_4 + a_4)b_5 + b_3a_5 \right.}$$

$$- \frac{a_5a_6b_3b_7}{\left| (b_5b_6 + a_6)b_7 + b_5a_7 \right.} - \cdots \tag{3.2.18}$$

The proofs of these results will not be given here, they can be found in the book of Perron (1954a).

One last important point is that the quantities $A_n$ and $B_n$ can be written in terms of *continuants* or determinants of tri-diagonal matrices, that is,

$$
A_n = K\left(\begin{matrix} a_1 & \cdots & a_n \\ b_0 \ b_1 & \cdots & b_n \end{matrix}\right) := \begin{vmatrix} b_0 & -1 & 0 & \cdots & 0 & 0 \\ a_1 & b_1 & -1 & \cdots & 0 & 0 \\ 0 & a_2 & b_2 & \cdots & 0 & 0 \\ \vdots & \vdots & \vdots & \ddots & \vdots & \vdots \\ 0 & 0 & 0 & \cdots & b_{n-1} & -1 \\ 0 & 0 & 0 & \cdots & a_n & b_n \end{vmatrix},
$$

$$(3.2.19)$$

$$
B_n = K\left(\begin{matrix} a_2 & \cdots & a_n \\ b_1 \ b_2 & \cdots & b_n \end{matrix}\right) := \begin{vmatrix} b_1 & -1 & 0 & \cdots & 0 & 0 \\ a_2 & b_2 & -1 & \cdots & 0 & 0 \\ 0 & a_3 & b_3 & \cdots & 0 & 0 \\ \vdots & \vdots & & \ddots & \vdots & \vdots \\ 0 & 0 & 0 & \cdots & b_{n-1} & -1 \\ 0 & 0 & 0 & \cdots & a_n & b_n \end{vmatrix}.
$$

The above expressions are consequences of the fact that the determinants satisfy the same recursive relations (3.2.2) and the same initial conditions (3.2.3) as $A_n$ and $B_n$. To see this, the determinants are expanded by the last row (or column). Note that we have already used this notation in Lemma 2.5.2 in order to describe the supporting polynomials $\underline{P}_m(x)$, $x\underline{Q}_m(x)$, $(x-1)\overline{P}_m(x)$ and $x(x-1)\overline{Q}_{m-1}(x)$. It will be shown in the subsequent discussion that these polynomials appear as partial denominators in the continued fraction expansion of the Stieltjes transform of the measures $d\mu$, $xd\mu$, $(1-x)d\mu$ and $x(1-x)d\mu$.

We conclude this section with some auxiliary results concerning continued fractions which will become important later.

**Lemma 3.2.3** *The continued fraction*

$$
-\frac{1}{\lvert 2b} - \frac{1}{\lvert 2b} - \frac{1}{\lvert 2b} - \cdots = w \qquad (3.2.20)
$$

*converges if and only if $b \in \mathbb{C}\backslash(-1,1)$. In the case of convergence*

$$
w = -b + \sqrt{b^2 - 1},
$$

*where the branch of the square root is defined such that $\lvert -b + \sqrt{b^2 - 1}\rvert < 1$.*

*Proof* Let $A_n/B_n$ denote the $n$th approximant of the continued fraction

(3.2.20). By (3.2.2) the denominator and numerator satisfy the difference equation

$$A_n = 2bA_{n-1} - A_{n-2} \quad (n \geq 1), \quad A_{-1} = 1, A_0 = 0,$$

$$B_n = 2bB_{n-1} - B_{n-2} \quad (n \geq 1), \quad B_{-1} = 0, B_0 = 1. \tag{3.2.21}$$

In order to solve the first of these equations, we put $A_n = (-s)^n$ and obtain

$$s^2 + 2bs + 1 = 0. \tag{3.2.22}$$

If $b \neq -1, 1$, this equation has two distinct solutions, say, $s_1$ and $s_2 = 1/s_1$. Then the general solution of the first line in (3.2.21) is given by

$$A_n = c_1(-s_1)^n + c_2(-s_2)^n$$

where $c_1$ and $c_2$ are determined by the initial conditions $A_{-1} = 1$, $A_0 = 0$. This yields

$$A_n = \frac{1}{s_2 - s_1} \{s_2(-s_1)^{n+1} - s_1(-s_2)^{n+1}\} \quad (n \geq -1),$$

and by a similar reasoning one obtains

$$B_n = \frac{1}{s_2 - s_1} \{(-s_1)^{n+1} - (-s_2)^{n+1}\} \quad (n \geq -1).$$

Therefore the $n$th approximant of (3.2.20) is given by

$$\frac{A_n}{B_n} = s_1 s_2 \frac{(-s_1)^n - (-s_2)^n}{(-s_2)^{n+1} - (-s_1)^{n+1}} \quad (n \geq 1), \tag{3.2.23}$$

which converges if and only if $|s_1| \neq |s_2|$. Observing (3.2.22), this is satisfied if and only if $b \in \mathbb{C} \setminus [-1, 1]$. Now, if $|s_1| < |s_2|$, it follows from (3.2.23) that

$$w = \lim_{n \to \infty} \frac{A_n}{B_n} = \lim_{n \to \infty} \frac{s_1 - s_2(\frac{s_1}{s_2})^{n+1}}{1 - (\frac{s_1}{s_2})^{n+1}} = s_1 \quad (n \geq 1),$$

which means that the continued fraction in (3.2.20) converges to the solution of (3.2.22) with smaller absolute value, that is,

$$w = s_1 = -b + \sqrt{b^2 - 1},$$

where the branch of the square root is defined by $|-b + \sqrt{b^2 - 1}| < 1$ (note that $s_1 = 1/s_2$). Finally, if $b = 1$, it is easy to see that $A_n = -n$, $B_n = n + 1$, and the continued fraction converges with limit $w = -1$. The case $b = -1$ is treated similarly with limit $w = 1$.  ∎

**Lemma 3.2.4**  *If $\rho_i > 0$, $1 \leq i \leq n$, then*

$$\frac{1|}{|1} - \frac{\rho_1|}{|1 + \rho_1} - \frac{\rho_2|}{|1 + \rho_2} - \cdots - \frac{\rho_n|}{|1 + \rho_n} = 1 + \sum_{\ell=1}^{n} \rho_1 \cdots \rho_\ell. \qquad (3.2.24)$$

*Proof*  The identity follows by noting that

$$m_k(w) = 1 + \rho_k w = \frac{1|}{|1} - \frac{\rho_k|}{|\rho_k} + \frac{1|}{|w}$$

and that both sides of (3.2.24) are given by $m_1 \circ m_2 \circ \cdots \circ m_n(1)$.  ∎

Note that since $\rho_i > 0$, the expression in (3.2.24) is increasing with $n$. Therefore, if $n \to \infty$, the limit is well defined if we include the possibility $+\infty$.

**Lemma 3.2.5**  *If $0 < g_i < 1$, $i \geq 1$, then*

$$S = \frac{1|}{|1} - \frac{g_1|}{|1} - \frac{(1 - g_1)g_2|}{|1} - \cdots = 1 + \sum_{\ell=1}^{\infty} \prod_{i=1}^{\ell} \frac{g_i}{(1 - g_i)}. \qquad (3.2.25)$$

*Proof*  Equation (3.2.25) follows from Lemma 3.2.4 and (3.2.11) by writing the continued fraction for $S$ as

$$\frac{1|}{|1} - \frac{\rho_1|}{|1 + \rho_1} - \frac{\rho_2|}{|1 + \rho_2} - \cdots,$$

where $\rho_i = g_i/(1 - g_i)$.  ∎

## 3.3  CONTINUED FRACTION EXPANSIONS OF THE STIELTJES TRANSFORM

It was indicated in the introductory Section 3.1 that the Stieltjes transform of a probability measure $\mu$ on the interval $[0, 1]$ was related to the canonical moments through the continued fraction expansion (3.1.3). This section develops some details regarding this expansion. Recall that the Stieltjes transform

$$S(z, \mu) = \int_0^1 \frac{d\mu(x)}{z - x}$$

is an analytic function of $z$ in the region $\mathbb{C}\setminus[0, 1]$.

**Theorem 3.3.1** *For any probability measure $\mu$ on the interval $[0, 1]$ and $z \in \mathbb{C} \backslash [0, 1]$, the Stieltjes transform of $\mu$ has the continued fraction expansion*

$$\int_0^1 \frac{d\mu(x)}{z - x} = \frac{1}{|z} - \frac{\zeta_1}{|1} - \frac{\zeta_2}{|z} - \cdots$$

$$= \frac{1}{|z - \zeta_1} - \frac{\zeta_1 \zeta_2}{|z - \zeta_2 - \zeta_3} - \frac{\zeta_3 \zeta_4}{|z - \zeta_4 - \zeta_5} - \cdots \cdots \qquad (3.3.1)$$

*The convergents of the continued fractions converge uniformly in any compact region that has a positive distance from the interval $[0, 1]$.*

Note that from (3.2.16) the second continued fraction is the even contraction of the first one. Both of the expansions on the right side will terminate if $\mu$ has finite support since in this case $p_{n+1} = 0$ or $1$ for some $n$. The measure in these cases will be either $\mu_n^-$ or $\mu_n^+$, and the corresponding canonical moment sequences will be given by $(p_1, \ldots, p_n, 0)$ or $(p_1, \ldots, p_n, 1)$, respectively. For these measures we have

$$\int_0^1 \frac{d\mu_n^-(x)}{z - x} = \frac{1}{|z} - \frac{\zeta_1}{|1} - \cdots - \frac{\zeta_n}{|\tau_n} \qquad (3.3.2)$$

and

$$\int_0^1 \frac{d\mu_n^+(x)}{z - x} = \frac{1}{|z} - \frac{\zeta_1}{|1} - \cdots - \frac{\zeta_n}{|\tau_n} - \frac{q_n}{|\tau_{n+1}}, \qquad (3.3.3)$$

where

$$\tau_n = \tau_n(z) = \begin{cases} z & n \text{ even,} \\ 1 & n \text{ odd.} \end{cases} \qquad (3.3.4)$$

*Proof of Theorem 3.3.1* Let $p_1, p_2, \ldots$ denote the canonical moments of the measure $\mu$ and $\mu_n^-$ be the measure corresponding to the truncated sequence $(p_1, \ldots, p_n, 0)$. The proof will consist of showing (3.3.2). Since $\mu_n^-$ converges weakly to $\mu$, the expansion (3.3.1) follows. The uniform convergence then follows, since the functions $g(x, z) = 1/(z - x)$ are equicontinuous for $z$ in any compact region that has a positive distance from $[0, 1]$. The left-hand side in (3.3.2) is a finite sum of equicontinuous functions and therefore also equicontinuous. Since it converges, the convergence must be uniform in any compact set with positive distance from $[0, 1]$ (see Royden, 1968, p. 178). Equations (3.3.2) and (3.3.3), with $n$ even and odd, constitute four equations. All four will follow provided that it is shown that (3.3.2) holds for $n = 2m - 1$. The other cases are derived by appropriately letting $p_n$ converge to 0 or 1. Thus, if (3.3.2) holds for

$\mu_{2m+1}$ and $p_1, \ldots, p_{2m+1}$, we let $p_{2m+1} \to 0$, obtaining (3.3.2) for $n = 2m$. To obtain (3.3.3), we let $p_n \to 1$ appropriately in (3.3.2).

To show (3.3.2) for $n = 2m - 1$, consider the polynomials $\underline{P}_k(x)$ orthogonal with respect to the measure $\mu$ with leading coefficient one. Let the right side of (3.3.2), or the convergents of (3.3.1), be denoted by

$$\frac{A_n(z)}{B_n(z)} = \frac{1}{|z} - \frac{\zeta_1}{|1} - \cdots - \frac{\zeta_n}{|\tau_n}. \tag{3.3.5}$$

From Lemma 2.5.2 and (3.2.19) it is clear that

$$B_{2m}(z) = K\begin{pmatrix} -\zeta_1 & \cdots & -\zeta_{2m-1} \\ z & 1 & \cdots & z & 1 \end{pmatrix} = \underline{P}_m(z), \qquad m \geq 0,$$

$$\tag{3.3.6}$$

$$B_{2m-1}(z) = K\begin{pmatrix} -\zeta_1 & \cdots & -\zeta_{2m-2} \\ z & 1 & \cdots & 1 & z \end{pmatrix} = z\underline{Q}_{m-1}(z), \qquad m \geq 1.$$

The even numerator polynomials are given by

$$A_{2k}(z) = P^{(1)}_{k-1}(z) = \int_0^1 \frac{\underline{P}_k(z) - \underline{P}_k(t)}{z - t} d\mu(t), \tag{3.3.7}$$

which follows from an even contraction [see the second line in (3.3.1)] and the fact that the polynomials $A_{2k}(z)$ satisfy the same recursion formula (see Corollary 2.3.4) as $\underline{P}_k(z)$ with initial conditions $A_0(z) := 0$ and $A_1(z) = 1$. Equation (3.3.7) holds for $k = 1, \ldots, m$ and any measure $\sigma$ with the same moments $c_k$ as $\mu$ for $k = 1, \ldots, 2m - 1$. Therefore

$$P^{(1)}_{m-1}(z) = \int_0^1 \frac{\underline{P}_m(z) - \underline{P}_m(t)}{z - t} d\mu_{2m-1}(t). \tag{3.3.8}$$

Since the support of $\mu_{2m-1}$ is the set of zeros of $\underline{P}_m(x)$, it follows that

$$\int_0^1 \frac{d\mu_{2m-1}(t)}{z - t} = \frac{P^{(1)}_{m-1}(z)}{\underline{P}_m(z)} = \frac{A_{2m}(z)}{B_{2m}(z)}. \tag{3.3.9}$$

Thus (3.3.2) holds for $n = 2m - 1$, which proves the assertion of the Theorem. ∎

The section will be concluded by giving the corresponding continued fraction expansions for the Stieltjes transform of the measures $x d\mu(x)$, $(1 - x)d\mu(x)$, and $x(1 - x)d\mu(x)$. It is also shown how the Stieltjes transform for a measure on the interval $[a, b]$ can be obtained from results already given, and an example is discussed.

**Theorem 3.3.2**  *For a given probability measure $\mu$ on the interval $[0, 1]$ and $z \in \mathbb{C}\backslash[0, 1]$, the following continued fraction expansions hold:*

1.
$$\int_0^1 \frac{d\mu(x)}{z - x} = \left.\frac{1}{z - \zeta_1}\right| - \left.\frac{\zeta_1 \zeta_2}{z - \zeta_2 - \zeta_3}\right| - \left.\frac{\zeta_3 \zeta_4}{z - \zeta_4 - \zeta_5}\right| - \cdots$$

2.
$$\int_0^1 \frac{x d\mu(x)}{z - x} = \left.\frac{\zeta_1}{z - \zeta_1 - \zeta_2}\right| - \left.\frac{\zeta_2 \zeta_3}{z - \zeta_3 - \zeta_4}\right| - \left.\frac{\zeta_4 \zeta_5}{z - \zeta_5 - \zeta_6}\right| - \cdots$$

3.
$$\int_0^1 \frac{(1 - x)d\mu(x)}{z - x} = \left.\frac{\gamma_1}{z - \gamma_2}\right| - \left.\frac{\gamma_2 \gamma_3}{z - \gamma_3 - \gamma_4}\right| - \cdots$$

4.
$$\int_0^1 \frac{x(1 - x)d\mu(x)}{z - x} = \left.\frac{q_1 p_1 q_2}{z - \gamma_2 - \gamma_3}\right| - \left.\frac{\gamma_3 \gamma_4}{z - \gamma_4 - \gamma_5}\right| - \cdots$$

*The convergents of the continued fractions converge uniformly on any compact region which has positive distance from the interval $[0, 1]$.*

*Proof*   The first equation is Theorem 3.3.1. Formulas in 2, 3, and 4 follow from Theorems 1.5.5, 1.5.4, and Corollary 1.5.6, respectively.   ∎

The continued fraction expansions for probability measures on $[a, b]$ can be derived from the corresponding expansions on $[0, 1]$. Thus, if $\mu_{ab}$ is a measure on $[a, b]$ and is transformed by the linear transformation $y = a + x(b - a)$ to $\mu$ on $[0, 1]$, then

$$\int_a^b \frac{d\mu_{ab}(y)}{z - y} = \frac{1}{b - a} \int_0^1 \frac{d\mu(x)}{w - x},$$

where $w = (z - a)/(b - a)$. Note that by Theorem 1.3.2 the measures $\mu$ and $\mu_{ab}$ have the same canonical moments. Using the expansion from Theorem 3.3.1 and making the equivalence operation (3.2.11) with $c_{2j-1} = b - a$ and $c_{2j} = 1$, we may arrive at

**Theorem 3.3.3** *If $\mu_{ab}$ is a probability measure on the interval $[a,b]$ and $z \in \mathbb{C}\backslash[a,b]$, then the Stieltjes transform of $\mu_{ab}$ has the continued fraction expansion*

$$\int_a^b \frac{d\mu_{ab}(y)}{z-y} = \frac{1|}{|z-a} - \frac{\zeta_1(b-a)|}{|1} - \frac{\zeta_2(b-a)|}{|z-a} - \cdots$$

$$= \frac{1|}{|z-a-\zeta_1(b-a)} - \frac{\zeta_1\zeta_2(b-a)^2|}{|z-a-(\zeta_2+\zeta_3)(b-a)}$$

$$- \frac{\zeta_3\zeta_4(b-a)^2|}{|z-a-(\zeta_4+\zeta_5)(b-a)} - \cdots$$

*The convergents of the continued fraction converge uniformly on any compact region that has positive distance from the interval $[a,b]$.*

**EXAMPLE 3.3.4** In this example we will use Theorem 3.3.3 for determining the canonical moments of the Beta distribution. On the interval $[0,1]$ the moments of this distribution are

$$c_n = \frac{1}{B(\alpha+1,\beta+1)}\int_0^1 x^{\beta+n}(1-x)^\alpha dx = \frac{(\beta+1)_n}{(\alpha+\beta+2)_n},$$

where $(x)_n = x(x+1).\ .\ .(x+n-1)$. Introducing the notation for the *Hypergeometric function*

$$_2F_1(a,b,c;z) = 1 + \frac{ab}{c}z + \frac{(a)_2(b)_2}{(c)_2}\frac{z^2}{2!} + \cdots,\tag{3.3.10}$$

it follows that the corresponding power series (3.1.2) for the Stieltjes transform of the Beta distribution is given by

$$z\int_0^1 \frac{w_{(\alpha,\beta)}(x)}{z-x}dx = z\sum_{n=0}^\infty \frac{c_n}{z^{n+1}}$$

$$= \sum_{n=0}^\infty \frac{(\beta+1)_n}{(\alpha+\beta+2)_n}\frac{1}{z^n}.\tag{3.3.11}$$

$$=_2F_1\left(\beta+1,1,\alpha+\beta+2;\frac{1}{z}\right).$$

In the following we will derive a continued fraction expansion for a Hypergeometric series of the form (3.3.10) which was originally found by Gauss in

(1812). It is easy to check that

$$_2F_1(a,b,c;z) =\,_2F_1(a,b+1,c+1;z) - \frac{a(c-b)}{c(c+1)}z\cdot_2F_1(a+1,b+1,c+2;z),$$

or equivalently

$$\frac{_2F_1(a,b+1,c+1;z)}{_2F_1(a,b,c;z)} = \left[1 - z\frac{a(c-b)}{c(c+1)}\frac{_2F_1(a+1,b+1,c+2;z)}{_2F_1(a,b+1,c+1;z)}\right]^{-1}. \quad (3.3.12)$$

Observing that

$$\frac{_2F_1(a+1,b+1,c+2;z)}{_2F_1(a,b+1,c+1;z)} = \frac{_2F_1(b+1,a+1,c+2;z)}{_2F_1(b+1,a,c+1;z)},$$

we obtain from (3.3.12),

$$\frac{_2F_1(b+1,a+1,c+2;z)}{_2F_1(b+1,a,c+1;z)}$$

$$= \left[1 - \frac{(b+1)(c+1-a)}{(c+1)(c+2)}z\cdot\frac{_2F_1(b+2,a+1,c+3;z)}{_2F_1(b+1,a+1,c+2;z)}\right]^{-1}.$$

Applying these relations successively yields the following continued fraction expansion (provided the right side converges):

$$\frac{_2F_1(a,b+1,c+1;z)}{_2F_1(a,b,c;z)} = \frac{1|}{|1} - \frac{\frac{a(c-b)z}{c(c+1)}|}{|1} - \frac{\frac{(b+1)(c+1-a)z}{(c+1)(c+2)}|}{|1} - \cdots$$

$$- \frac{\frac{(b+n)(c-a+n)z}{(c+2n-1)(c+2n)}|}{|1} - \frac{\frac{(a+n)(c-b+n)z}{(c+2n)(c+2n+1)}|}{|1} - \cdots.$$

By (3.3.11), putting $a = \beta+1$, $b = 0$, $c = \alpha+\beta+1$ and replacing $z$ by $1/z$ gives the continued fraction expansion for the Stieltjes transform of the Beta distribution:

$$\frac{1}{z}\cdot_2F_1\left(\beta+1,1,\alpha+\beta+2;\frac{1}{z}\right) =$$

$$\frac{1|}{|z} - \frac{\frac{\beta+1}{\alpha+\beta+2}|}{|1} - \frac{\frac{\alpha+1}{(\alpha+\beta+2)(\alpha+\beta+3)}|}{|z} - \cdots - \frac{\frac{(\alpha+n)n}{(\alpha+\beta+2n)(\alpha+\beta+2n+1)}|}{|z}$$

$$- \frac{\frac{(\alpha+\beta+n+1)(\beta+n+1)}{(\alpha+\beta+2n+1)(\alpha+\beta+2n+2)}|}{|1} \cdots$$

Comparing this expansion with Theorem 3.3.3 yields

$$\zeta_{2n-1} = \frac{(\beta + n)(\alpha + \beta + n)}{(2n - 1 + \alpha + \beta)(2n + \alpha + \beta)},$$

$$\zeta_{2n} = \frac{(\alpha + n)n}{(2n + \alpha + \beta)(2n + 1 + \alpha + \beta)}, \qquad n = 1, 2, \ldots.$$

By a straightforward induction we now obtain for the canonical moments of the Beta distribution:

$$p_{2n-1} = \frac{\beta + n}{2n + \alpha + \beta}, \quad p_{2n} = \frac{n}{2n + 1 + \alpha + \beta}, \qquad n = 1, 2, \ldots.$$

## 3.4 MORE SYMMETRY OR DUALITY

In Section 1.3 of Chapter 1 some basic symmetry or invariance properties of the canonical moments were presented. For example, if the probability measure $\mu$ on $[0, 1]$ has canonical moments $p_1, p_2, \ldots$ and the measure $\mu$ is transformed to $\nu$ on $[0, 1]$ via the transformation $y = 1 - x$, the resulting canonical moments were then $q_1, p_2, q_3, p_4, \ldots$, that is, the odd $p_i$ change to $q_i$. An interesting question is how to relate the measure with canonical moments $q_1, q_2, q_3, \ldots$ to the original measure corresponding to $p_1, p_2, p_3, \ldots$. The answer to this question is given in Theorem 3.4.2 below and is a consequence of a careful analysis of the finite expansions for the Stieltjes transforms of $\mu_n^-$ and $\mu_n^+$ in Equations (3.3.2) and (3.3.3). It was indicated that these expansions can be written as the ratio of various orthogonal polynomials. Since the $p_i$ values will be all switched to $q_i$, there will be a need of showing the explicit dependence of the polynomials $\underline{P}_n$, $\underline{Q}_n$, $\overline{P}_n$, and $\overline{Q}_n$ on the canonical moment sequence $p = (p_1, p_2, \ldots)$. This dependence will be explicitly given by writing $\underline{P}_n(x) = \underline{P}_n(x, p)$, and so on. This will allow us to use $\underline{P}_n(x, q)$, which is defined of course using the same recursion formula in Corollary 2.3.4, part 1, by replacing $p_i$ by $q_i$.

**Theorem 3.4.1** *If $n < N(c)$,*

$$S_n^-(z, p) := S(z, \mu_n^-) = \begin{cases} \dfrac{\overline{Q}_{m-1}(z, q)}{\underline{P}_m(z, p)} & n = 2m - 1, \\[4mm] \dfrac{\overline{P}_m(z, q)}{z \underline{Q}_m(z, p)} & n = 2m, \end{cases}$$

$$S_n^+(z, p) := S(z, \mu_n^+) = \begin{cases} \dfrac{\underline{P}_m(z, q)}{z(z - 1)\overline{Q}_{m-1}(z, p)} & n = 2m - 1, \\[4mm] \dfrac{\underline{Q}_m(z, q)}{(z - 1)\overline{P}_m(z, p)} & n = 2m. \end{cases}$$

*Proof* The Stieltjes transforms $S(z, \mu_n^-)$ and $S(z, \mu_n^+)$ are given in equations (3.3.2) and (3.3.3), respectively. These are written as the ratio of determinants using equation (3.2.19). The denominators are then written using Lemma 2.5.2, and the numerators are expressed using Lemmas 2.5.3, 2.5.4, 2.5.5, and Theorem 2.5.6. ∎

With the aid of Theorem 3.4.1 the Stieltjes transform of the measure corresponding to the sequence $(q_1, q_2, \ldots, q_n, 0)$ or $(q_1, q_2, \ldots, q_n, 1)$ can be expressed in terms of the transforms involving the $p_i$. In the next theorem the notation from Theorem 3.4.1 is used. In addition $S(z, p)$ will denote the Stieltjes transform of the measure with canonical moments $p = (p_1, p_2, \ldots)$.

**Theorem 3.4.2** *If* $n < N(c)$,

$$z(z - 1)S_n^-(z, q)S_n^+(z, p) = 1, \tag{3.4.1}$$

$$z(z - 1)S(z, q)S(z, p) = 1. \tag{3.4.2}$$

*Proof* Equation (3.4.1) is immediate from Theorem 3.4.1. The second equation is obtained by taking a limit as $n$ goes to infinity. ∎

**EXAMPLE 3.4.3** Consider the sequence of canonical moments $p = (p, p, \ldots)$. The continued fraction expansion for the Stieltjes transform of the corresponding measure on the interval $[0, 1]$ is obtained from Theorem 3.3.1:

$$S(z, p) = \frac{1}{|z - p} - \frac{p^2 q}{|z - 2pq} - \frac{p^2 q^2}{|z - 2pq} - \cdots$$

$$= \frac{1}{|z - p} - \frac{p}{|2b} - \frac{1}{|2b} - \frac{1}{|2b} - \cdots,$$

where $b = (z - 2pq)/2pq$. An application of Lemma 3.2.3 shows that this continued fraction converges whenever

$$b = \frac{z - 2pq}{2pq} \in \mathbb{C} \setminus (-1, 1),$$

and the limit is given by

$$S(z, p) = \left[ z - p + p \left\{ -\frac{z - 2pq}{2pq} + \sqrt{\frac{(z - 2pq)^2}{4p^2 q^2} - 1} \right\} \right]^{-1}$$

$$= \frac{(2q - 1)z - \sqrt{z(z - 4pq)}}{2pz(1 - z)}, \tag{3.4.3}$$

where the branch of the square root is defined by

$$\left| -\frac{(z - 2pq)}{2pq} + \sqrt{\frac{(z - 2pq)^2}{4p^2q^2} - 1} \right| < 1. \tag{3.4.4}$$

The Stieltjes transform of the measure corresponding to the sequence $q = (q, q, \ldots)$ is either obtained by switching $p$ and $q$ in (3.4.3) or by an application of Theorem 3.4.2. This gives

$$S(z, q) = \left[ z(z - 1)S(z, p) \right]^{-1} = \frac{2p}{-(2q - 1)z + \sqrt{z(z - 4pq)}}$$

$$= \frac{(2p - 1)z - \sqrt{z(z - 4pq)}}{2qz(1 - z)}.$$

The measures corresponding to these two transforms will be given in Example 3.6.8 below.


## 3.5  CANONICAL MOMENTS AND ROOTS OF POLYNOMIALS

The finite canonical moment sequences and the measures $\mu$ with finite support are in one-to-one correspondence. The measures corresponding to $(p_1, \ldots, p_n, 0)$ and $(p_1, \ldots, p_n, 1)$ were denoted by $\mu_n^-$ and $\mu_n^+$. The two measures $\mu_n^-$ and $\mu_n^+$ both have precisely $n$ "free" parameters. In Section 1.2 it was shown that the roots of $\mu_n^-$ and $\mu_n^+$ could be expressed by (1.2.8) and (1.2.9) or

and $(3.5.1)$

Consider, for example, $\mu_{2m-1}^-$. The support in this case has $m$ points $0 < t_1 < t_2 < \ldots < t_m < 1$. These points together with the $m - 1$ "free" weights $\lambda_1, \lambda_2, \ldots, \lambda_m$ $(\sum_{i=1}^{m} \lambda_i = 1)$ give $2m - 1 = n$ parameters. The $n$ variables in $(p_1, \ldots, p_{2m-1}, 0)$ are in a one-to-one correspondence with $(t_1, \ldots, t_m, \lambda_1, \ldots, \lambda_{m-1})$. The other cases are entirely similar.

Now note that in each case, $n = 2m$ or $2m - 1$, the total number of interior zeros is precisely $n$. Thus, if $n = 2m - 1$, there are $2m - 1$ quantities $t_1, \ldots, t_m$, and $s_1, \ldots, s_{m-1}$. These are the roots of the polynomials $\underline{P}_m(x)$ and $\overline{Q}_{m-1}(x)$, respectively. If $n = 2m$, there are $2m$ quantities $t_1, \ldots, t_m$ and $s_1, \ldots, s_m$ that are the roots of $\underline{Q}_m(x)$ and $\overline{P}_m(x)$. The following theorem shows precisely how the interior roots of the upper and lower principal representations determine the canonical moments $(p_1, \ldots, p_n)$.

**Theorem 3.5.1**

$$\frac{(z-1)\overline{P}_m(z)}{z\underline{Q}_m(z)} = 1 - \frac{q_{2m}}{\big|\ z\ } - \frac{\zeta_{2m}}{\big|\ 1\ } - \cdots - \frac{\zeta_1}{\big|\ z\ }, \qquad (3.5.2)$$

$$\frac{z(z-1)\overline{Q}_{m-1}(z)}{\underline{P}_m(z)} = z - \frac{q_{2m-1}}{\big|\ 1\ } - \frac{\zeta_{2m-1}}{\big|\ z\ } - \cdots - \frac{\zeta_1}{\big|\ z\ }. \qquad (3.5.3)$$

*Proof*   The proof is based on considering the expansion (3.3.3), Lemma 2.5.2, and Lemma 3.2.2. For equation (3.5.2) we have

$$\frac{(z-1)\overline{P}_m(z)}{z\underline{Q}_m(z)} = \frac{K\left(\begin{array}{ccccc} -\zeta_1 & \cdots & -\zeta_{2m} & -q_{2m} \\ z & 1 & \cdots & z & 1 \end{array}\right)}{K\left(\begin{array}{cccc} -\zeta_1 & \cdots & -\zeta_{2m} \\ z & 1 & \cdots & z \end{array}\right)}$$

$$= 1 - \frac{q_{2m}}{\big|\ z\ } - \frac{\zeta_{2m}}{\big|\ 1\ } - \cdots - \frac{\zeta_1}{\big|\ z\ }.$$

The other case is very similar and therefore omitted. ∎

In each case the mapping under consideration is "onto" as well as one-to-one. By this we mean, for example, that any set of $2m - 1$ points $0 < t_1 < s_1 < t_2 < \ldots < s_{m-1} < t_m < 1$ determines a moment point $p_1, \ldots, p_{2m-1}$. To see this, let $\lambda_1, \ldots, \lambda_m$ denote possible weights for $t_1, \ldots, t_m$ and $\mu_0, \mu_1, \ldots, \mu_m$ weights for $s_0 = 0 < s_1 < \ldots < s_m = 1$. Then the requirement is

$$\sum_{i=0}^{m} \mu_i s_i^k - \sum_{i=1}^{m} \lambda_i t_i^k = \begin{cases} 0 & \text{if } k = 0, 1, \ldots, 2m - 1, \\ \\ c & \text{if } k = 2m, \end{cases}$$

where $c$ is potentially $c_{2m}^+ - c_{2m}^-$. There are $2m + 1$ equations in $2m + 1$ unknowns. If the $s_i$ and $t_i$ are written in order, we find that for $c > 0$ the corresponding $\mu_i$ and $\lambda_i$ are positive. The value of $c$ is chosen so that $\sum_{i=1}^{m} \lambda_i = \sum_{i=0}^{m} \mu_i = 1$. These weights determine the unique moments given by the expressions

$$c_k = \sum_{i=0}^{m} \mu_i s_i^k = \sum_{i=1}^{m} \lambda_i t_i^k \quad k = 0, 1, \ldots, 2m - 1.$$

A simple calculation on the ratio of the polynomials in either (3.5.2) or (3.5.3) shows that in both cases $q_n$ is the sum of all the roots in the numerator minus the sum of all the roots in the denominator. Referring to (3.5.1), let

$$I_i = \begin{cases} s_i - t_{i-1}, & i = 1, \ldots, m+1, \; n = 2m, \\ s_i - t_i, & i = 1, \ldots, m+1, \; n = 2m + 1, \end{cases}$$

denote the length of the interval between successive roots where we start with a $t_i$ value and go to the next $s_j$ value. The corresponding intervals $(t_{i-1}, s_i)$ $(n = 2m)$ or $(t_i, s_i)$ $(n = 2m + 1)$ are called the *lower canonical intervals*. The remaining intervals are called *upper canonical intervals*. The reason for this nomenclature is that if one adds another coordinate $p_{n+1}$, the upper (lower) principal representation of $(p_1, \ldots, p_n, p_{n+1})$ has roots in each of the upper (lower) canonical intervals.

**Corollary 3.5.2**    *If $n \leq N(c)$ and $n = 2m$ or $2m + 1$, then*

$$q_n = \sum_{i=1}^{m+1} I_i.$$

Corollary 3.5.2 and expressions for the next term in Equation (3.5.2) are given in Skibinsky (1986).

**EXAMPLE 3.5.3**    If $n = 1$ $(m = 0)$, then $s_0 = 0 < t_1 < s_1 = 1$, $p_1 = t_1$ and $q_1 = s_1 - t_1 = 1 - t_1$. The interval $(p_1, 1)$ is a lower canonical interval, and $(0, p_1)$ is an upper canonical interval. The moment point $(p_1, p_2)$ has a lower principal representation supported on 0 and $t_1'$ for some $t_1' \in (p_1, 1)$ and an upper principal representation supported on $s_1'$ and 1 for some $s_1' \in (0, p_1)$. This follows from Theorem 2.2.3 and Corollary 2.3.4. Similarly for $n = 2$ $(m = 1)$, $t_0 = 0$, $t_1 = p_1 + q_1 p_2 = p_2 + q_2 p_1$, $s_1 = p_1 q_2$, and $s_2 = 1$. Then

$$I_1 + I_2 = s_1 - t_0 + s_2 - t_1 = q_2,$$

the lower canonical intervals are $(0, p_1 q_2)$ and $(p_2 + p_1 q_2, 1)$, and the upper canonical interval is $(p_1 q_2, p_2 + p_1 q_2)$.

The proof of Theorem 3.5.1 used Lemma 3.2.2 and the expansion of the Stieltjes transform for $\mu_n^+$ in (3.3.3). If the measure $\mu_n^-$ and its expansion (3.3.2) is used instead the following theorem is obtained:

**Theorem 3.5.4**

$$\frac{z\underline{Q}_m(z)}{\underline{P}_m(z)} = z - \frac{\zeta_{2m}}{|1} - \frac{\zeta_{2m-1}}{|z} - \cdots - \frac{\zeta_1}{|z}, \tag{3.5.4}$$

$$\frac{\underline{P}_m(z)}{z\underline{Q}_{m-1}(z)} = 1 - \frac{\zeta_{2m-1}}{|z} - \frac{\zeta_{2m-2}}{|1} - \cdots - \frac{\zeta_1}{|z} \tag{3.5.5}$$

Note that, for example, if $n = 2m$, then (3.5.4) expresses $\zeta_1, \zeta_2, \ldots, \zeta_{2m}$ or $p_1, \ldots, p_{2m}$ in terms of the roots of the lower principal representations for $(p_1, \ldots, p_{2m})$ and $(p_1, \ldots, p_{2m-1})$. The interpretation for $\zeta_n$ in Theorem 3.5.4 is the same as for $q_n$ in Theorem 3.5.1. Thus $\zeta_{2m}$ is the sum of the zeros of the polynomial $z\underline{Q}_m(z)$ minus the sum of the zeros of of the polynomial $\underline{P}_m(z)$. For example, if $m = 1$, then $\underline{Q}_1(z) = z - (\zeta_1 + \zeta_2)$ and $\underline{P}_1(z) = z - \zeta_1$, and $\zeta_2$ is the difference of the roots.

## 3.6   INVERSION OF THE STIELTJES TRANSFORM

The problem of determining the measure $\mu$ and some of its properties from the Stieltjes transform will be considered in this section. A few brief remarks about the case where $\mu$ has finite support will be mentioned before discussing the general inversion formula.

If $\mu$ has finite support, then

$$S(z) = S(z, \mu) = \int \frac{d\mu(x)}{z - x} = \sum_{i=1}^{n} \frac{\lambda_i}{z - x_i}, \tag{3.6.1}$$

where the support points of the measure $\mu$ are denoted by $x_1, x_2, \ldots, x_n$ with corresponding weights $\lambda_1, \lambda_2, \ldots, \lambda_n$. There are therefore essentially two problems in determining the measure $\mu$, namely finding the support points $x_k$ and determining the weights $\lambda_k$.

The support points, in the present situation, will be simple poles of the function $S(z, \mu)$. For example, if $S(z) = z/(z^2 - 1)$, then one can perform a simple partial fraction decomposition to give

$$S(z) = \frac{z}{z^2 - 1} = \frac{1}{2}\left(\frac{1}{z - 1} + \frac{1}{z + 1}\right).$$

The corresponding measure $\mu$ then has mass $\frac{1}{2}$ at the points $x = \pm 1$.

Generally the transform of a measure with finite support will appear as the ratio of two polynomials as in Theorem 3.4.1 where the polynomials have no common zeros. The zeros of the denominator determine the support of $\mu$. If $x_k$ is known to be a support point, then, by the partial fraction expansion (3.6.1), the corresponding weight is given by

$$\lambda_k = \lim_{z \to x_k} (z - x_k) S(z, \mu). \tag{3.6.2}$$

If the transform is written as

$$S(z, \mu) = \frac{P_{n-1}^{(1)}(z)}{P_n^*(z)}, \tag{3.6.3}$$

where $P_n^*(z) = \prod_{i=1}^n (z - x_i)$, then $x_1, \ldots, x_n$ provide the support of $\mu$ and the weights $\lambda_k$ in (3.6.1) or (3.6.2) can alternatively be written as

$$\lambda_k = \frac{P_{n-1}^{(1)}(x_k)}{\frac{d}{dx} P_n^*(x)|_{x=x_k}}, \tag{3.6.4}$$

since in this case (3.6.2) quickly reduces to (3.6.4).

The case where the canonical moments are given will be of some importance. Recall that the canonical moments must be calculated with reference to some basic interval. If the interval is $[a, b]$ and $\xi$ is supported at $n$ points (i.e., $\zeta_{2n-1}\zeta_{2n} = 0$), then by Theorem 3.3.3 the Stieltjes transform has an expansion

$$\int_a^b \frac{d\mu(x)}{z - x} = \frac{1}{\left| z - a - (b-a)\zeta_1 \right.} - \frac{(b-a)^2\zeta_1\zeta_2}{\left| z - a - (b-a)(\zeta_2 + \zeta_3) \right.} - \cdots$$
$$- \frac{(b-a)^2\zeta_{2n-3}\zeta_{2n-2}}{\left| z - a - (b-a)(\zeta_{2n-2} + \zeta_{2n-1}) \right.}.$$

It follows from (3.2.2) and (3.2.3) that the numerator and denominator polynomials in (3.6.3) satisfy the same recursion formula with different initial conditions. Thus, if

$$W_{k+1}(x) = (x - a - (b-a)(\zeta_{2k} + \zeta_{2k+1})) W_k(x)$$
$$- (b-a)^2 \zeta_{2k-1}\zeta_{2k} W_{k-1}(x) \tag{3.6.5}$$

$(\zeta_0 = 0)$, then

$$P_k^*(x) = W_k(x) \qquad \text{when } W_{-1}(x) \equiv 0, \ W_0(x) \equiv 1, \tag{3.6.6}$$

and

$$P_k^{(1)}(x) = W_{k+1}(x) \qquad \text{when } W_0(x) \equiv 0, \ W_1(x) \equiv 1. \tag{3.6.7}$$

**Theorem 3.6.1** *Let $\mu$ denote a measure on the interval $[a, b]$ supported on $n$ points with canonical moments $p_1, p_2, \ldots$. Then the support of $\mu$ consists of the zeros of $P_n^*(x)$ determined from (3.6.6), and the weights $\lambda_k$ are given by (3.6.4) where $P_k^{(1)}(x)$ is determined by (3.6.7).*

**EXAMPLE 3.6.2** In this example we consider the moment space $M_n$ for $n = 1, 2, 3$ and the lower principal representations of moment points. The measure $\mu_1^-$ corresponding to the sequence $(p_1, 0)$ on the interval $[0, 1]$ has mass 1 at the point $p_1 \in (0, 1)$. The measure $\mu_2^-$ corresponding to $(p_1, p_2, 0)$ has two support points and is obtained from Theorem 3.6.1. We calculate $P_2^*$ and $P_2^{(1)}$ by (3.6.5), (3.6.6), and (3.6.7) and obtain

$$P_2^*(x) = x^2 - x(\zeta_1 + \zeta_2),$$

$$P_1^{(1)}(x) = x - \zeta_2.$$

Observing (3.6.4), we find that $\mu_2^-$ is supported at the points 0 and $p_1 + q_1 p_2$ with masses proportional to $q_1 p_2$ and $p_1$, respectively. Similarly the measure $\mu_3^-$ corresponding to $(p_1, p_2, p_3, 0)$ on the interval $[0, 1]$ puts masses proportional to $-\gamma + \sqrt{\alpha^2 - \beta}$ and $\gamma + \sqrt{\alpha^2 - \beta}$ at the points $\alpha - \sqrt{\alpha^2 - \beta}$ and $\alpha + \sqrt{\alpha^2 - \beta}$, where $2\alpha = \zeta_1 + \zeta_2 + \zeta_3$, $\beta = \zeta_1 \zeta_3$ and $2\gamma = \zeta_1 - \zeta_2 - \zeta_3$. Upper principal measures can be treated similarly and are discussed in Section 4.2.

**EXAMPLE 3.6.3** Consider the interval $[-1, 1]$ and the sequence of canonical moments

$$p_i \equiv \tfrac{1}{2}, \qquad i = 1, \ldots, 2n - 1, \quad p_{2n} = 0.$$

The representing measure corresponds to a lower principal representation supported at $n$ points in $(-1, 1)$. Observing (3.6.5) (with $[a, b] = [-1, 1]$), we obtain the recursive relation

$$W_{k+1}(x) = x W_k(x) - \tfrac{1}{4} W_{k-1}(x)$$

whenever $k \geq 2$. From (3.6.6) and (3.6.7) the initial conditions for $P_k^*(x)$ and $P_k^{(1)}(x)$ are

$$W_1(x) = x, \quad W_2(x) = x^2 - \tfrac{1}{2},$$

$$W_1(x) = 1, \quad W_2(x) = x,$$

respectively. Comparing this relation with (2.1.7) and (2.1.8), it follows by induction that the polynomials $P_n^*(x)$ and $P_n^{(1)}(x)$ are the monic versions of the Chebyshev polynomials of the first and second kind, that is,

$$P_n^*(x) = \frac{1}{2^{n-1}} T_n(x),$$

$$P_{n-1}^{(1)}(x) = \frac{1}{2^{n-1}} U_{n-1}(x).$$

From the trigonometric representation (2.1.5) we obtain for the support points

$$x_k = \cos\left(\frac{2k-1}{2n}\pi\right), \qquad k = 1, \ldots, n,$$

and (2.1.5) and (2.1.6) yield $dT_n(x)/dx = nU_{n-1}(x)$. Consequently

$$\mu(\{x_k\}) = \frac{U_{n-1}(x_k)}{\frac{d}{dx} T_n(x)|_{x=x_k}} = \frac{1}{n}, \qquad k = 1, \ldots, n,$$

and $\mu$ is the uniform distribution on the points $\{\cos((2k-1)\pi/2n)\}_{k=1}^n$.

If the transform is given by some more analytical expression, then more general inversion formula are needed. Roughly speaking the measure $\mu$ is carried by the imaginary part of the Stieltjes transform $S(z, \mu)$. Note that if $z = u - iv$, then

$$\frac{1}{z-x} = \frac{1}{u-x-iv} = \frac{u-x+iv}{(u-x)^2 + v^2}$$

and

$$\frac{1}{\pi} \operatorname{Im} S(z, \mu) = \int_{-\infty}^{\infty} \frac{v}{\pi[(u-x)^2 + v^2]} d\mu(x)$$

$$= \int_{-\infty}^{\infty} \frac{1}{v} K\left(\frac{u-x}{v}\right) d\mu(x).$$

(3.6.8)

Here

$$K(y) = \frac{1}{\pi(1+y^2)}$$

is the *Cauchy density*. The inversion of the Stieltjes transform relies on the fact that the function

$$K_v(u-x) = \frac{1}{v} K\left(\frac{u-x}{v}\right)$$

converges to a point mass or Dirac measure at $u$ as $v$ approaches zero. Note that $K$ is symmetric about zero and decreases to zero as $|y| \to \infty$. The following properties can be shown by standard analysis:

1. $\int_{-\infty}^{\infty} K_v(u - x)dx = 1$.

2. $\lim_{v \to 0} \int_{|u-x| \le \delta} K_v(u - x)dx = 1 \ (\delta > 0)$.

3. $\lim_{v \to 0} \max_{|u-x| > \delta} K_v(u - x) = 0 \ (\delta > 0)$.

Note also that $\overline{S(z)} = S(\bar{z})$ from the definition of $S(z)$ in (3.6.1) and hence that

$$\text{Im } S(u - iv) = \frac{1}{2i}[S(u - iv) - S(u + iv)]. \tag{3.6.9}$$

**Theorem 3.6.4** *If $d\mu(x) = f(x)dx$ in a neighborhood of $u_0$ and $f(x)$ is continuous at $u_0$, then*

$$\lim_{v \to 0} \frac{1}{2\pi i}[S(u_0 - iv) - S(u_0 + iv)] = \lim_{v \to 0} \frac{1}{\pi} \text{Im} S(u_0 - iv) = f(u_0).$$

*Proof* By the discussion in (3.6.8) we have

$$\frac{1}{\pi} \text{Im } S(u_0 - iv) = \int_{-\infty}^{\infty} K_v(u_0 - x)d\mu(x)$$

$$\tag{3.6.10}$$

$$= \int_A K_v(u_0 - x)f(x)dx + \int_{A^c} K_v(u_0 - x)d\mu(x),$$

where $A = \{x \in \mathbb{R} \mid |u_0 - x| \le \delta\}$ and $\delta$ is sufficiently small. The second integral converges to zero by property 3 of the function $K_v$. For the remaining term we obtain, using property 1,

$$\int_A K_v(u_0 - x)f(x)dx - f(u_0) = \int_A K_v(u_0 - x)(f(x) - f(u_0))dx$$

$$- f(u_0) \int_{A^c} K_v(u_0 - x)dx.$$

Thus the first integral in (3.6.10) converges to $f(u_0)$ by the continuity assumption on $f$ and property 2. ∎

Isolated mass points can usually be determined by the singularities of $S(z)$ and the previous discussion; however, some care should be exercised because a given support point may be an accumulation point of other mass points and singular parts of $\mu$ may be present. The mass points of $\mu$ can always be simply determined by the following:

**Theorem 3.6.5**   *For every $u_0 \in (-\infty, \infty)$,*

$$\lim_{v \to 0} v\mathrm{Im}\, S(u_0 - iv) = \lim_{v \to 0} \int_{-\infty}^{\infty} \frac{v^2}{(x - u_0)^2 + v^2} \, d\mu(x) = \mu(\{u_0\}).$$

*Proof*   The proof uses the fact that the function

$$\frac{v^2}{(x - u_0)^2 + v^2}$$

has value one at $x = u_0$ and otherwise converges to zero as $v \to 0$. Since it is bounded and $\mu$ is a finite measure, the result follows from the dominated convergence theorem.   ∎

To describe the general inversion formula of a Stieltjes transform, let

$$\bar{\mu}(s) = \tfrac{1}{2}\left\{\mu((-\infty, s]) + \mu((-\infty, s))\right\} \tag{3.6.11}$$

be the cumulative distribution of $\mu$, "normalized" at points of discontinuity.

**Theorem 3.6.6**

$$\bar{\mu}(t) - \bar{\mu}(s) = \frac{1}{\pi} \lim_{v \to 0^+} \mathrm{Im} \int_{s-iv}^{t-iv} S(z)\,dz$$

$$= \frac{1}{\pi} \lim_{v \to 0^+} \mathrm{Im} \int_{s}^{t} S(u - iv)\,du$$

$$= \frac{1}{2\pi i} \lim_{v \to 0^+} \int_{s}^{t} (S(u - iv) - S(u + iv))\,du.$$

*Proof*   From (3.6.8) we have

$$\frac{1}{\pi} \mathrm{Im} \int_{s}^{t} S(u - iv)\,du = \int_{s}^{t} \int_{-\infty}^{\infty} K_v(u - x)\,d\mu(x)\,du$$

$$= \int_{-\infty}^{\infty} \int_{s}^{t} K_v(u - x)\,du\,d\mu(x),$$

where the interchange of integrals is permissible since the integrand is positive.

By properties of $K$ the inner integral

$$g_v(x) = \int_s^t K_v(u-x)du = \frac{1}{\pi}\left[\arctan\left(\frac{t-x}{v}\right) - \arctan\left(\frac{s-x}{v}\right)\right]$$

is bounded by one and if $v \to 0^+$, then $g_v$ converges to

$$g(x) = \begin{cases} 0, & x \notin [s,t], \\ \frac{1}{2}, & x = s \text{ or } t, \\ 1, & x \in (s,t). \end{cases}$$

Since $\mu$ has finite measure the result follows again from the dominated convergence theorem. ∎

Usually the mass points of $\mu$ can be separated out using Theorem 3.6.5, and the density part can be determined immediately from the imaginary part of $S(u)$. A precise result in this regard is given by

**Theorem 3.6.7**   *If $S(z,\mu)$ can be extended from the lower half plane to a continuous function in a neighborhood of $u_0$ on the real line, then $\mu$ has a continuous density in a neighborhood of $u_0$ given by*

$$f(u) = \frac{1}{\pi} \operatorname{Im} S(u,\mu).$$

*Proof*   With some abuse of notation, we let $S(u)$ denote the value of the extension for $u$ in a neighborhood of $u_0$. The assumption states that $S(z)$ is continuous in some region $z = u - iv$, where $|u - u_0| \le \delta$ and $0 \le v \le \delta$ for some $\delta$. Since $S(z)$ is continuous, it follows that

$$\frac{1}{\pi} \operatorname{Im} S(u - iv)$$

is bounded and converges to $(1/\pi) \operatorname{Im} S(u)$ as $v \to 0^+$. By the dominated convergence theorem and Theorem 3.6.6,

$$\overline{\mu}(t) - \overline{\mu}(s) = \lim_{v \to 0^+} \int_s^t \frac{1}{\pi} \operatorname{Im} S(u - iv)du$$

$$= \int_s^t \frac{1}{\pi} \operatorname{Im} S(u)du$$

for any $s$ and $t$ in $(u_0 - \delta, u_0 + \delta)$. Therefore $\mu$ has a density in the interval $(u_0 - \delta, u_0 + \delta)$ given by $(1/\pi) \operatorname{Im} S(u)$. ∎

**EXAMPLE 3.6.8**   Consider the measure $\mu^{(p)}$ corresponding to the sequence of canonical moments $p = (p, p, p, \ldots)$. It is shown in Example 3.4.3 that the Stieltjes transform is given by

$$S(z, p) = \frac{(2q - 1)z - \sqrt{z(z - 4pq)}}{2pz(1 - z)},$$

where the branch of the square root is determined by (3.4.4). For every $u_0 \in (0, 4pq)$ the transform $S(z, p)$ can be extended from the lower half-plane to a continuous function in a neighborhood of $u_0$. This extension uses the square root for which $\mathrm{Im}(\sqrt{\cdots}) < 0$. It now follows from Theorem 3.6.7 that $\mu^{(p)}$ has a density in $(0, 4pq)$ given by

$$\frac{1}{\pi} \, \mathrm{Im} \; S(u, p) = \frac{1}{2p\pi} \frac{\sqrt{u(4pq - u)}}{u(1 - u)}, \qquad u \in (0, 4pq). \qquad (3.6.12)$$

By Theorem 3.6.5 mass points can only occur at $u = 0$ or $u = 1$. Because

$$\lim_{v \to 0} v \; \mathrm{Im} \; S(-iv, p) = 0,$$

there is no jump at 0. For a discussion of the mass at the remaining point 1, we remark that for sufficiently small $v$, (3.4.4) defines the branch of the square root in $S(1 - iv, p)$ as $\mathcal{R}(\sqrt{\cdots}) > 0$. Therefore Theorem 3.6.5 gives for the discrete part of the measure $\mu^{(p)}$,

$$\mu^{(p)}(\{1\}) = \begin{cases} \dfrac{2p - 1}{p} & \text{if } \; p \geq \dfrac{1}{2}, \\[2mm] 0 & \text{if } \; p < \dfrac{1}{2}. \end{cases} \qquad (3.6.13)$$

Note that in the case $p = \frac{1}{2}$, (3.6.12) and (3.6.13) give a further proof that on the interval $(0, 1)$ the arc-sine distribution has canonical moments equal to $\frac{1}{2}$.

CHAPTER 4

# Special Sequences of Canonical Moments

## 4.1  INTRODUCTION

In the previous chapters we discussed the relationships between the canonical moments and orthogonal polynomials, continued fractions, and Stieltjes transforms. These results were mainly used to derive the canonical moment sequences corresponding to various probability measures on compact intervals. We will now address in more detail the converse problem of identifying the measure corresponding to a given sequence of canonical moments. Problems of this type will become important in the applications in the following chapters, where numerous solutions of optimal design problems are obtained in terms of "optimal" sequences of canonical moments. Many of the proofs are somewhat technical, and the reader might consider omitting them on a first reading. However, the following list of results is in no way complete, and the proofs and technical details become important if similar moment problems have to be solved and cannot be reduced to the cases considered in this chapter.

Our approach is based on the relation between the canonical moments and the continued fraction expansions of the Stieltjes transform in Chapters 2 and 3. It is pointed out in Theorem 3.6.1 that the roots and weights of the measure corresponding to a terminating sequence of canonical moments can be found by calculating the zeros and evaluating certain orthogonal polynomials that appear in the continued fraction expansions of the Stieltjes transform. In many cases these polynomials can be connected to classical orthogonal polynomials, and the corresponding measures (more precisely, the support points and weights) can be found explicitly.

The identification of measures with an infinite sequence of canonical moments is more complicated because this task requires the calculation of the Stieltjes transform from its corresponding continued fraction expansion and the application of some kind of Stieltjes-Perron inversion formula as described in Section 3.6. Problems of this type will become useful in the calculation of the limit distribution of sequences of measures given by their moments.

We finally remark, once more, that on every interval $[a, b]$ there exists exactly one probability measure $\mu_{a,b}$ corresponding to a given sequence of canonical moments and that by Theorem 1.3.2 measures on different intervals with the same sequence of canonical moments are related by linear transformations. In the following sections we will mainly work on the intervals $[0, 1]$ and $[-1, 1]$. The results can easily be transferred to other intervals by a linear transformation. The specific choice of the interval $[0, 1]$ or $[-1, 1]$ is motivated by our intention to present the proofs in the most transparent form. Because our arguments require some basic facts about special functions, it turns out that some proofs are easier to understand on the interval $[0, 1]$ while others are more transparent on the interval $[-1, 1]$.

## 4.2  SIMPLE SEQUENCES

The simplest case in the identification of a measure supporting a given sequence of canonical moments is probably the determination of the probability measure $\sigma_1^+$ on the interval $[0, 1]$ corresponding to the sequence $(p_1, 1)$ where $p_1 \in (0, 1)$. By the discussion of Section 1.3 the measure $\sigma_1^+$ is the upper principal representation of the moment point $p_1 \in \text{Int } M_1$, and the support points are given by 0 and 1. The corresponding masses are easily identified as $\sigma_1^+(1) = 1 - \sigma_1^+(0) = p_1$.

Next we consider the measure $\sigma_2^+$ corresponding to the terminating sequence $(p_1, p_2, 1)$ $(p_i \in (0, 1)$, $i = 1, 2)$, which is the upper principal representation of the associated moment point $(p_1, p_1(p_1 + q_1 p_2)) \in \text{Int } M_2$. Theorem 3.4.1 yields for the Stieltjes transform of $\sigma_2^+$:

$$\int_0^1 \frac{d\sigma_2^+(x)}{z - x} = \frac{Q_1(z, q)}{(z - 1)\overline{P}_1(z, p)},$$

where

$$\underline{Q}_1(z, q) = z - q_1 - p_1 q_2,$$

$$\overline{P}_1(z, p) = z - p_1 q_2,$$

by Corollary 2.3.4. Following the discussion at the beginning of Section 3.6, the support points of the measure $\sigma_2^+$ are $p_1 q_2$ and 1, while the corresponding masses are obtained via Theorem 3.6.1 as

$$\sigma_2^+(1) = 1 - \sigma_2^+(p_1 q_2) = \frac{\underline{Q}_1(1, q)}{\frac{d}{dz}\left[(z - 1)\overline{P}_1(z, p)\right]\big|_{z=1}}$$

$$= \frac{\underline{Q}_1(1, q)}{\overline{P}_1(1, p)} = \frac{p_1 p_2}{1 - p_1 q_2}$$

(4.2.1)

(note that in the notation of Theorem 3.6.1 we have $P_2^*(x) = (x - 1)\overline{P}_1(x, \boldsymbol{p})$, $P_1^{(1)}(x) = \underline{Q}_1(x, \boldsymbol{q})$).

Finally we consider the measure $\sigma_3^+$ corresponding to the terminating sequence $(p_1, p_2, p_3, 1)$ ($p_i \in (0, 1)$, $i = 1, 2, 3$). From Theorem 3.4.1 and Corollary 2.3.4 we have

$$\int_0^1 \frac{d\sigma_3^+(x)}{z - x} = \frac{\underline{P}_2(z, \boldsymbol{q})}{z(z - 1)\overline{Q}_1(z, \boldsymbol{p})}, \tag{4.2.2}$$

where

$$\overline{Q}_1(z, \boldsymbol{p}) = z - p_1 q_2 - p_2 q_3,$$

$$\underline{P}_2(z, \boldsymbol{q}) = z^2 - z(q_1 + p_1 q_2 + p_2 q_3) + q_1 p_2 q_3.$$

Thus the support points of (the upper principal representation) $\sigma_3^+$ are 0, 1 and $p_1 q_2 + p_2 q_3$. The corresponding masses can be obtained by similar arguments as in (4.2.1); for example, we have from (4.2.2) for the mass at the origin

$$\sigma_3^+(0) = \frac{\underline{P}_2(0, \boldsymbol{q})}{-\overline{Q}_1(0, \boldsymbol{p})} = \frac{q_1 p_2 q_3}{p_1 q_2 + p_2 q_3}.$$

We summarize these results in the following:

**Theorem 4.2.1**
1. *The probability measure on the interval* $[0, 1]$ *corresponding to the sequence* $(p_1, 1)$ *has masses* $q_1$ *and* $p_1$ *at the points* 0 *and* 1.
2. *The probability measure on the interval* $[0, 1]$ *corresponding to the sequence* $(p_1, p_2, 1)$ *has masses* $q_1/(1 - p_1 q_2)$ *and* $p_1 p_2/(1 - p_1 q_2)$ *at the points* $p_1 q_2$ *and* 1.
3. *The probability measure on the interval* $[0, 1]$ *corresponding to the sequence* $(p_1, p_2, p_3, 1)$ *has masses*

$$\frac{q_1 p_2 q_3}{p_1 q_2 + p_2 q_3}, \quad \frac{q_1 p_1 q_2}{(p_1 q_2 + p_2 q_3)(q_1 q_2 + p_2 p_3)}, \quad \frac{p_1 p_2 p_3}{q_1 q_2 + p_2 p_3},$$

*at the points* $0, p_1 q_2 + p_2 q_3$, *and* 1, *respectively.*

**Corollary 4.2.2**
1. *The probability measure on the interval* $[-1, 1]$ *corresponding to the sequence* $(\frac{1}{2}, p_2, \frac{1}{2}, 1)$ *has masses* $p_2/2, q_2, p_2/2$ *at the points* $-1, 0, 1$, *respectively.*
2. *The probability measure on the interval* $[-1, 1]$ *corresponding to the sequence* $(\frac{1}{2}, p_2, 1/2, p_4, \frac{1}{2}, 1)$ *has masses* $\alpha, \frac{1}{2} - \alpha, \frac{1}{2} - \alpha, \alpha$ *at the points* $-1, -\sqrt{p_2 q_4}$, $\sqrt{p_2 q_4}, 1$, *respectively, where* $\alpha = p_2 p_4/(2(1 - p_2 q_4))$.

3. *The probability measure on the interval* $[-1,1]$ *corresponding to the sequence* $(\frac{1}{2}, p_2, \frac{1}{2}, p_4, \frac{1}{2}, p_6, \frac{1}{2}, 1)$ *has masses* $\alpha_1, \alpha_2, 1 - 2\alpha_1 - 2\alpha_2, \alpha_2, \alpha_1$ *at the points* $-1, -\sqrt{p_2q_4 + p_4q_6}, 0, \sqrt{p_2q_4 + p_4q_6}, 1,$ *respectively, where*

$$\alpha_1 = \frac{p_2p_4p_6}{2(q_2q_4 + p_4p_6)}, \quad \alpha_2 = \frac{p_2q_2q_4}{2(p_2q_4 + p_4q_6)(q_2q_4 + p_4p_6)}.$$

*Proof* We will only show part 1, all other cases are treated similarly. If $\mu^{(s)}$ denotes the measure on the interval $[-1,1]$ corresponding to the sequence $(\frac{1}{2}, p_2, \frac{1}{2}, 1)$, then by Theorem 1.3.5, the measure $\mu$ on the interval $[0,1]$ corresponding to $(p_2, 1)$ is related to $\mu^{(s)}$ by the transformation

$$\mu^{(s)}([-x, x]) = \mu([0, x^2]) \qquad x \in [0, 1].$$

The assertion is now an immediate consequence of Theorem 4.2.1. ∎

## 4.3 MEASURES WITH "NEARLY" EQUAL WEIGHTS

In this section we will discuss terminating sequences of canonical moments whose corresponding probability measures have "nearly" equal masses on their support points. We will start the discussion with an investigation of some reversing properties of the sequence of canonical moments corresponding to the Beta distribution with density on $[0,1]$ proportional to $x^\beta(1-x)^\alpha$. From Example 1.3.6 we observe that this measure has canonical moments for $j \geq 1$,

$$p_{2j} = p_{2j}(\alpha, \beta) = \frac{j}{2j + 1 + \alpha + \beta},$$

$$(4.3.1)$$

$$p_{2j-1} = p_{2j-1}(\alpha, \beta) = \frac{\beta + j}{2j + \alpha + \beta}.$$

Now, consider the first $n$ canonical moments of this sequence $(p_1, \ldots, p_n)$ and the corresponding upper and lower principal representation associated with the sequences $(p_1, \ldots, p_n, 1)$ and $(p_1, \ldots, p_n, 0)$, respectively. By Theorem 2.5.1 the measures associated with the sequences $(q_n, \ldots, q_1, 1)$ and $(p_1, \ldots, p_n, 1)$ have the same support and the measures associated with $(p_n, \ldots, p_1, 0)$ and $(p_1, \ldots, p_n, 0)$ have the same support. The following results show that for the Beta distribution these measures have essentially uniform weight on all support points and that this property actually characterizes the Beta distribution.

The proof of the following Theorem 4.3.1 will be given on the interval $[0, 1]$, since this makes several steps in the proof slightly easier. The polynomials, orthogonal to $x^\beta(1-x)^\alpha$ on $[0, 1]$, are called Jacobi polynomials on the interval $[0, 1]$. The normalized version of degree $k$ with leading coefficient 1 will be denoted by $\hat{P}_k^{(\alpha,\beta)}(x)$. By convention the same polynomials on $[-1, 1]$ with a different normalization are denoted by $P_k^{(\alpha,\beta)}(x)$ in Example 2.1.2. It follows from (2.1.15) that the monic orthogonal polynomials with respect to the Beta distribution on the interval $[0, 1]$ are given by

$$\hat{P}_n^{(\alpha,\beta)}(x) = \frac{n!\Gamma(\alpha+\beta+n+1)}{\Gamma(\alpha+\beta+2n+1)} P_n^{(\alpha,\beta)}(2x-1)$$

(4.3.2)

$$= \frac{\Gamma(\alpha+n+1)}{\Gamma(\alpha+\beta+2n+1)} \sum_{j=0}^{n} \binom{n}{j} \frac{\Gamma(\alpha+\beta+n+j+1)}{\Gamma(\alpha+j+1)} (x-1)^j.$$

**Theorem 4.3.1** *For $n \in \mathbb{N}$ and the canonical moments in (4.3.1), let $\rho_n^+$ denote the upper principal representation of $(q_n, q_{n-1}, \ldots, q_1)$ and $\rho_n^-$ denote the lower principal representation of $(p_n, p_{n-1}, \ldots, p_1)$ on the interval $[0, 1]$.*

1. *$\rho_{2m-1}^+$ has equal masses at the zeros of the $(m-1)$th Jacobi polynomial*

$$\hat{P}_{m-1}^{(\alpha+1,\beta+1)}(x),$$

   *while the masses at the points 0 and 1 are $(\beta+1)$ and $(\alpha+1)$ times bigger than the masses at the interior support points.*

2. *$\rho_{2m-1}^-$ has equal masses at the zeros of the $m$th Jacobi polynomial*

$$\hat{P}_m^{(\alpha,\beta)}(x).$$

3. *$\rho_{2m}^+$ has equal masses at the zeros of the $m$th Jacobi polynomial*

$$\hat{P}_m^{(\alpha+1,\beta)}(x),$$

   *while the remaining mass at the point 1 is $(\alpha+1)$ times bigger than the masses at the interior points.*

4. *$\rho_{2m}^-$ has equal masses at the zeros of the $m$th Jacobi polynomial*

$$\hat{P}_m^{(\alpha,\beta+1)}(x),$$

   *while the remaining mass at the point 0 is $(\beta+1)$ times bigger than the masses at the interior points.*

The results of the preceding theorem can be summarized in the following display:

| Sequence | Support polynomial | Mass at 0 | Mass at 1 |
|---|---|---|---|
| $(p_{2m-1}, \ldots, p_1, 0)$ | $\hat{P}_m^{(\alpha,\beta)}(x)$ | $0$ | $0$ |
| $(p_{2m}, \ldots, p_1, 0)$ | $\hat{P}_m^{(\alpha,\beta+1)}(x)$ | $\dfrac{\beta+1}{m+\beta+1}$ | $0$ |
| $(q_{2m}, \ldots, q_1, 1)$ | $\hat{P}_m^{(\alpha+1,\beta)}(x)$ | $0$ | $\dfrac{\alpha+1}{m+\alpha+1}$ |
| $(q_{2m-1}, \ldots, q_1, 1)$ | $\hat{P}_{m-1}^{(\alpha+1,\beta+1)}(x)$ | $\dfrac{\beta+1}{m+\alpha+\beta+1}$ | $\dfrac{\alpha+1}{m+\alpha+\beta+1}$ |

where the canonical moments are given by (4.3.1) and the interior support points of the principal representation are given by zeros of the polynomial in the second column. The masses at these points are all equal. If there are masses at the points 0 or 1, these are $\beta+1$ or $\alpha+1$ times bigger than the masses at the remaining roots.

*Proof of Theorem 4.3.1.* All four cases are somewhat similar, and we restrict ourselves to the proofs of cases 1 and 2.

For case 1 we are concerned with the measure $\rho_{2m-1}^+$ corresponding to the sequence

$$q_{2m-1}(\alpha, \beta), \ldots, q_1(\alpha, \beta), \quad 1. \tag{4.3.3}$$

By Theorem 2.5.1 the support of $\rho_{2m-1}^+$ is the same as the support of the measure corresponding to

$$p_1(\alpha, \beta), \ldots, p_{2m-1}(\alpha, \beta), \quad 1.$$

Now the polynomials $\hat{P}_n^{(\alpha+1,\beta+1)}(x)$ are orthogonal with respect to the measure $x^{\alpha+1}(1-x)^{\beta+1}dx$, and Theorem 2.2.3 part 4, and Theorem 2.2.1 show that the support consists of the roots of

$$x(x-1)\hat{P}_{m-1}^{(\alpha+1,\beta+1)}(x). \tag{4.3.4}$$

This establishes the correct support.

To determine the associated weights, we first recall from Theorem 3.4.1 that the Stieltjes transform for a given sequence $p_1, \ldots, p_{2m-1}, 1$ is given by

$$S_{2m-1}^+(z,\boldsymbol{p}) = \frac{\underline{P}_m(z,\boldsymbol{q})}{z(z-1)\overline{Q}_{m-1}(z,\boldsymbol{p})}.$$

The sequence $(p_1, \ldots, p_{2m-1}, 1)$ under consideration is the sequence (4.3.3). The above discussion has established that the denominator is (4.3.4). For the numerator, using the sequence (4.3.3), we have to find $\underline{P}_m$ for the sequence

$$p_{2m-1}(\alpha, \beta), \ldots, p_1(\alpha, \beta).$$

Now the zeros of the polynomial $\underline{P}_m$ correspond to the support of the lower representation for $n = 2m - 1$, and hence by Theorem 2.5.1 the numerator corresponds to the lower principal representation of the sequence

$$p_1(\alpha, \beta), \ldots, p_{2m-1}(\alpha, \beta),$$

which is $\hat{P}_m^{(\alpha,\beta)}(x)$, by Theorems 2.2.1 and 2.2.3, part 1. The Stieltjes transform of the upper principal representation $\rho_{2m-1}^+$ is therefore

$$S(z, \rho_{2m-1}^+) = \frac{\hat{P}_m^{(\alpha,\beta)}(z)}{z(z-1)\hat{P}_{m-1}^{(\alpha+1,\beta+1)}(z)}. \tag{4.3.5}$$

By Theorem 3.6.1 the weights at the support points $x_i$ are given by

$$\rho_{2m-1}^+(x_i) = \frac{\hat{P}_m^{(\alpha,\beta)}(x_i)}{\frac{d}{dx}\left[x(x-1)\hat{P}_{m-1}^{(\alpha+1,\beta+1)}(x)\right]\Big|_{x=x_i}}. \tag{4.3.6}$$

To evaluate this, we need two further results. The first is that for the monic Jacobi polynomials on the interval $[0, 1]$:

$$\frac{d}{dx}\left[x(x-1)\hat{P}_{m-1}^{(\alpha+1,\beta+1)}(x)\right] = (m + \alpha + \beta + 1)\hat{P}_m^{(\alpha,\beta)}(x) \tag{4.3.7}$$
$$- (x(\alpha + \beta) - \beta)\hat{P}_{m-1}^{(\alpha+1,\beta+1)}(x).$$

The second is that if $p_1, p_2, \ldots$ correspond to a measure $d\mu$ on $[0, 1]$ with orthogonal polynomials $\{\underline{P}_k\}_{k \geq 0}$, then

$$\underline{P}_k(0) = (-1)^k \zeta_1 \zeta_3 \cdots \zeta_{2k-1}. \tag{4.3.8}$$

The last named result (4.3.8) follows by an induction from Corollary 2.3.4. Equation (4.3.7) is a consequence of Rodrigues's formula, which states that the monic Jacobi polynomials can be represented as

$$\hat{P}_n^{(\alpha,\beta)}(x) = f_n^{(\alpha,\beta)} x^{-\beta}(1-x)^{-\alpha}\frac{d^n}{dx^n}(x^{n+\beta}(1-x)^{n+\alpha}), \tag{4.3.9}$$

where

$$f_n^{(\alpha,\beta)} = \frac{(-1)^n}{(\alpha + \beta + n + 1)_n}.$$

The representation (4.3.9) can easily be obtained from (2.1.15) by transforming the interval $[-1, 1]$ onto $[0, 1]$.

With the aid of (4.3.7) and (4.3.8), the weights in (4.3.6) can be evaluated. All of the roots in $(0, 1)$, or the roots of $\hat{P}_{m-1}^{(\alpha+1,\beta+1)}(x)$, have equal weight $1/(m + \alpha + \beta + 1)$. Using (4.3.1), (4.3.8), and a small amount of algebra shows that the weight at zero is $(\beta + 1)/(m + \alpha + \beta + 1)$. The corresponding weight at $x = 1$ must be $(\alpha + 1)/(m + \alpha + \beta + 1)$.

The proof of case 2 follows arguments similar to case 1. Here we want to determine the measure $\rho_{2m-1}^-$ corresponding to

$$p_{2m-1}(\alpha, \beta), \ldots, p_1(\alpha, \beta), \quad 0. \tag{4.3.10}$$

By Theorem 2.5.1 the support points are the same as for the reversed sequence, so Theorem 2.2.1 and 2.2.3 show that the support consists of the zeros of the $m$th orthogonal polynomial with resect to Beta distribution, i.e. $\hat{P}_m^{(\alpha,\beta)}(x)$.

By Theorem 3.4.1 the Stieltjes transform of the lower principal representation of $p_1, \ldots, p_{2m-1}$ is given by

$$S_{2m-1}^-(z, \boldsymbol{p}) = \frac{\overline{Q}_{m-1}(z, \boldsymbol{q})}{\underline{P}_m(z, \boldsymbol{p})}.$$

From the above discussion the denominator is $\hat{P}_m^{(\alpha,\beta)}(x)$. For the numerator the polynomial $\overline{Q}_{m-1}$ corresponds to

$$q_{2m-1}(\alpha, \beta), \ldots, q_1(\alpha, \beta). \tag{4.3.11}$$

Now $\overline{Q}_{m-1}$ corresponds to the upper principal representation of (4.3.11) which, by Theorem 2.5.1, is the same as the upper principal representation of

$$p_1(\alpha, \beta), \ldots, p_{2m-1}(\alpha, \beta).$$

By the discussion in the proof of part 1, this is $\hat{P}_{m-1}^{(\alpha+1,\beta+1)}(x)$. Therefore the Stieltjes transform of $\rho_{2m-1}^-$ is given by

$$S(z, \rho_{2m-1}^-) = \frac{\hat{P}_{m-1}^{(\alpha+1,\beta+1)}(x)}{\hat{P}_m^{(\alpha,\beta)}(x)}. \tag{4.3.12}$$

For the monic Jacobi polynomials on the interval $[0, 1]$, it follows from (4.3.2) that

$$\frac{d}{dx}\hat{P}_m^{(\alpha,\beta)}(x) = m\hat{P}_{m-1}^{(\alpha+1,\beta+1)}(x).$$

Using this, the equality of the weights can be obtained from (4.3.12) and Theorem 3.6.1. ∎

Note that $\rho_{2m-1}^-$ $(m \in \mathbb{N})$ is the lower principal representation of the "truncated" and "reversed" sequence $(p_{2m-1}, \ldots, p_2, p_1)$. It turns out that the property that "$\rho_{2m-1}^-$ has equal weights at its support points" for all $m$ actually characterizes the Beta distribution. To see this, we consider a measure $\mu$ with canonical moments $p_1, p_2, \ldots$ and for $m \in \mathbb{N}$ let $\rho_{2m-1}^-$ denote the measure corresponding to the terminated and reversed sequence $(p_{2m-1}, \ldots, p_1)$. We are looking for all measures $\mu$ on the interval $[0, 1]$ such that, for all $m \in \mathbb{N}$, $\rho_{2m-1}^-$ puts equal masses on its $m$ support points:

$$\rho_{2m-1}^-(\{x\}) = \frac{1}{m} \qquad \forall x \in \text{ supp}(\rho_{2m-1}^-). \tag{4.3.13}$$

This condition has a nice geometric interpretation. Recall the definition of the moment space in Definition 1.1.1:

$$M_{2m} = \{(c_1, \ldots, c_{2m}) | c_i = \int_0^1 x^i d\mu, \ i = 1, \ldots, 2m\}.$$

It is shown in Section 1.2 that for each $(c_1, \ldots, c_{2m-1}) \in M_{2m}$ the point $(c_1, \ldots, c_{2m-1}, c_{2m}^-)$ is a boundary point of $M_{2m}$ corresponding to the "lower principal representation" of $c_1, \ldots, c_{2m-1}$ or of $(p_1, \ldots, p_{2m-1})$. Let $\underline{P}_m(x)$ denote the $m$th orthogonal polynomial corresponding to $c_1, \ldots, c_{2m-1}$. By Corollary 2.2.4 the hyperplane supporting $M_{2m}$ at $(c_1, \ldots, c_{2m-1}, c_{2m}^-)$ is determined by $\underline{P}_m^2(x)$, and the corresponding face of $M_{2m}$ has extreme points $(x_i, x_i^2, \ldots, x_i^{2m})$, $i = 1, \ldots, m$. The measure $\rho_{2m-1}^-$, which puts equal masses on its support points, can be viewed as the "center" of this face.

**Theorem 4.3.2** *A probability measure on a compact interval is a Beta distribution if and only if it satisfies* (4.3.13) *for all* $m \in \mathbb{N}$.

*Proof* The direct part is established in Theorem 4.3.1. For the converse part consider an (infinite) measure $\mu$ on the interval $[0, 1]$ with corresponding sequence $p_1, p_2, \ldots$. Define $\mu_{2m-1}^-$ and $\rho_{2m-1}^-$ as the lower principal representations of the sequences $\boldsymbol{p}_{2m-1} = (p_1, \ldots, p_{2m-1})$ and $\boldsymbol{p}_{2m-1}^R = (p_{2m-1}, \ldots, p_1)$, respectively $(m \in \mathbb{N})$. The Stieltjes transform of the measure corresponding to the reversed sequence and its continued fraction expansion are obtained from Theorem 3.3.1 and 3.4.1 as

$$\Phi(z) = \int_0^1 \frac{d\rho_{2m-1}^-(x)}{z - x} = \frac{\overline{Q}_{m-1}(z, \boldsymbol{q}_{2m-1}^R)}{\underline{P}_m(z, \boldsymbol{p}_{2m-1}^R)}$$

$$= \frac{1}{\lfloor z \rfloor} - \frac{p_{2m-1}}{\lfloor 1 \rfloor} - \frac{\gamma_{2m-1}}{\lfloor z \rfloor} - \cdots - \frac{\gamma_2}{\lfloor 1 \rfloor} \tag{4.3.14}$$

$$= \frac{z^{m-1} - (\sum_{j=2}^{2m-1} \gamma_j) z^{m-2} + (\sum_{i=2}^{2m-1} \sum_{j=i+2}^{2m-1} \gamma_i \gamma_j) z^{m-3} - \cdots}{\prod_{j=1}^m (z - x_j)},$$

where $\gamma_j = q_j p_{j-1}$ $(j \geq 2)$ and $x_1, \ldots, x_m$ denote the support points of $\rho_{2m-1}^-$. Note that the last identity in (4.3.14) follows from the determinant representation of the partial numerator and denominator of a continued fraction in (3.2.19). Alternatively, it can be obtained from the representation in the first line and Corollary 2.3.4.

On the other hand, $\rho_{2m-1}^-$ puts equal weights on its support points, and it follows that

$$\Phi(z) = \frac{1}{m} \sum_{i=1}^{m} \frac{1}{z - x_i}$$

$$= \frac{\frac{1}{m}\left[ m\, z^{m-1} - (m-1)\left(\sum_{i=1}^{m} x_i\right) z^{m-2} + (m-2)\left(\sum_{i<j} x_i x_j\right) z^{m-3} \cdots \right]}{\prod_{j=1}^{m}(z - x_j)}.$$

$$(4.3.15)$$

Because the measures $\mu_{2m-1}^-$ and $\rho_{2m-1}^-$ have the same support (Theorem 2.5.1), Theorem 2.2.3, and Corollary 2.3.4 show

$$\prod_{j=1}^{m}(z - x_j) = \underline{P}_m(z, \mathbf{p}_{2m-1})$$

$$= z^m - \left(\sum_{j=1}^{2m-1} \zeta_j\right) z^{m-1} + \left(\sum_{i=1}^{2m-1} \sum_{j=i+2}^{2m-1} \zeta_i \zeta_j\right) z^{m-2} - \cdots,$$

which yields

$$\sum_{j=1}^{m} x_j = \sum_{j=1}^{2m-1} \zeta_j, \quad \sum_{i=1}^{m} \sum_{j=i+1}^{m} x_i x_j = \sum_{i=1}^{2m-1} \sum_{j=i+2}^{2m-1} \zeta_i \zeta_j.$$

By a combination of (4.3.14) and (4.3.15) and a comparison of the coefficients of $z^{m-2}$ and $z^{m-3}$ in the polynomials of the numerators, we now obtain the following equations for all $m \geq 2$:

$$m\left(\sum_{j=2}^{2m-1} \gamma_j\right) = (m-1)\left(\sum_{j=1}^{2m-1} \zeta_j\right),$$

$$(4.3.16)$$

$$m\left(\sum_{i=2}^{2m-1} \sum_{j=i+2}^{2m-1} \gamma_i \gamma_j\right) = (m-2)\left(\sum_{i=1}^{2m-1} \sum_{j=i+2}^{2m-1} \zeta_i \zeta_j\right).$$

To simplify these equations, we note that

$$\sum_{j=1}^{2m-1} \zeta_j = \sum_{j=1}^{m} x_j = p_{2m-1} + \sum_{j=2}^{2m-1} \gamma_j,$$

$$\sum_{i=1}^{2m-1}\sum_{j=i+2}^{2m-1} \zeta_i \zeta_j = \sum_{i=1}^{m}\sum_{j=i+1}^{m} x_i x_j = \sum_{i=2}^{2m-1}\sum_{j=i+2}^{2m-1} \gamma_i \gamma_j + p_{2m-1}\sum_{j=2}^{2m-2} \gamma_j,$$

which follow readily from (4.3.14) by writing the denominator in terms of $\gamma_j$ and using (3.2.19). This gives the system

$$(m-1)p_{2m-1} = \sum_{j=2}^{2m-1} \gamma_j,$$

$$(m-2)p_{2m-1}\left(\sum_{j=2}^{2m-2} \gamma_j\right) = 2\sum_{i=2}^{2m-1}\sum_{j=i+2}^{2m-1} \gamma_i \gamma_j, \qquad m \ge 2.$$

(4.3.17)

We will now simplify (4.3.17), using the fact that the equations must hold for all $m \ge 2$. This yields for $m-1$,

$$\sum_{j=2}^{2m-3} \gamma_j = (m-2)p_{2m-3},$$

$$2\sum_{i=2}^{2m-3}\sum_{j=i+2}^{2m-3} \gamma_i \gamma_j = (m-3)p_{2m-3}(\gamma_{2m-4} + (m-3)p_{2m-5}),$$

and (4.3.17) reduces to $(m \ge 3)$

$$p_{2m-1}[(m-1)+p_{2m-2}] = p_{2m-2} + \gamma_{2m-2} + (m-2)p_{2m-3},$$

$$(m-2)p_{2m-1}[(m-1)+p_{2m-2}]$$

$$= 2(m-2)p_{2m-2} + [2q_{2m-2} + (m-3)][\gamma_{2m-4} + (m-3)p_{2m-5}].$$

(4.3.18)

We now prove successively (for $n \ge 2$) that the solutions of the equation (4.3.17) or equivalently (4.3.18) are given by

$$p_{2n-1} = \frac{(n-1)p_2 + \gamma_2}{(2n-3)p_2 + 1}, \qquad p_{2n-2} = \frac{(n-1)p_2}{(2n-4)p_2 + 1}$$

(4.3.19)

$(n = 2, 3, \dots )$. In the case $n = 2$ we obtain from (4.3.17) (note that this case gives only one equation for $p_3$)

$$p_3 = \gamma_2 + (1 - p_3)p_2$$

which is (4.3.19) for $n = 2$ [the second representation in (4.3.19) is obvious for $n = 2$]. Now assume that the representation (4.3.19) holds for $n = 1, \ldots, m - 1$, and consider (4.3.18) for $n = m$. By straightforward calculations (using the induction hypothesis) we obtain

$$\gamma_{2m-4} + (m - 3)p_{2m-5} = (m - 2)p_{2m-5}\frac{(2m - 7)p_2 + 1}{(2m - 6)p_2 + 1}. \tag{4.3.20}$$

Equating the two equations of (4.3.18), solving with respect to $p_{2m-2}$, and using (4.3.20) yields

$$
\begin{aligned}
p_{2m-2}&\left[1 + p_{2m-3} - 2p_{2m-5}\frac{(2m - 7)p_2 + 1}{(2m - 6)p_2 + 1}\right]\\
&= (m - 1)\left[p_{2m-3} - p_{2m-5}\frac{(2m - 7)p_2 + 1}{(2m - 6)p_2 + 1}\right].
\end{aligned}
\tag{4.3.21}
$$

Finally we observe the representations

$$p_{2m-3} - p_{2m-5}\frac{(2m - 7)p_2 + 1}{(2m - 6)p_2 + 1} = \frac{p_2[(m - 3)p_2 + (1 - \gamma_2)]}{[(2m - 5)p_2 + 1][(2m - 6)p_2 + 1]},$$

$$1 - p_{2m-5}\frac{(2m - 7)p_2 + 1}{(2m - 6)p_2 + 1} = \frac{(m - 3)p_2 + (1 - \gamma_2)}{(2m - 6)p_2 + 1}$$

(which follow from the induction hypothesis) and obtain from (4.3.21)

$$p_{2m-2}[(m - 3)p_2 + 1 - \gamma_2][(2m - 5)p_2 + 1 + p_2]$$
$$= (m - 1)p_2 [(m - 3)p_2 + 1 - \gamma_2].$$

This reduces to

$$p_{2m-2} = \frac{(m - 1)p_2}{(2m - 4)p_2 + 1},$$

and the first equation in (4.3.18) can be simplified [using (4.3.20)], that is,

$$p_{2m-1} = \frac{(m - 1)p_2 + \gamma_2}{(2m - 3)p_2 + 1}.$$

This shows that the solution of the Equations (4.3.18) is given by (4.3.19). Because every probability measure on $[0,1]$ that satisfies (4.3.13) for all $m \in \mathbb{N}$ must also satisfy the Equations (4.3.18), we have shown that the corresponding sequence of canonical moments $p_1, p_2, \ldots$ of such a probability measure is determined by

$$p_{2m-1} = \frac{(m-1)p_2 + q_2 p_1}{(2m-3)p_2 + 1}, \quad p_{2m-2} = \frac{(m-1)p_2}{(2m-4)p_2 + 1} \quad (m \geq 2).$$

If we replace the "free" parameters $p_1, p_2 \in (0, 1)$ by

$$p_1 = \frac{\beta + 1}{\alpha + \beta + 2}, \quad p_2 = \frac{1}{\alpha + \beta + 3} \quad (a, \beta > -1),$$

we obtain

$$p_{2m-2} = \frac{m-1}{2m-1+\alpha+\beta}, \quad p_{2m-1} = \frac{\beta + m}{2m + \alpha + \beta}.$$

It was shown in Example 1.5.3 that these are the canonical moments of the Beta distribution, which proves the assertion. ∎

As a special case of Theorem 4.3.1 consider the sequence of canonical moments

$$p_{2i} = \frac{m - i + z}{2(m-i) + z}, \quad i = 1, \ldots, m - 1,$$

$$p_{2i-1} = \frac{1}{2}, \quad i = 1, \ldots, m, \qquad (4.3.22)$$

$$p_{2m} = 1,$$

where $z > -1$ is a given number. This is obtained by truncating the sequence of canonical moments of the Beta distribution in (4.3.1) at $2m - 1$ and considering the upper principal representation of the reversed sequence $(q_{2m-1}, \ldots, q_1)$ for $\alpha = \beta = (z - 1)/2$.

It was already pointed out in Section 1.6 that this sequence with $z = 1$ plays an important role in the maximization of the Hankel determinant $\underline{H}_{2m}$ with respect to all probability measures on the interval $[0, 1]$. Sequences of the general form (4.3.22) turn out to be useful for finding the probability measures $\xi$, which maximize weighted products of Hankel determinants of the form

$$\prod_{j=1}^{m} \underline{H}_{2j}^{\beta_j} = \begin{vmatrix} c_0 & c_1 \\ c_1 & c_2 \end{vmatrix}^{\beta_1} \begin{vmatrix} c_0 & c_1 & c_2 \\ c_1 & c_2 & c_3 \\ c_2 & c_3 & c_4 \end{vmatrix}^{\beta_2} \cdots \begin{vmatrix} c_0 & \cdots & c_m \\ \vdots & & \vdots \\ c_m & \cdots & c_{2m} \end{vmatrix}^{\beta_m},$$

where $c_i = \int x^i d\xi(x)$ denotes the $i$th moment of $\xi$ $(i = 1, \ldots, 2m)$ and $\beta_1, \ldots, \beta_m$ denote nonnegative numbers. These results will have applications in the construction of discrimination and model robust designs (see Chapter 6) and in the solution of certain extremal problems for polynomials in approximation theory. The sequence (4.3.22) also appears in the construction of efficient designs for multivariate polynomial regression models as described in Section 5.8.

**Corollary 4.3.3** *The probability measure $\xi^z$ on the interval $[-1, 1]$ corresponding to the sequence of canonical moments in (4.3.22) is supported at the zeros of the polynomial*

$$(x^2 - 1)C_{m-1}^{(z/2+1)}(x).$$

*The masses of $\xi^z$ at the zeros of $C_{m-1}^{(z/2+1)}(x)$ are all equal to $1/(m+z)$, while the masses of $\xi^z$ at the boundary points are given by $(z+1)/(2(m+z))$.*

*Proof* To obtain the probability measure corresponding to (4.3.22) on the interval $[-1, 1]$ analytically, recall from Example 2.1.2 and (2.1.18) that the ultraspherical polynomials $C_n^{(\lambda)}(x)$ are orthogonal on the interval $[-1, 1]$ with respect to the measure $(1 - x^2)^{\lambda-1/2}dx$ and are proportional to the Jacobi polynomials $P_n^{(\lambda-1/2,\lambda-1/2)}(x)$. The result in the corollary then follows from part 1 of Theorem 4.3.1 transferred to the interval $[-1, 1]$. ∎

It may be worthwhile mentioning a couple of special cases in Corollary 4.3.3. For $z = 1$ the corresponding measure has equal weight $1/(m+1)$ at the zeros of the polynomial

$$(x^2 - 1)C_{m-1}^{(3/2)}(x) = (x^2 - 1)P'_m(x),$$

where $P_m(x)$ denotes the $m$th Legendre polynomial orthogonal with respect to the Lebesgue measure on the interval $[-1, 1]$. The above identity follows from (2.1.24) and (2.1.20). For $z = 0$ we obtain masses proportional to

$$1 : 2 : \ldots : 2 : 1$$

at the zeros of the polynomial

$$(x^2 - 1)C_{m-1}^{(1)}(x) = (x^2 - 1)U_{m-1}(x)$$

[see (2.1.19)], where $U_n(x)$ is the Chebyshev polynomial of the second kind defined in (2.1.6). The support points are called *Chebyshev points* and can be found explicitly from the trigonometric representation (2.1.6) as $x_j = \cos(\pi(m-j)/m)$, $j = 0, \ldots, m$.

There are numerous results with a similar structure that can be used for maximizing Hankel determinants corresponding to the functions

$$x^k, x^{2k}, x^{3k}, \ldots$$

(see Dette, 1992). Moreover properties of this type can be used for characterizing sieved ultraspherical polynomials (see Al-Salam, Allaway, and Askey, 1984) just as Theorem 4.3.2 characterizes the Jacobi polynomials (see Dette, 1995e).

## 4.4 PARTITIONED SEQUENCES

In this section we discuss a technique for the identification of the measure corresponding to a given (terminating) sequence of canonical moments that partitions into two parts.

The partitioning will make use of an elementary lemma regarding the continuant expressions for the orthogonal polynomial. Let

$$K(r,n) = K\begin{pmatrix} a_{r+1} & & \cdots & a_n \\ b_r & b_{r+1} & \cdots & b_n \end{pmatrix}$$

denote the continuant defined in (3.2.19), where $K(n,n) = b_n$, $K(n,j) = 1$ if $j < n$.

**Lemma 4.4.1**  *If $1 \le r < n$,*

$$K(1,n) = K(1,r)K(r+1,n)$$
$$+ a_{r+1}K(1,r-1)K(r+2,n).$$

*Proof*  The expression follows from *Laplace's expansion* of $K(1,n)$, using the first $r$ rows in the expansion. ∎

**Theorem 4.4.2**  *For every $k \in \{1, \ldots, n-1\}$ the support of the probability measure $\xi$ on the interval $[-1,1]$ corresponding to the sequence of canonical moments*

$$\tfrac{1}{2}, p_2, \tfrac{1}{2}, p_4, \tfrac{1}{2}, \ldots, \tfrac{1}{2}, p_{2n-2}, \tfrac{1}{2}, 1$$

*is given by the zeros of the polynomial*

$$(x^2 - 1)\{G_{n-k-1}(x)H_k(x) - p_{2k}q_{2k+2}G_{n-k-2}(x)H_{k-1}(x)\},$$

*where $G_{n-k-1}(x)$ and $H_k(x)$ are the polynomials with leading coefficient 1 which* *give the interior support points of the measures corresponding to the sequences*

$$\tfrac{1}{2}, p_{2k+2}, \tfrac{1}{2}, p_{2k+4}, \tfrac{1}{2}, \ldots, \tfrac{1}{2}, p_{2n-2}, \tfrac{1}{2}, 1$$

*and*

$$\tfrac{1}{2}, p_2, \tfrac{1}{2}, p_4, \tfrac{1}{2}, \ldots, \tfrac{1}{2}, p_{2k}, \tfrac{1}{2}, 1,$$

*respectively* $(G_0(x) = 1)$.

*Proof* By Theorem 2.2.3 and Corollary 2.3.6 the support of $\xi$ consists of $\pm 1$ and the zeros of $\overline{S}_{n-1}(x)$. The polynomial $\overline{S}_{n-1}(x)$ can be expressed, for $n \geq 2$, as

$$\overline{S}_{n-1}(x) = K \begin{pmatrix} & -p_2 q_4 & \cdots & -p_{2n-4}q_{2n-2} \\ x & x & \cdots & x \end{pmatrix},$$

since $\overline{S}_{n-1}(x)$ and the continuant satisfy the same recursion formula and initial conditions $(\overline{S}_0(x) = 1, \overline{S}_1(x) = x)$.

Now applying Lemma 4.4.1, it follows that

$$\overline{S}_{k-1}(x) = H_k(x) G_{n-k-1}(x) - p_{2k}q_{2k+2} H_{k-1}(x) G_{n-k-2}(x),$$

where

$$H_k(x) = K \begin{pmatrix} & -p_2 q_4 & \cdots & -p_{2k-2}q_{2k} \\ x & x & \cdots & x \end{pmatrix}$$

and

$$G_{n-k-1}(x) = K \begin{pmatrix} & -p_{2k+2}q_{2k+4} & \cdots & -p_{2n-4}q_{2n-2} \\ x & x & \cdots & x \end{pmatrix}.$$

The statements in the theorem regarding $H_k(x)$ and $G_{n-k-1}(x)$ follow from the comments regarding $\overline{S}_{n-1}(x)$ in the first paragraph. ∎

**Theorem 4.4.3** *For $0 \leq r \leq n-1$ the probability measure $\xi^{(r)}$ on the interval $[-1,1]$ corresponding to the sequence of canonical moments*

$$p_{2j-1} = \frac{1}{2}, \qquad j = 1, \ldots, n,$$

$$p_{2j} = \frac{1}{2}, \qquad j = 1, \ldots, r,$$

$$p_{2j} = \frac{n-j+1}{2(n-j)+1}, \qquad j = r+1, \ldots, n,$$

*is supported at the $n+1$ zeros $-1 = x_0 < x_1 < \ldots < x_{n-1} < x_n = 1$ of the polynomial*

$$(x^2 - 1)\{P'_{n-r}(x)U_r(x) - P'_{n-r-1}(x)U_{r-1}(x)\},$$

*where $P_k(x)$ and $U_k(x)$ denote the kth Legendre and kth Chebyshev polynomial of the second kind, respectively. The masses at the support points are given by*

$$\xi^{(r)}(x_j) = \frac{2}{2n + 1 + U_{2r}(x_j)}, \qquad j = 0, \ldots, n . \tag{4.4.1}$$

*Proof* At this point we only prove the statement regarding the support points; the assertion (4.4.1) will be shown in the proof of Theorem 5.6.1 in Chapter 5. The case $r = 0$ is treated by Corollary 4.3.3 and the discussion following its proof. For $r \geq 1$ an application of Theorem 4.4.2 (with $k = r$) yields that the interior support points can be obtained by looking at the monic polynomials $H_r(x)$ and $G_{n-r-1}(x)$ supporting the sequences

$$p_j^{(1)} = \tfrac{1}{2}, \qquad j = 1, \ldots, 2r + 1, \qquad p_{2r+2}^{(1)} = 1 \tag{4.4.2}$$

and

$$p_{2j}^{(2)} = p_{2j+2r} = \frac{n - r - j + 1}{2(n - r - j) + 1}, \qquad j = 1, \ldots, n - r, \tag{4.4.3}$$

$(p_{2j-1}^{(2)} = \tfrac{1}{2}, j = 1, \ldots, n - r)$, respectively. Applying Corollary 4.3.3 twice (with $z$, $n$ replaced by $0$, $r+1$ and $1$, $n-r$, respectively), it follows that $H_r(x)$ and $G_{n-r-1}(x)$ are the monic versions of ultraspherical polynomials. More precisely,

$$H_r(x) \quad = \hat{C}_r^{(1)}(x) \quad = 2^{-r}C_r^{(1)}(x),$$

$$G_{n-r-1}(x) \quad = \hat{C}_{n-r-1}^{(3/2)}(x) \quad = \frac{\Gamma(\tfrac{3}{2})\Gamma(n-r)}{\Gamma(n-r+\tfrac{1}{2})2^{(n-r-1)}} \; C_{n-r-1}^{(3/2)}(x),$$

where we use (2.3.19) for the last equalities. From (2.1.19), (2.1.20), and (2.1.24), we have $C_r^{(1)}(x) = U_r(x)$, $C_k^{(3/2)}(x) = \frac{d}{dx}C_{k+1}^{(1/2)}(x) = P'_{k+1}(x)$, and Theorem 4.4.2 shows that the measure $\xi^{(r)}$ is supported at the zeros of the polynomial

$$(x^2 - 1)\left\{G_{n-r-1}(x)H_r(x) - \frac{1}{2}\frac{n-r-1}{2(n-r)-1}G_{n-r-2}(x)H_{r-1}(x)\right\}$$

$$= \frac{1}{2^{n-1}}\frac{\Gamma(\tfrac{3}{2})\Gamma(n-r)}{\Gamma(n-r+\tfrac{1}{2})}(x^2 - 1)\{P'_{n-r}(x)U_r(x) - P'_{n-r-1}(x)U_{r-1}(x)\},$$

which proves the assertion of the theorem.                                        ∎

**Theorem 4.4.4**  *For any $\beta \in (-1, 1)$ the probability measure $\xi^{(\beta)}$ on the interval $[-1, 1]$ corresponding to the sequence of canonical moments*

$$p_{2i-1} = \frac{1}{2}, \qquad i = 1, \ldots, n,$$

$$p_{2i} = \frac{1}{2}, \qquad i = 1, \ldots, n-2, \qquad (4.4.4)$$

$$p_{2n-2} = \frac{1+\beta}{2}, \qquad p_{2n} = 1$$

*$(n \geq 2)$ is supported at the zeros $-1 = x_0 < x_1 < \ldots < x_{n-1} < x_n = 1$ of the polynomial*

$$(x^2 - 1)\{U_{n-1}(x) + \beta U_{n-3}(x)\},$$

*where $U_k(x)$ denotes the kth Chebyshev polynomial of the second kind $(U_{-1}(x) = 0)$. The weights at the support points are given by*

$$\xi^{(\beta)}(\mp 1) = \frac{1+\beta}{2[n + \beta(n-2)]},$$

$$\xi^{(\beta)}(x_j) = \left[ n - 1 - \frac{(1+\beta)U_{n-2}(x_j)}{U_n(x_j) + \beta U_{n-2}(x_j)} \right]^{-1}, \qquad j = 1, \ldots, n-1.$$

*Proof*  In view of Theorem 4.4.2 we have to find the polynomials $H_k(x)$, $G_1(x)$ which vanish at the roots of the measures corresponding to the sequences $p_j = \frac{1}{2}$ $(j \leq 2k+1)$, $p_{2k+2} = 1$ $(k = n-2, n-1)$ and the sequence $(\frac{1}{2}, (1+\beta)/2, \frac{1}{2})$. By the arguments given in the proof of Theorem 4.4.3, we find that

$$H_k(x) = \frac{1}{2^k} U_k(x),$$

and from Corollary 4.2.2 we have $G_1(x) = x$. Thus we obtain from Theorem 4.4.2 $(k = n-2)$ that $\xi^{(\beta)}$ is supported at the zeros of the polynomial

$$(x^2 - 1)Q_{n-1}(x) = \frac{1}{2^{n-2}}(x^2 - 1)\left\{ xU_{n-2}(x) - \frac{1-\beta}{2}U_{n-3}(x) \right\}$$

$$= \frac{1}{2^{n-1}}(x^2 - 1)\{U_{n-1}(x) + \beta U_{n-3}(x)\}, \qquad (4.4.5)$$

where we used the recursive relation for the Chebyshev polynomials of the second kind $U_{-1}(x) = 0$, $U_0(x) = 1$,

$$U_{k+1}(x) = 2xU_k(x) - U_{k-1}(x), \qquad k \geq 0, \tag{4.4.6}$$

(see Example 2.1.1). This proves the assertion regarding the support points. For the calculation of the corresponding weights, we apply Theorem 3.6.1 and find for the polynomial in the numerator of (3.6.4) ($\zeta_{2n+1} = 0$) that

$$P_n^{(1)}(x) = K \begin{pmatrix} -4\zeta_3\zeta_4 & \cdots & -4\zeta_{2n-1}\zeta_{2n} \\ x+1-2(\zeta_2+\zeta_3) & \cdots & x+1-2(\zeta_{2n}+\zeta_{2n+1}) \end{pmatrix}$$

$$= K \begin{pmatrix} -\dfrac{1}{4} & \cdots & -\dfrac{1+\beta}{4} & -\dfrac{1-\beta}{2} \\ x & x & \cdots & x & x & x \end{pmatrix}.$$

Here the first equality follows because the continuant and the polynomial $P_n^{(1)}$ satisfy the same recursive relations and initial conditions $P_0^{(1)}(x) = 1$, $P_1^{(1)}(x) = x+1-2(\zeta_2+\zeta_3) = x$. The second equality is a consequence of (4.4.4). By expanding this determinant successively with respect to last row, $P_n^{(1)}$ can be found recursively from $P_n^{(1)}(x) = C_n(x)$, $C_0(x) = 1$, $C_1(x) = x$, and

$$C_{k+1}(x) = x\,C_k(x) - \tfrac{1}{4}C_{k-1}(x), \qquad 1 \leq k \leq n-3,$$

$$C_{n-1}(x) = x\,C_{n-2}(x) - \tfrac{1}{4}(1+\beta)C_{n-3}(x), \tag{4.4.7}$$

$$C_n(x) = x\,C_{n-1}(x) - \tfrac{1}{2}(1-\beta)C_{n-2}(x).$$

It is easy to see that the first line of (4.4.7) is just the recursion for the monic version of the Chebyshev polynomials of the second kind:

$$C_k(x) = \frac{1}{2^k}U_k(x), \qquad k = 0, \ldots, n-2.$$

An application of (4.4.6) and (4.4.7) now yields

$$C_{n-1}(x) = \frac{1}{2^{n-1}}\{2xU_{n-2}(x) - (1+\beta)U_{n-3}(x)\}$$

$$= \frac{1}{2^{n-1}}\{U_{n-1}(x) - \beta U_{n-3}(x)\},$$

$$P_n^{(1)}(x) = C_n(x) = \frac{1}{2^{n-1}}\{x[U_{n-1}(x) - \beta U_{n-3}(x)] - (1-\beta)U_{n-2}(x)\}.$$

By Theorem 3.6.1 and (4.4.5) the weights at the support points

$-1 = x_0 < x_1 < \ldots < x_{n-1} < x_n = 1$ can be calculated as

$$\xi^{(\beta)}(x_j) = \frac{P_n^{(1)}(x_j)}{\frac{d}{dx}(x^2-1)Q_{n-1}(x)\big|_{x=x_j}}$$

$$= \frac{x_j\left[U_{n-1}(x_j) - \beta U_{n-3}(x_j)\right] - (1-\beta)U_{n-2}(x_j)}{2x_j\left[U_{n-1}(x_j) + \beta U_{n-3}(x_j)\right] + (x_j^2-1)\left[U'_{n-1}(x_j) + \beta U'_{n-3}(x_j)\right]}.$$

(4.4.8)

For $j = 0, n$ we thus have

$$\xi^{(\beta)}(-1) = \xi^{(\beta)}(1) = \frac{1+\beta}{2[n+\beta(n-2)]},$$

where we have used $U_{n-1}(1) = (-1)^{n-1}U_{n-1}(-1) = n$ which follows from (4.4.6) and a simple induction.

To show the assertion for the remaining support points, we need the identity

$$(x^2-1)U'_k(x) = kxU_k(x) - (k+1)U_{k-1}(x) \qquad (k \geq 1), \qquad (4.4.9)$$

which follows from the trigonometric representations (2.1.5) and (2.1.6) and from well-known results for the trigonometric functions. The remaining support points $x_1, \ldots, x_{n-1}$ satisfy

$$U_{n-1}(x_j) + \beta U_{n-3}(x_j) = 0, \qquad j = 1, \ldots, n-1, \qquad (4.4.10)$$

and we obtain from (4.4.6), (4.4.8), and (4.4.9),

$$\xi^{(\beta)}(x_j) = \frac{2x_j U_{n-1}(x_j) - (1-\beta)U_{n-2}(x_j)}{2x_j U_{n-1}(x_j) - nU_{n-2}(x_j) - \beta(n-2)U_{n-4}(x_j)}$$

$$= \frac{U_n(x_j) + \beta U_{n-2}(x_j)}{U_n(x_j) - (n-1)U_{n-2}(x_j) - \beta(n-2)U_{n-4}(x_j)}.$$

Now an application of (4.4.6) and (4.4.10) shows that

$$\xi^{(\beta)}(x_j) = \frac{U_n(x_j) + \beta U_{n-2}(x_j)}{(n-1)\left[U_n(x_j) + \beta U_{n-2}(x_j)\right] - (1+\beta)U_{n-2}(x_j)}$$

$$= \left[(n-1) - \frac{(1+\beta)U_{n-2}(x_j)}{U_n(x_j) + \beta U_{n-2}(x_j)}\right]^{-1}, \qquad j = 1, \ldots, n-1,$$

as desired. ∎

## 4.5  INFINITE SEQUENCES OF CANONICAL MOMENTS

In the preceding sections we identified probability measures corresponding to some terminating sequences of canonical moments. In this case the calculations become relatively straightforward because the Stieltjes transform of the measure is a rational function and the masses and weights can be found by discussing the partial fraction expansion of this function. For infinite sequences of canonical moments, this technique does not work any longer, and there are several methods for this situation described in Section 3.6. Usually the calculations have to be performed in two steps. The Stieltjes transform is first identified from its continued fraction expansion, and the Stieltjes inversion formula of Section 3.6 is then applied in order to find the corresponding distribution.

We start our investigations with the sequence

$$p_{2j} = \frac{j}{2j+2}, \qquad j \geq 1, \tag{4.5.1}$$

$$p_{2j-1} = \frac{1}{2}, \qquad j \geq 1 .$$

It was already shown in Example 1.3.6 that the measure on the interval $[-1, 1]$ corresponding to this sequence is the distribution $\mu^U$ with density $2(1 - x^2)^{1/2}/\pi$. To demonstrate the calculation of the distribution corresponding to an infinite sequence of canonical moments, we will now go backward and identify this density from the sequence of canonical moments in (4.5.1). For definiteness we consider the interval $[-1, 1]$ and note that by Theorem 3.3.3 and (4.5.1) the continued fraction expansion of the Stieltjes transform of $\mu^U$ is given by

$$S(z, \mu^U) = \int_{-1}^{1} \frac{d\mu^U(x)}{z - x} = \frac{1}{|z|} - \frac{\frac{1}{4}}{|z|} - \frac{\frac{1}{4}}{|z|} - \frac{\frac{1}{4}}{|z|} - \cdots$$

$$= 2\left(z - \sqrt{z^2 - 1}\right) \tag{4.5.2}$$

$(z \in \mathbb{C} \setminus [-1, 1])$. Here the branch of the square root of is defined such that

$$\left| z - \sqrt{z^2 - 1} \right| < 1, \tag{4.5.3}$$

and the equality in (4.5.2) follows from Lemma 3.2.3. For every $u_0 \in (-1, 1)$, $S(u_0, \mu)$ can be extended from the lower half-plane to a continuous function in a neighborhood of $u_0$. The branch of the square root in this expansion satisfies $\mathrm{Im}\sqrt{(\cdots)} < 0$, and Theorem 3.6.7 shows that $\mu^U$ has a density $f$ on $(-1, 1)$ given by

$$f(x) = \frac{1}{\pi} \mathrm{Im}\ S(x, \mu^U) = \frac{2}{\pi}\sqrt{1 - x^2}, \qquad x \in (-1, 1).$$

As a second example consider the sequence corresponding to the arc-sine measure

$$p_j = \tfrac{1}{2} \qquad (j \geq 1),$$

which was already discussed in Example 3.6.8 on the interval $[0, 1]$. Theorem 3.3.3 shows that the transform of the corresponding distribution $\mu^T$ has the continued fraction expansion

$$S(z, \mu^T) = \int_{-1}^{1} \frac{d\mu^T(x)}{z - x} = \frac{1|}{|z} - \frac{\frac{1}{2}|}{|z} - \frac{\frac{1}{4}|}{|z} - \frac{\frac{1}{4}|}{|z} - \cdots$$

$$= \left[ z - \frac{1}{2} S(z, \mu^U) \right]^{-1}$$

where $S(z, \mu^U)$ is defined in (4.5.2). Thus we obtain from (4.5.2) that

$$S(z, \mu^T) = \frac{1}{\sqrt{z^2 - 1}}, \tag{4.5.4}$$

where the sign of the square root is defined by (4.5.3). Now $S(u_0, \mu^T)$ can be extended from the lower half-plane to a continuous function in a neighborhood of every $u_0 \in (-1, 1)$, and therefore $\mu^T$ is absolute continuous with density

$$f(x) = \frac{1}{\pi} \operatorname{Im} S(x, \mu^T) = \frac{1}{\pi} \frac{1}{\sqrt{1 - x^2}}, \qquad x \in (-1, 1).$$

**Theorem 4.5.1** *The Stieltjes transform of the probability measure $\xi^{g,h}$ on the interval $[-1, 1]$ corresponding to the sequence of canonical moments*

$$p_{4j-2} = g \in (0, 1), \qquad j \geq 1,$$

$$p_{4j} = h \in (0, 1), \qquad j \geq 1, \tag{4.5.5}$$

$$p_{2j-1} = \tfrac{1}{2}, \qquad j \geq 1,$$

*is given by*

$$S(z, \xi^{g,h}) = \frac{1}{2h} \frac{(1 - 2h)z^2 + (h - g) - \sqrt{(z^2 - \eta)^2 - 4\mu}}{z(1 - z^2)},$$

*where*

$$\eta = g(1 - h) + h(1 - g),$$

$$\mu = g(1 - g)h(1 - h), \tag{4.5.5}$$

*and the branch of the square root is defined such that*

$$\left| \frac{z^2 - \eta}{2\sqrt{\mu}} - \sqrt{\frac{(z^2 - \eta)^2}{4\mu} - 1} \right| < 1 . \tag{4.5.7}$$

*Proof*  The continued fraction expansion of $S(z, \xi^{g,h})$ is given by

$$S(z, \xi^{g,h}) = \frac{1|}{|z} - \frac{g|}{|z} - \frac{(1-g)h|}{|z} - \frac{(1-h)g|}{|z} - \frac{(1-g)h|}{|z} - \cdots$$

$$= \frac{z|}{|z^2 - g} - \frac{g(1-g)h|}{|z^2 - \eta} - \frac{\mu|}{|z^2 - \eta} - \frac{\mu|}{|z^2 - \eta} - \cdots,$$

where the last identity follows from the even contraction in (3.2.17) and the equivalence transformation (3.2.11). Using once more (3.2.17), (3.2.11), and (4.5.2), we find that

$$H(z) = \frac{1|}{|z^2 - \eta} - \frac{\mu|}{|z^2 - \eta} - \frac{\mu|}{|z^2 - \eta} - \cdots$$

$$= \frac{1/(2\sqrt{\mu})|}{|(z^2 - \eta)/(2\sqrt{\mu})} - \frac{\frac{1}{4}|}{|(z^2 - \eta)/(2\sqrt{\mu})} - \frac{\frac{1}{4}|}{|(z^2 - \eta)/(2\sqrt{\mu})} - \cdots$$

$$= \frac{1}{\sqrt{\mu}} \left( \frac{z^2 - \eta}{2\sqrt{\mu}} - \sqrt{\frac{(z^2 - \eta)^2}{4\mu} - 1} \right),$$

where the branch of the square root is defined by (4.5.7). Thus it follows that

$$S(z, \xi^{g,h}) = \frac{z}{z^2 - g - g(1-g)hH(z)}$$

$$= \frac{1}{2h} \frac{(1-2h)z^2 + (h-g) - \sqrt{(z^2 - \eta)^2 - 4\mu}}{z(1-z^2)}, \tag{4.5.8}$$

where the branch of the square root is defined by (4.5.7).    ∎

The identification of the corresponding probability measure is more complicated because this measure may have an absolute continuous and a discrete part. This property usually depends on the (relative) size of the parameters $g$ and $h$.

**Theorem 4.5.2**    *The probability measure $\xi^{g,h}$ on the interval $[-1, 1]$ corresponding to the sequence of canonical moments (4.5.5) has an absolute continuous component* *given by*

$$\frac{d\xi^{g,h}}{dx}(x) = \frac{1}{2\pi h}\frac{\sqrt{4\mu - (x^2 - \eta)^2}}{|x|(1 - x^2)}I\{|x^2 - \eta| < 2\sqrt{\mu}\}\ .$$

*If $g + h > 1$, $\xi^{g,h}$ has mass $(g + h - 1)/2h$ at the points $-1$ and $1$. If $h > g$, $\xi^{g,h}$ has mass $(h - g)/h$ at the point $0$. If $g + h \leq 1$ and $h \leq g$, there are no masses at the points $-1, 1$, and $0$, respectively.*

*Proof* The assertion regarding the absolute continuous part follows from Theorem 3.6.7 because $S(z, \xi^{g,h})$ can be extended from the lower half-plane to a continuous function in a neighborhood of any $u_0 \in (-1, 1) \setminus \{0\}$. Jumps of $\xi^{g,h}$ are only possible at the poles of $S(z, \xi^{g,h})$. We investigate the situation at $z = 0$, the other cases are treated similarly. If $g = h$, it is straightforward to show that $S(z, \xi^{g,h})$ has no pole at $z = 0$ and that consequently $\xi^{g,h}(\{0\}) = 0$ in this case (see Theorem 3.6.5). Observing the definition of $\eta$ and $\mu$ in (4.5.6), we see that $h \neq g$ if and only if $\eta^2 > 4\mu$. For $z = -iv$ and sufficiently small $v$, this determines the sign of the square root in (4.5.7) to satisfy $\mathcal{R}(\sqrt{\ldots}) < 0$. Now Theorem 3.6.5 yields

$$\xi^{g,h}(\{0\}) = \lim_{v \to 0} \text{Im}\ S\{v\ S(-iv, \xi^{g,h})\}$$

$$= \lim_{v \to 0} \text{Im}\ \left\{\frac{(2h - 1)v^2 + (h - g) + \sqrt{(v^2 + \eta)^2 - 4\mu}}{-i2h(1 + v^2)}\right\}$$

$$= \frac{h - g + |h - g|}{2h} = \begin{cases} \dfrac{(h - g)}{h} & \text{if}\ \ h > g, \\[2mm] 0 & \text{if}\ \ h \leq g, \end{cases}$$

which proves the assertion.    ∎

CHAPTER 5

# Canonical Moments and Optimal Designs—First Applications

## 5.1 INTRODUCTION

In the previous chapters we explained parts of the theoretical background of canonical moments. Canonical moments were defined geometrically, and the connections to Hankel determinants, orthogonal polynomials, and continued fractions were illustrated. In the following two chapters we discuss an important area of application of these results which is in the construction of optimum designs for polynomial regression models. In many cases the problem of deciding how to collect observations in a most efficient way is equivalent to the problem of maximizing functions of Hankel determinants over the moment space $M_n$. These determinants can be expressed in terms of canonical moments, and the optimization problem usually reduces to an "elementary" exercise in calculus. We have already demonstrated this approach in Section 1.6.

Section 5.2 starts with a short introduction to the linear model, parameter estimation, and the univariate and multivariate polynomial regression model. The theory of exact and approximate designs is explained in Sections 5.3 and 5.4 and (elementary) applications of canonical moments to the solution of various design problems in polynomial regression models are illustrated in Sections 5.5 to 5.8.

## 5.2 LINEAR MODELS

A large part of statistics involves itself with what has become known as the *linear regression model*:

$$Y = \boldsymbol{f}(x)^T \boldsymbol{\theta} + \varepsilon, \tag{5.2.1}$$

where $\boldsymbol{\theta} = (\theta_1, \ldots, \theta_k)^T$ is a vector of unknown parameters, $\boldsymbol{f}(x) = (f_1(x), \ldots, f_k(x))^T$ is a vector of real-valued, linearly independent, continuous regres-

128

sion functions and $\varepsilon$ is a random error term with mean $E(\varepsilon) = 0$ and variance $\mathrm{Var}(\varepsilon) = \sigma^2 > 0$. The interpretation of (5.2.1) is that $Y$ is the result of a measurement at a point $x \in \mathcal{X}$ which is the sum of the expectation, *the deterministic mean effect* $f(x)^T \theta$, and an additive error term $\varepsilon$. $Y$ is called the *response* at the point $x \in \mathcal{X}$. In general, the relationship between $x$ and $Y$ would have $f(x)^T \theta$ replaced by some arbitrary unknown function $g(x)$. For convenience this function $g(x)$ is assumed to be in the finite dimensional *linear* space of functions generated by $f_1, f_2, \ldots, f_k$.

The set $\mathcal{X}$ of all possible points where observations can be taken is called the *design space* and is assumed to be a compact subset of some Euclidean space. The variance of the random term $\varepsilon$ in (5.2.1) (which subsumes quite different sources of error) is assumed to be independent of the specific point $x$, where the response $Y$ is observed. This assumption is referred to in the literature as the *homoscedastic assumption*. For point estimation the two moment assumptions on the random variable

$$E(Y|x) = f(x)^T \theta, \quad \mathrm{Var}(Y|x) = \sigma^2 > 0, \tag{5.2.2}$$

are adequate. However, for the problem of constructing confidence regions and testing hypotheses, it is usually assumed that the response $Y$ at the point $x$ is normally distributed with mean $f(x)^T \theta$ and variance $\sigma^2 > 0$, i.e.

$$Y \sim \mathcal{N}(f(x)^T \theta, \ \sigma^2). \tag{5.2.3}$$

We assume that the experimenter can take $n$ observations

$$Y_j = f(x_j)^T \theta + \varepsilon_j \quad (j = 1, \ldots, n) \tag{5.2.4}$$

at experimental conditions $x_1, \ldots, x_n \in \mathcal{X}$. The $x_i$ values are not necessarily distinct, namely repeated observations at some $x_i$ are allowed; however; all observations are assumed to be uncorrelated:

$$E(\varepsilon_i \varepsilon_j) = \begin{cases} \sigma^2 & \text{if } i = j, \\ 0 & \text{else.} \end{cases} \tag{5.2.5}$$

If the different responses and errors are collected in vectors $Y = (Y_1, \ldots, Y_n)^T$ and $\varepsilon = (\varepsilon_1, \ldots, \varepsilon_n)^T$, then (5.2.2) and (5.2.5) can be conveniently written in matrix form

$$Y = X\theta + \varepsilon,$$

where

$$X = (f(x_1), \ldots, f(x_n))^T \in \mathbb{R}^{n \times k}$$

denotes the $n \times k$ *design matrix*. The expectation and the dispersion (matrix) of the random vector $Y$ are given by (note (5.2.5))

$$E(Y) = X\boldsymbol{\theta}, \quad D(Y) = \sigma^2 I_n, \tag{5.2.6}$$

where $I_n$ denotes the $n \times n$ identity matrix. Following Pukelsheim (1993) we call (5.2.6) the *linear model with moment assumptions*. If additionally the normality assumption (5.2.3) holds for $Y_1, \ldots, Y_n$, we obtain

$$Y \sim \mathcal{N}(X\boldsymbol{\theta}, \sigma^2 I_n) \tag{5.2.7}$$

and call (5.2.7) the *linear model with normality assumption*.

An important part of the statistical analysis of the functional relationship (5.2.2) is the estimation of the unknown parameters $\boldsymbol{\theta}, \sigma^2$ from the observed data $Y = (Y_1, \ldots, Y_n)^T$. We restrict our considerations to linear unbiased estimates for $\boldsymbol{\theta}$, that is,

$$\hat{\boldsymbol{\theta}}_L = LY, \tag{5.2.8}$$

where $L \in \mathbb{R}^{k \times n}$ is a given $k \times n$ matrix and

$$E[\hat{\boldsymbol{\theta}}_L] = LX\boldsymbol{\theta} = \boldsymbol{\theta} \qquad \text{for all } \boldsymbol{\theta} \in \mathbb{R}^k. \tag{5.2.9}$$

In order to compare different linear unbiased estimators, we introduce the concept of *Loewner ordering*.

**Definition 5.2.1** *The set of all nonnegative (positive) definite $k \times k$ matrices is denoted by $NND(k)$ $(PD(k))$, the set of all symmetric matrices is denoted by $SYM(k)$. Define for $A, B \in SYM(k)$,*

$$\begin{aligned} A \geq B \quad &\text{if and only if} \quad A - B \in NND(k), \\ A > B \quad &\text{if and only if} \quad A - B \in PD(k). \end{aligned} \tag{5.2.10}$$

*The partial ordering defined on $SYM(k)$ is called the Loewner ordering.*

Note that the dispersion matrix of a linear estimator (5.2.8) is nonnegative definite, i.e.

$$D(\hat{\boldsymbol{\theta}}_L) = D(LY) = \sigma^2 LL^T \geq 0.$$

This dispersion matrix can be minimized (in the Loewner ordering) with respect to all linear unbiased estimators $\hat{\boldsymbol{\theta}}_L$ for $\boldsymbol{\theta}$.

**Theorem 5.2.2** *For the linear model with moment assumptions (5.2.6) where rank $(X) = k$, the estimator*

$$\hat{\theta}^{GM} = (X^T X)^{-1} X^T Y \qquad (5.2.11)$$

*is the best linear unbiased estimator (BLUE) with respect to the Loewner ordering, that is,*

$$\sigma^2 (X^T X)^{-1} = D(\hat{\theta}^{GM}) \leq D(\hat{\theta}_L) \qquad (5.2.12)$$

*for all linear unbiased estimators $\hat{\theta}_L$ for $\theta$.*

*Proof* From (5.2.9) we obtain $LX = I$ for any $L$ for which $E[LY] = \theta$ for all $\theta \in \mathbb{R}^k$. Then, since

$$((X^T X)^{-1} X^T - L)((X^T X)^{-1} X^T - L)^T \geq 0,$$

it follows that

$$(X^T X)^{-1} - LX(X^T X)^{-1} - (X^T X)^{-1} X^T L^T + LL^T \geq 0.$$

Since $LX = I$, we have

$$D(\hat{\theta}_L) = \sigma^2 LL^T \geq \sigma^2 (X^T X)^{-1}. \qquad \blacksquare$$

In many cases the experimenter is particularly interested in inference about certain linear combinations of the unknown parameters, say, $z_j^T \theta$, where $z_j \in \mathbb{R}^k$ $(j = 1, \ldots, s)$. Let $K = (z_1, \ldots, z_s)$ denote a $k \times s$ matrix of rank $s \leq k$, and consider the *parameter subsystem* $K^T \theta \in \mathbb{R}^s$. $K^T \theta$ is called *estimable* if and only if there exists a linear unbiased estimator for $K^T \theta$, or equivalently if the *range inclusion*

$$\text{range}(K) \subseteq \text{range}(X^T) \qquad (5.2.13)$$

is satisfied. The best linear unbiased estimate (with respect to the Loewner ordering) for the parameter subsystem is given by

$$\hat{\theta}^K = K^T (X^T X)^- X^T Y, \qquad (5.2.14)$$

where $(X^T X)^-$ denotes a *generalized inverse* of the matrix $X^T X$, that is, any matrix $G$ that satisfies $(X^T X) G (X^T X) = X^T X$. The (minimum) dispersion matrix of this estimator is given by

$$D(\hat{\theta}^K) = \sigma^2 K^T (X^T X)^- K. \qquad (5.2.15)$$

It can be shown that under the range inclusion (5.2.13), the estimator $\hat{\theta}^K$ and its dispersion matrix $D(\hat{\theta}^K)$ do not depend on the specific choice of the generalized inverse of $X^T X$. The estimators $\hat{\theta}^{GM}$ and $\hat{\theta}^K$ are called the *Gauss-Markov or best linear unbiased (BLUE) estimators* for the full parameter vector $\theta$ and the parameter subsystem $K^T \theta$, respectively.

**EXAMPLE 5.2.3** If $f(x) = (1, x, \ldots, x^m)^T$ denotes the vector of monomials, then (5.2.1) reduces to the *univariate polynomial regression model* of degree $m$. Here $k = m + 1$ and

$$Y = \sum_{j=0}^{m} \theta_j x^j + \varepsilon. \tag{5.2.16}$$

If observations are taken at the points $x_1, \ldots, x_n$, then the design matrix $X$ in (5.2.6) is given by

$$X = \begin{bmatrix} 1 & x_1 & \cdots & x_1^m \\ 1 & x_2 & \cdots & x_2^m \\ \vdots & \vdots & & \vdots \\ 1 & x_n & \cdots & x_n^m \end{bmatrix}.$$

This matrix has rank $m + 1$ if and only if there are at least $m + 1$ different points among $x_1, \ldots, x_n$. The covariance matrix of the Gauss-Markov estimator for a parameter subsystem $K^T \theta$ ($K \in \mathbb{R}^{m+1 \times s}$) is given by

$$\sigma^2 K^T (X^T X)^- K,$$

where

$$\frac{1}{n} X^T X = \begin{bmatrix} 1 & c_1 & c_2 & \cdots & c_m \\ c_1 & c_2 & c_3 & \cdots & c_{m+1} \\ \vdots & \vdots & \vdots & & \vdots \\ c_m & c_{m+1} & c_{m+2} & \cdots & c_{2m} \end{bmatrix}. \tag{5.2.17}$$

and $c_j = \frac{1}{n} \sum_{i=1}^{n} x_i^j, j = 0, \ldots, 2m$.

**EXAMPLE 5.2.4** In many practical situations the response $Y$ depends on more than one explanatory variable. For example, the growth rate of plants is often modeled as a function of temperature and fertilizer. An obvious generalization of the univariate polynomial regression to models with more than one factor is the *q-way mth degree polynomial*. The design space is a subset of $\mathbb{R}^q$, and the model contains

$$N_{q,m} = \binom{m+q}{q}$$

regression functions of the form

$$\prod_{j=1}^{q} t_j^{m_j}, \tag{5.2.18}$$

where $(t_1, \ldots, t_q)^T \in \mathcal{X} \subseteq \mathbb{R}^q$ and the exponents in (5.2.18) are nonnegative integers such that

$$\sum_{j=1}^{q} m_j \leq m. \tag{5.2.19}$$

The simplest case where $m = 1$ gives a *linear regression in q variables*

$$Y = \alpha_0 + \sum_{i=1}^{q} \alpha_i t_i + \varepsilon.$$

The *multivariate quadratic regression* or *q-way second degree model* is

$$Y = \alpha_0 + \sum_{i=1}^{q} \alpha_i t_i + \sum_{i=1}^{q} \sum_{j=i}^{q} \alpha_{i,j} t_i t_j + \varepsilon.$$

Finally the general form of the *q-way mth degree polynomial* is given by

$$Y = \alpha_0 + \sum_{i=1}^{q} \alpha_i t_i + \sum_{1 \leq i_1 \leq i_2 \leq q} \alpha_{i_1, i_2} t_{i_1} t_{i_2} + \cdots \tag{5.2.20}$$

$$+ \sum_{1 \leq i_1 \leq \ldots \leq i_m \leq q} \alpha_{i_1, \ldots, i_m} \prod_{j=1}^{m} t_{i_j} + \varepsilon.$$

If observations are taken at $n$ different points $t^{(1)}, \ldots, t^{(n)} \in \mathbb{R}^q$ $[t^{(j)} = (t_{j1}, \ldots, t_{jq})^T]$, then the entries in the matrix $X^T X$ are of the form

$$\sum_{j=1}^{n} \prod_{i=1}^{q} t_{ji}^{m_i} \prod_{i=1}^{q} t_{ji}^{m_i'} \tag{5.2.21}$$

where $m_i$ and $m_i'$ are nonnegative integers $(i = 1, \ldots, q)$ such that $\sum_{i=1}^{q} m_i \leq m$ and $\sum_{i=1}^{q} m_i' \leq m$.

We will conclude this section with a brief discussion of the problem of testing hypotheses in linear models. Consider the linear model with normality assumption (5.2.7), and assume that the experimenter is interested in a hypothesis of the form

$$H_0: K^T \theta = 0$$

where $K \in \mathbb{R}^{k \times s}$ is a given matrix of rank $s \leq k$. To estimate $K^T \theta$, we assume that the range inclusion (5.2.13) is satisfied. If $n > \text{rank}(X)$, the $F$-test for testing $H_0: K^T \theta = 0$ rejects $H_0$ for large values of the test statistic

$$F = \frac{n - \text{rank}(X)}{\text{rank}(K)} \cdot \frac{(\hat{\theta}^K)^T (K^T (X^T X)^- K)^- \hat{\theta}^K}{Y^T (I_n - X(X^T X)^- X^T) Y}, \qquad (5.2.22)$$

where $\hat{\theta}^K = K^T (X^T X)^- X^T Y$. Under the normality assumption (5.2.7), $F$ has a noncentral $F$-distribution with numerator degrees of freedom $\text{rank}(K)$, denominator degrees of freedom $n - \text{rank}(X)$, and noncentrality parameter

$$\frac{1}{\sigma^2} (K^T \theta)^T (K^T (X^T X)^- K)^- (K^T \theta). \qquad (5.2.23)$$

We also note that the power function of the $F$-test for the hypothesis $H_0 : K^T \theta = 0$ is an increasing function of the noncentrality parameter.

**EXAMPLE 5.2.5** Consider the univariate polynomial regression model (5.2.16). If the experimenter wants to find out if a polynomial regression of degree $m$ or $m - 1$ is appropriate for describing the relation between the response and the explanatory variable, he could perform an $F$-test for testing $H_0: K^T \theta = \theta_m = 0$, where $K = e_m = (0, \ldots, 0, 1)^T \in \mathbb{R}^{m+1}$ denotes the $(m+1)$th unit vector. It can be shown that the range inclusion (5.2.13) is satisfied if and only if there are at least $m + 1$ different points among $x_1, \ldots, x_n$ (i.e., rank $(X) = m + 1$). Since $\hat{\theta}_m^{e_m} = e_m^T \hat{\theta}^{GM}$, the test statistic of the $F$-test for the hypotheses $H_0: \theta_m = 0$ is given by ($n > m + 1$),

$$F = \frac{(\hat{\theta}_m^{e_m})^2 (e_m^T (X^T X)^{-1} e_m)^{-1}}{\frac{1}{n-m-1} Y^T (I_n - X(X^T X)^{-1} X^T) Y} \qquad (5.2.24)$$

and has a noncentral $F$-distribution with numerator degrees of freedom 1, denominator degrees of freedom $n - m - 1$, and noncentrality parameter

$$\frac{1}{\sigma^2} \theta_m^2 (e_m^T (X^T X)^{-1} e_m)^{-1}. \qquad (5.2.25)$$

The power function of the $F$-test is an increasing function of the noncentrality parameter. Consequently the power of the $F$-test for the hypotheses $H_0 : \theta_m = 0$

could be improved by decreasing

$$e_m^T (X'X)^{-1} e_m$$

with respect to the choice of the points $x_1, \ldots, x_n$ in the design space $\mathcal{X}$. This minimization is a typical problem of optimum experimental design.

## 5.3 OPTIMUM EXPERIMENTAL DESIGNS

Consider the linear regression model with the moment assumptions (5.2.6)

$$E[Y] = X\theta, \quad D(Y) = \sigma^2 I_n,$$

and recall that the design matrix $X$ is given by

$$X = (f(x_1), \ldots, f(x_n))^T \in \mathbb{R}^{n \times k}.$$

Assume that the distinct points among $x_1, \ldots, x_n$ are the points $x_1, \ldots, x_\ell$ ($\ell \le n$), and let $n_i$ denote the number of times the particular point $x_i$ occurs among $x_1, \ldots, x_n$ ($i = 1, \ldots, \ell$). By this procedure one obtains a probability measure $\xi_{(n)}$ on the design space $\mathcal{X}$ with finite support $\{x_1, \ldots, x_\ell\}$ and mass $n_i/n$ at the point $x_i$ ($i = 1, \ldots, \ell$). We call any probability measure with finite support and masses that are multiples of $1/n$ an *exact design* for sample size $n$ and summarize the information of such a measure in the matrix

$$\xi_{(n)} = \begin{pmatrix} x_1 & \cdots & x_\ell \\ \dfrac{n_1}{n} & \cdots & \dfrac{n_\ell}{n} \end{pmatrix}.$$

The first row of this matrix gives the points in the design space $\mathcal{X}$ where observations have to be taken, and the second row tells the experimenter how many observations have to be taken at these points. With this notation the matrix $X^T X$ can be written as a Stieltjes integral

$$X^T X = \sum_{j=1}^{n} f(x_j) f^T(x_j) = \sum_{j=1}^{\ell} n_j f(x_j) f^T(x_j) = n \sum_{j=1}^{\ell} \frac{n_j}{n} f(x_j) f^T(x_j)$$

$$= n \int_{\mathcal{X}} f(x) f^T(x) d\xi_{(n)}(x) =: n M(\xi_{(n)}). \tag{5.3.1}$$

Let $K \in \mathbb{R}^{k \times s}$ denote a given matrix of rank $s \le k$, and consider the problem of estimating the parameter subsystem $K^T \theta$. Since

$$\mathrm{range}(X^T X) = \mathrm{range}(X^T)$$

the parameter subsystem $K^T\theta$ is estimable by an exact design $\xi_{(n)}$ if and only if

$$\text{range}(K) \subseteq \text{range}(M(\xi_{(n)})). \tag{5.3.2}$$

For the exact design $\xi_{(n)}$ which satisfies (5.3.2), the BLUE for $K^T\theta$ (see Theorem 5.2.2) is given by

$$\hat{\theta}^K = K^T(X^TX)^-X^TY = \frac{1}{n}K^TM^-(\xi_{(n)})X^TY$$

with minimum dispersion matrix

$$D(\hat{\theta}^K) = \frac{\sigma^2}{n}K^TM^-(\xi_{(n)})K.$$

Since this matrix depends on the design $\xi_{(n)}$, it is reasonable to investigate if the dispersion matrix of the BLUE for the parameter subsystem $K^T\theta$ can be further minimized by an appropriate choice of the exact design $\xi_{(n)}$. To this end we define $\mathcal{A}_n(K)$ as the set of all exact designs for sample size $n \in \mathbb{N}$ that satisfy the range inclusion (5.3.2). Since the dispersion matrix does not depend on the vector of unknown parameters and because of the simple way it depends on $\sigma^2/n$, the problem of an optimum exact design is: Choose an exact design for sample size $n$ $\xi_{(n)} \in \mathcal{A}_n(K)$ to make

$$C_K(M(\xi_{(n)})) := (K^TM^-(\xi_{(n)})K)^{-1} \tag{5.3.3}$$

large in some sense.

In general there does not exist a design $\xi_{(n)}^* \in \mathcal{A}_n(K)$ that maximizes $C_K(M(\xi_{(n)}))$ in the very strong sense of Loewner ordering

$$C_K(M(\xi_{(n)}^*)) \geq C_K(M(\xi_{(n)})) \text{ for all } \xi_{(n)} \in \mathcal{A}_n(K).$$

Such a $\xi_{(n)}^*$ is likely to exist only in exceptional circumstances (e.g., $K \in \mathbb{R}^k$), and usually optimum designs are determined by maximizing appropriate functions of the "information matrix" $C_K(M(\xi_{(n)}))$, which are called *optimality criteria* or *information functions*. Usually information functions are defined as real valued, positively homogeneous, nonconstant, upper semicontinuous, isotonic, and concave functions on the set of nonnegative definite $s \times s$ matrices $NND(s)$ (see Pukelsheim, 1993). Throughout this text we restrict ourselves to the most popular class of information functions, *Kiefer's $\phi_p$-criteria*.

**Definition 5.3.1**  *Let $p \in [-\infty, 1]$; the pth matrix mean*

$$\phi_p: NND(s) \longrightarrow \mathbb{R}$$

*is defined through*

$$\phi_p(C) = \begin{cases} \left(\dfrac{1}{s}\ trace\ C^p\right)^{1/p} & \text{if } p \neq -\infty, 0, \\ (\det C)^{1/s} & \text{if } p = 0, \\ \lambda_{min}(C) & \text{if } p = -\infty, \end{cases}$$

*for positive definite $C \in PD(s)$ and*

$$\phi_p(C) = \begin{cases} \left(\dfrac{1}{s}\ trace\ C^p\right)^{1/p} & \text{if } p \in (0, 1], \\ 0 & \text{if } p \in [-\infty, 0], \end{cases}$$

*for nonnegative definite matrices $C \in NND(s)$. An exact $\phi_p$-optimal design $\xi^*_{(n)}$ for the parameter subsystem $K^T\boldsymbol{\theta}$ maximizes the function*

$$\phi_p(C_K(M(\xi_{(n)})))$$

*in the set of designs $\mathcal{A}_n(K)$ for which $K^T\boldsymbol{\theta}$ is estimable.*

There are a few optimality criteria that enjoy particular popularity, and they will be explained in the following. For $p = 0$, Definition 5.3.1 gives the *D-optimality criterion*

$$\phi_0(C_K(M(\xi_{(n)}))) = (\det(K^T M^-(\xi_{(n)})K))^{-1/s}. \tag{5.3.4}$$

In order to explain the statistical motivation for the use of the *D*-optimality criterion, we consider the linear model with normality assumption (5.2.7) and a matrix $K \in \mathbb{R}^{k \times s}$ of full rank. If $n > \text{rank}(X)$, a $(1 - \alpha)$-confidence ellipsoid for the vector $K^T\boldsymbol{\theta}$ is given by the set of $\mu \in \mathbb{R}^s$ satisfying

$$(\hat{\boldsymbol{\theta}}^K - \mu)^T (K^T(M^-(\xi_{(n)})K)^{-1}(\hat{\boldsymbol{\theta}}^K - \mu) \leq \frac{s\hat{\sigma}^2}{n} F_{s, n-\text{rank}(X), 1-\alpha}, \tag{5.3.5}$$

where $F_{n,m,1-\alpha}$ denotes the $(1 - \alpha)$- quantile of the *F*-distribution, with $n$ numerator degrees of freedom and $m$ denominator degrees of freedom, and $\hat{\sigma}^2 = (n - \text{rank}(X))^{-1} Y^T (I_n - X(X^T X)^- X^T) Y$ is the common estimator of the variance. The volume of this ellipsoid is proportional to

$$\{\det(K^T M^-(\xi_{(n)})K)\}^{1/2},$$

and therefore a *D*-optimum design $\xi^D_{(n)}$ (for $K^T\boldsymbol{\theta}$) minimizes the volume of the ellipsoid of concentration for the vector $K^T\boldsymbol{\theta}$ with respect to the choice of designs $\xi_{(n)} \in \mathcal{A}_n(K)$. Note that if $K = I_k$, then a *D*-optimum design $\xi_{(n)}$ maximizes $\det(M(\xi_{(n)}))$.

If $p = -1$, we obtain from Definition 5.3.1 the *A-optimality criterion*

$$\phi_{-1}(C_K(M(\xi_{(n)}))) = \left\{ \frac{1}{s} \text{trace} \ (K^T M^-(\xi_{(n)})K) \right\}^{-1}$$

$$= \left[ \frac{1}{s} \sum_{j=1}^{s} z_j^T M^-(\xi_{(n)})z_j \right]^{-1},$$

where $z_1, \ldots, z_s$ denote the columns of the $k \times s$ matrix $K$. By (5.2.15) the term $z_j^T M^-(\xi_{(n)})z_j$ is proportional to the variance of the Gauss-Markov estimator of the linear combination $z_j^T \theta$, and therefore an $A$-optimal design $\xi_{(n)}^A$ minimizes the sum of the variances of the Gauss-Markov estimators of the linear combinations $z_j^T \theta \ (j = 1, \ldots, s)$.

In the case $p = -\infty$ Definition 5.3.1 gives the *E-optimality criterion*

$$\phi_{-\infty}(C_K(M(\xi_{(n)}))) = \lambda_{\min}((K^T M^-(\xi_{(n)})K)^{-1})$$

$$= \{\lambda_{\max}(K^T M^-(\xi_{(n)})K)\}^{-1}$$

$$= \{ \max_{z \in \mathbb{R}^s, ||z||=1} z^T K^T M^-(\xi_{(n)})Kz \}^{-1},$$

which has the following minimax motivation: For a vector $z \in \mathbb{R}^s$ the variance of the Gauss-Markov estimator for the linear combination $z^T K^T \theta$ is proportional to $z^T K^T M^-(\xi_{(n)})Kz$. Therefore an $E$-optimal design $\xi_{(n)}^E$ minimizes the worst possible variance among all linear combinations $z^T K^T \theta$ with a $z$ vector of Euclidean norm 1.

**EXAMPLE 5.3.2** Consider a *univariate quadratic regression* model on the design space $\mathcal{X} = [-1, 1]$,

$$Y = \theta_0 + \theta_1 x + \theta_2 x^2 + \varepsilon, \qquad x \in [-1, 1]. \tag{5.3.6}$$

If the experimenter is interested in the estimation of the highest coefficient $\theta_i$ or in a test to see whether the quadratic term is present in this model (see Example 5.2.6), we put $K = e_2 = (0, 0, 1)^T$. Let $\xi_{(n)}$ denote an exact design of sample size $n$, then the matrix $M(\xi_{(n)})$ defined in (5.3.1) is given by

$$M(\xi_{(n)}) = \int_{-1}^{1} f(x)f^T(x)d\xi_{(n)}(x) = \begin{bmatrix} 1 & c_1 & c_2 \\ c_1 & c_2 & c_3 \\ c_2 & c_3 & c_4 \end{bmatrix},$$

where $f(x) = (1, x, x^2)^T$ denotes the vector of regression functions and $c_j = \int_{-1}^{1} x^j d\xi_{(n)}(x)$ denotes the $j$th moment of $\xi_{(n)}$. It is easy to see that the parameter $\theta_2 = K^T\theta$ is estimable [i.e., $e_2 \in$ range $(M(\xi_{(n)}))$] if and only if $\xi_{(n)}$ has at least three support points. For these designs the dispersion "matrix" of the Gauss-Markov estimator for $\theta_2 = e_2^T\theta$ is proportional to

$$\{C_K(M(\xi_{(n)}))\}^{-1} = e_2^T M^{-1}(\xi_{(n)})e_2 = \frac{c_2 - c_1^2}{|M(\xi_{(n)})|} \in \mathbb{R}^+. \tag{5.3.7}$$

An optimal design for $\theta_2$ maximizes $C_K(M(\xi_{(n)}))$ and therefore minimizes the variance of the Gauss-Markov estimator for the parameter $\theta_2$. The maximization of $C_K(M(\xi_{(n)}))$ in the set of all exact designs with nonsingular matrix $M(\xi_{(n)})$ is somewhat complicated, and we only state the result, which is due to Krafft and Schaefer (1995).

If $n = 4p + q$, $q \in \{0, 1, 3\}$ and $p \geq 1$ (or $p = 0$ and $q = 3$), the optimal design $\xi_{(n)}^*$ for the estimation of the parameter $\theta_2$ is unique and given by

$$\xi_{(4p)}^* = \begin{pmatrix} -1 & 0 & 1 \\ \dfrac{1}{4} & \dfrac{1}{2} & \dfrac{1}{4} \end{pmatrix} \quad (n = 4p),$$

$$\xi_{(4p+1)}^* = \begin{pmatrix} -1 & 0 & 1 \\ \dfrac{p}{4p+1} & \dfrac{2p+1}{4p+1} & \dfrac{p}{4p+1} \end{pmatrix} \quad (n = 4p + 1),$$

$$\xi_{(4p+3)}^* = \begin{pmatrix} -1 & 0 & 1 \\ \dfrac{p+1}{4p+3} & \dfrac{2p+1}{4p+3} & \dfrac{p+1}{4p+3} \end{pmatrix} \quad (n = 4p + 3).$$

In the case $n = 4p + 2$ the situation is more complicated, and there are two (exact) optimal designs, namely

$$\xi_{(4p+2)}^* = \begin{pmatrix} -1 & x_0(p) & 1 \\ \dfrac{p}{4p+2} & \dfrac{2p+1}{4p+2} & \dfrac{p+1}{4p+2} \end{pmatrix} \quad (n = 4p + 2) \tag{5.3.8}$$

and its reflection at the point 0. Here $x_0(p)$ is the real root of the cubic polynomial

$$(2p + 1)^2 x^3 - 3(2p + 1)x^2 + (20p^2 + 20p + 3)x - 2p - 1. \tag{5.3.9}$$

## 5.4 APPROXIMATE OPTIMAL DESIGNS

Example 5.3.2 indicates that in general, the determination of optimal exact designs is an extremely hard problem, which in many cases of practical interest is an intractable integer optimization problem. The situation is analogous to the much simpler problem of maximizing a real-valued function $f$ on the integers where no calculus techniques are applicable. A commonly used device for this simpler problem is to extend the function $f$ to the real axis, determine its maximum and argue that the maximum, of the function over the integers will occur at an integer that is close to the point where the function over the real axis attains its maximum. We will use for the design problems of Section 5.3 a similar concept, which is called the *approximate theory of optimum design* and is due to Kiefer.

**Definition 5.4.1** *An approximate design is a probability measure on the design space $\mathcal{X}$ with finite support. The set of all approximate designs is denoted by $\Xi$.*

A convenient notation for an approximate design is the matrix form

$$\xi = \begin{pmatrix} x_1 & \cdots & x_\ell \\ w_1 & \cdots & w_\ell \end{pmatrix} \tag{5.4.1}$$

where $x_1, \ldots, x_\ell$ denote the support points and $w_1, \ldots, w_\ell$ the corresponding weights. Obviously an exact design for the sample size $n$ is also an approximate one, but the converse is in general not true because the weights in (5.4.1) are not necessarily multiples of $1/n$. The interpretation of an approximate design is that the weights $w_j$ represent the relative proportion of total observations that should be taken at $x_j$ ($j = 1, \ldots, \ell$). Very often an approximate design is called a *design for an infinite sample size* because it arises from the exact design of sample size $n$ when $n$ tends to infinity. However, for a finite sample size $n$ the numbers $w_j n$ are not necessarily integers, and an optimal approximate design has to be approximated by an exact design for sample size $n$ using appropriate rounding procedures (see Example 5.4.5).

**Definition 5.4.2** *Let $\xi \in \Xi$ denote an approximate design; the matrix*

$$M(\xi) = \int_{\mathcal{X}} f(x) f^T(x) d\xi(x) = \sum_{j=1}^{\ell} w_j f(x_j) f^T(x_j)$$

*is called a moment matrix. For a given matrix $K \in \mathbb{R}^{k \times s}$ of full rank $s \le k$, we call*

$$\mathcal{A}(K) = \{\xi \in \Xi \mid \operatorname{range}(K) \subseteq \operatorname{range}(M(\xi))\}$$

*the set of all feasible designs and define for $\xi \in \mathcal{A}(K)$,*

$$C_K(M(\xi)) = (K^T M^-(\xi) K)^{-1},$$

*as the information matrix of the approximate design $\xi \in \mathcal{A}(K)$ for the parameter subsystem $K^T \theta$.*

The interpretation of the preceding definition is relatively obvious. If $\xi$ is an approximate design and $\xi_{(n)}$ an approximation of $\xi$ by an exact design of sample size $n$, then the moment and the information matrix of $\xi_{(n)}$ are expected to be "close" to the corresponding matrices of the approximate design $\xi$. Consequently, if an optimality criterion can be maximized in the class $\mathcal{A}(K)$ at the point $\xi^*$ and $\xi^*_{(n)} \in \mathcal{A}_n(K)$ is an approximation of $\xi^*$ by an exact design of sample size $n$, then $\xi^*_{(n)}$ should be an efficient exact design for estimating the parameter subsystem $K^T\theta$. In the following text all designs are understood in the approximate sense, and this is not mentioned explicitly. Exact designs for finite sample sizes can be found by apportionment methods from the optimal approximate designs (see Pukelsheim and Rieder, 1993, or Fedorov, 1972).

**Definition 5.4.3** *Let* $-\infty \le p < 1$; *a design* $\xi \in \Xi$ *is called* $\phi_p$-*optimal for estimating* $K^T\theta$ *if and only if* $\xi \in \mathcal{A}(K)$ *and* $\xi$ *maximizes* $\phi_p(C_K(M(\xi)))$ *in the class of all feasible designs* $\mathcal{A}(K)$.

**Theorem 5.4.4** *The set of all feasible designs is convex. For every* $p \in [-\infty, 1)$ *and for every* $K \in \mathbb{R}^{k \times s}$ *of full rank* $s \le k$, *the function*

$$\phi_p \circ C_K : \begin{cases} \mathcal{A}(K) \to \mathbb{R}, \\ \xi \to \phi_p(C_K(M(\xi))), \end{cases}$$

*is concave.*

The proof of the second part in Theorem 5.4.4 relies on the fact that the $\phi_p$-criteria in Definition 5.3.1 are concave functions of the eigenvalues of the matrix $C_K$. For example, if $p = -\infty$, $K = I_k$, we obtain the $E$-criterion

$$\phi_E(\xi) = \lambda_{\min}(M(\xi)).$$

The concavity of $\phi_E$ as a function on the set $\mathcal{A}_K = \{\xi \| M(\xi)\| \neq 0\}$ follows from the representation

$$\lambda_{\min}(A) = \min_{\|z\|=1} z^T A z.$$

A proof of the general statement in Theorem 5.4.4 can be found in the monograph of Pukelsheim (1993).

**EXAMPLE 5.4.5** Consider the quadratic regression Example 5.3.2 where $\theta = (\theta_0, \theta_1, \theta_2)^T$, $f(x) = (1, x, x^2)^T$ and $\mathcal{X} = [-1, 1]$, and assume that the experimenter is interested in the estimation of the highest coefficient $\theta_2$ (i.e., $e_2 = (0, 0, 1)^T$). It is easy to see that $\mathcal{A}(K) = \{\xi \in \Xi \| M(\xi)\| \neq 0\}$, and observing

(5.3.7), it follows that an optimal design $\xi$ for $\theta_2$ maximizes

$$C_{e_2}(M(\xi)) = \frac{|M(\xi)|}{c_2 - c_1^2} = \frac{\underline{H}_4(\xi)}{\underline{H}_2(\xi)}, \tag{5.4.2}$$

where $c_j = \int_{-1}^{1} x^j d\xi(x)$ denotes the $j$th moment of the design $\xi \in \mathcal{A}(K)$ and $\underline{H}_{2m}(\xi) = |(c_{i+j})_{i,j=0}^m|$ denote the Hankel determinants formed by the moments of the measure of $\xi$. By Remark 2.3.7 the ratio in (5.4.2) can be expressed in terms of canonical moments of $\xi$, that is,

$$C_{e_2}(M(\xi)) = 2^4 \prod_{j=1}^{2} \zeta_{2j-1}\zeta_{2j} = 2^4 \left( \prod_{j=1}^{3} q_j p_j \right) p_4.$$

A maximization in terms of canonical moments yields $p_j = \frac{1}{2} (j = 1, 2, 3), p_4 = 1$. The design corresponding to this sequence is optimal for the estimation of $\theta_2$ and can be identified by an application of Corollary 4.2.2:

$$\xi^* = \begin{pmatrix} -1 & 0 & 1 \\ \frac{1}{4} & \frac{1}{2} & \frac{1}{4} \end{pmatrix}.$$

This means that the experimenter should take $\frac{1}{4}$ of the observations at the points $-1$ and $1$ and $\frac{1}{2}$ of the observations at the point $0$. If the sample size $n$ is not a multiple of 4, a rounding procedure is applied to produce an exact design for the sample size $n$. In order to compare (exact) designs obtained by this procedure with the optimal exact designs $\xi_{(n)}^*$ of Example 5.3.2, we apply the following (simple) apportionment method. If $n_0$ is the closest integer to $n/4$, we use the exact design

$$\tilde{\xi}_{(n)} = \begin{pmatrix} -1 & 0 & 1 \\ \dfrac{n_0}{n} & 1 - \dfrac{2n_0}{n} & \dfrac{n_0}{n} \end{pmatrix}$$

as approximation of the optimal design $\xi^*$ (if there are two integers with the same distance to $n/4$, we define $n_0$ as the smaller one). Whenever $n \neq 4p + 2$, the design $\tilde{\xi}_{(n)}$ coincides with the optimal exact design $\xi_{(n)}^*$ of Example 5.3.2.

As pointed out in Section 5.3 the power of the $F$-test for the hypothesis $H_0 : \theta_2 = 0$ is an increasing function of $C_{e_2}(M(\xi))$. Therefore it is reasonable to compare the performance of the two designs $\xi_{(4p+2)}^*$ and $\tilde{\xi}_{(4p+2)}$ by the ratio

$$r(\tilde{\xi}_{(4p+2)}, \xi_{(4p+2)}^*) = \frac{C_{e_2}(M(\tilde{\xi}_{(4p+2)}))}{C_{e_2}(M(\xi_{(4p+2)}^*))}.$$

The following table contains these ratios and the solution $x_0(p)$ of the equation

(5.3.9) for different values of $p$:

| $p$ | 1 | 2 | 3 | 4 | 5 |
|---|---|---|---|---|---|
| $n$ | 6 | 10 | 14 | 18 | 22 |
| $x_0(p)$ | 0.0707 | 0.0408 | 0.0289 | 0.0224 | 0.0183 |
| $r(\tilde{\xi}_{(4p+2)}, \xi^*_{(4p+2)})$ | 0.9327 | 0.9759 | 0.9877 | 0.9925 | 0.9950 |

As expected, there appear only minor differences between the exact design $\tilde{\xi}_{(n)}$ (constructed by an approximation to the optimal approximate design) and the optimum exact design $\xi^*_{(n)}$ (constructed by integer optimization). In this case the approximate design approach provides an efficient solution of the exact design problem.

The preceding example served as a motivation for the concept of approximate design theory. It also introduced a method for comparing designs by forming ratios of the criterion values of a given design and a best design.

**Definition 5.4.6** *The $\phi_p$-efficiency of $\xi \in A(K)$ for estimating $K^T\theta$ is defined as*

$$\text{eff}_p(\xi) = \frac{\phi_p(C_K(M(\xi)))}{\sup_{\eta \in A(K)} \phi_p(C_K(M(\eta)))}.$$

*The $\phi_p$-efficiency of a design $\xi_1 \in A(K)$ with respect to a design $\xi_2 \in A(K)$ is defined as*

$$r_p(\xi_1, \xi_2) = \frac{\phi_p(C_K(M(\xi_1)))}{\phi_p(C_K(M(\xi_2)))}.$$

Note that $\text{eff}_p(\xi) \in [0, 1]$, while $r_p(\xi_1, \xi_2) \in [0, \infty]$. Obviously $\xi_1$ is called *better* than $\xi_2$ with respect to the $\phi_p$-optimality if and only if $r_p(\xi_1, \xi_2) > 1$. The efficiency of a design $\xi$ with respect to the $\phi_p$-optimal design $\xi^*$ is precisely the $\phi_p$-efficiency of $\xi$.

An important advantage of the consideration of the approximate design problem is that it "replaces" the integer optimization problem by a convex optimization problem. More precisely, consider the set of all moment matrices

$$\mathcal{M}(\Xi) = \{M(\xi)|\xi \in \Xi\}, \tag{5.4.3}$$

which is obviously a convex subset of the nonnegative definite matrices $NND(k)$. Because the design space $X$ is compact and the regression functions are continuous, it follows that

$$\mathcal{F} = \{f(x)f^T(x)|x \in \mathcal{X}\} \tag{5.4.4}$$

is a compact subset of $NND(k)$. By the definition of the moment matrix $M(\xi)$, the set $\mathcal{M}(\Xi)$ is the convex hull of $\mathcal{F}$ and therefore also compact. Thus the determination of a $\phi_p$-optimal design is first of all a maximization problem of a concave function on a convex and compact subset in $NND(k)$. The second step is the determination of a design $\xi \in \mathcal{A}(K)$ corresponding to the optimal moment matrix found in the first step. Thus, by introducing the concept of approximate designs, the powerful duality theory of convex analysis becomes applicable. These methods provide useful characterizations for the optimal moment matrices and designs, which are called *equivalence theorems*. For the purposes of Chapter 6 we first state these results for the very general problem of maximizing the function $\phi_p$ on a convex and compact subset of the nonnegative definite matrices and then specialize to the $\phi_p$-optimal design problem of Definition 5.4.3. For the $D$-optimality criterion where $p = 0$, the corresponding statement and its proof will be illustrated in Theorem 5.5.2 below where the results are fairly simple. The proof of the following general results can be found in Pukelsheim (1993).

**Theorem 5.4.7** *Let* $\mathcal{M}$ *denote a convex and compact subset of* $NND(k)$, $p \in (-\infty, 1)$, *and define*

$$\mathcal{M}_K = \{M \in \mathcal{M}|\text{ range } K \subset \text{ range } M\}. \tag{5.4.5}$$

*If* $\mathcal{M}_K \neq \phi$ *there exists a matrix* $M \in \mathcal{M}_K$ *that maximizes the function* $\phi_p(C_K(M))$ *over the set* $\mathcal{M}_K$.

*A matrix* $M \in \mathcal{M}_K$ *maximizes* $\phi_p(C_K(M))$ *over the set* $\mathcal{M}_K$ *if and only if there exists a generalized inverse* $G$ *of* $M$ *such that the inequality*

$$\text{trace } AGK(C_K(M))^{p+1}K^TG^T \leq \text{ trace}(C_K(M))^p \tag{5.4.6}$$

*holds for all* $A \in \mathcal{M}$. *If* $M \in \mathcal{M}_K$ *maximizes* $\phi_p(C_K(M))$, *equality is obtained in* (5.4.6) *for* $A = M$.

**Theorem 5.4.8** *Let* $\mathcal{M}$ *denote a convex and compact subset of* $NND(k)$; *if* $\mathcal{M}_K \neq \phi$ *there exists a matrix* $M \in \mathcal{M}_K$ *that maximizes*

$$\phi_{-\infty}(C_K(M)) = \lambda_{\min}((K^TM^-K)^{-1})$$

*over the set* $\mathcal{M}_K$. *A matrix* $M \in \mathcal{M}_K$ *maximizes* $\phi_{-\infty}(C_K(M))$ *over the set* $\mathcal{M}_K$ *if and only if there exists a generalized inverse* $G$ *of* $M$ *and a* $s \times s$ *matrix*

$$E \in \mathcal{E} := Co(\{zz^T|z \in \mathbb{R}^s, \ ||z|| = 1, \ \lambda_{\min}(C_K(M))z = C_K(M)z\})$$

*such that the inequality*

$$\text{trace } AGKC_K(M)EC_K(M)K^T G^T \leq \lambda_{\min}(C_K(M)) \qquad (5.4.7)$$

*holds for all $A \in \mathcal{M}$. If $M$ maximizes $\phi_{-\infty}(C_K(M))$, equality holds in (5.4.7) for $A = M$.*

As a first application consider the set $\mathcal{M}(\Xi)$ of all possible moment matrices corresponding to designs on $\mathcal{X}$. Obviously the set $\mathcal{M}_K$ defined in (5.4.5) coincides with the set of all feasible designs in Definition 5.4.2. Since $\mathcal{M}(\Xi)$ is the convex hull of the set $\mathcal{F}$ in (5.4.4), a straightforward application of Theorems 5.4.7 and 5.4.8 yields the following equivalence theorem for $\phi_p$-optimal designs:

**Theorem 5.4.9** *For $p \in [-\infty, 1)$ there exists a $\phi_p$-optimal design for $K^T\theta$.*
1. *If $p \in (-\infty, 1)$, a design $\xi^* \in \mathcal{A}(K)$ is $\phi_p$-optimal for $K^T\theta$ if and only if there exists a generalized inverse $G$ of the moment matrix $M(\xi^*)$ such that the inequality*

$$f^T(x)GK(C_K(M(\xi^*)))^{p+1}K^T G^T f(x) \leq \text{trace}(C_K(M(\xi^*)))^p \qquad (5.4.8)$$

   *holds for all $x \in \mathcal{X}$. Moreover there is equality in (5.4.8) for every support point of the $\phi_p$-optimal design $\xi^*$.*
2. *A design $\xi^* \in \mathcal{A}(K)$ is $\phi_{-\infty}$-optimal for $K^T\theta$ if and only if there exists a generalized inverse $G$ of the moment matrix $M(\xi^*)$ and a $s \times s$ matrix $E \in \mathcal{E}$ such that the inequality*

$$f^T(x)GKC_K(M(\xi^*))EC_K(M(\xi^*))K^T G^T f(x) \leq \lambda_{\min}(C_K(M(\xi^*))) \qquad (5.4.9)$$

   *holds for all $x \in \mathcal{X}$. Moreover there is equality in (5.4.9) for every support point of the $\phi_{-\infty}$-optimal design $\xi^*$.*

Roughly speaking, Theorem 5.4.9 provides a "checking condition" for $\phi_p$-optimality of a given design $\xi^*$. If the inequality (5.4.8) $(-\infty < p < 1)$ or (5.4.9) $(p = -\infty)$ is satisfied for all $x \in \mathcal{X}$, then the design $\xi^*$ is $\phi_p$-optimal for $K^T\theta$. There are two difficulties with the application of these inequalities. The first one is that the generalized inverse $G$ of $M(\xi)$ is not necessarily unique. Even if the design $\xi^*$ is $\phi_p$-optimal for $K^T\theta$, the inequality (5.4.8) or (5.4.9) might not be true for any generalized inverse of $M(\xi^*)$. However, in the estimation problem of the full parameter vector $\theta$ and in many other problems of practical interest, the set of feasible designs satisfies $\mathcal{A}(K) = \{\xi \|M(\xi)\| \neq 0\}$ and $G$ is just the inverse of $M(\xi)$. The second difficulty appears for the $\phi_{-\infty}$-optimality criterion, where the matrix $E \in \mathcal{E}$ is not uniquely determined if the multiplicity of the minimum eigenvalue of $C_K(M(\xi^*))$ is larger than 1. This reflects the well-known fact that in general, the $E$-optimality criterion is not differentiable. Note that the same comments apply to the optimization problem considered in Theorems 5.4.7 and 5.4.8.

## 5.5 *D*- AND *G*-OPTIMAL DESIGNS FOR UNIVARIATE POLYNOMIAL REGRESSION

It was already mentioned in Section 5.3 that the most popular optimality criterion is the *D*-criterion which is obtained from Kiefer's $\phi_p$-criteria for the choice $p = 0$. One reason for this popularity is that the *D*-optimal criterion depends only on the linear space spanned by the regression functions $f_1, f_2, \ldots, f_k$ in (5.2.1) and not on the particular basis used.

**Theorem 5.5.1**  *Let $f(x)$ denote the vector of regression functions in the linear model (5.2.1), and let $g(x) = Af(x)$, where A is a nonsingular $k \times k$ matrix. The D-optimal design in the linear regression model*

$$Y = \rho^T g(x) + \varepsilon \qquad (5.5.1)$$

*coincides with the D-optimal design in the linear model (5.2.1).*

*Proof*  The proof is an immediate consequence of the identity

$$|M_g(\xi)| = |\int_{\mathcal{X}} g(x)g^T(x)d\xi(x)|$$

$$= |A \int_{\mathcal{X}} f(x)f^T(x)d\xi(x)A^T| = |A|^2 |M_f(\xi)|,$$

which is obvious from the definition of the moment matrices in the linear models (5.5.1) and (5.2.1).  ∎

By Theorem 5.4.9 ($p = 0$, $K = I_k$, $C_K(M) = M$), it follows that a design $\xi^D$ is *D*-optimal if and only if

$$f^T(x)M^{-1}(\xi^D)f(x) \leq k \qquad \text{for all } x \in \mathcal{X}. \qquad (5.5.2)$$

Since there is equality in (5.5.2) for every support point of $\xi^D$, this inequality is equivalent to

$$G(\xi^D) = \max_{x \in \mathcal{X}} f^T(x)M^{-1}(\xi^D)f(x) = k. \qquad (5.5.3)$$

Now for any design $\xi$ on $\mathcal{X}$ it follows that

$$k = \text{trace } (M(\xi)M^{-1}(\xi)) = \sum_{j=1}^{\ell} \xi(x_j)f^T(x_j)M^{-1}(\xi)f(x_j) \leq G(\xi),$$

and therefore (5.5.3) is equivalent to

$$\xi^D \quad \text{minimizes} \quad G(\xi) = \max_{x \in \mathcal{X}} f^T(x) M^{-1}(\xi) f(x) \qquad (5.5.4)$$

Note that the existence of a $D$-optimal design follows from Theorem 5.4.9, and therefore a design minimizing (5.5.4) has to satisfy (5.5.3).

A design $\xi^*$ is called $G$-*optimal for the parameter* $\boldsymbol{\theta}$ if $|M(\xi)| > 0$ and $\xi^*$ minimizes

$$G(\xi) = \max_{x \in \mathcal{X}} f^T(x) M^{-1}(\xi) f(x).$$

**Theorem 5.5.2**   *A design* $\xi^*$ *with* $|M(\xi^*)| \neq 0$ *is $G$-optimal for the parameter* $\boldsymbol{\theta}$ *if and only if* $\xi^*$ *is $D$-optimal, that is, if and only if (5.5.2) or (5.5.3) is satisfied.*

Theorem 5.5.2 is the earliest "Equivalence Theorem" in design theory, and it was proved by Kiefer and Wolfowitz in 1960. The $G$-optimal designs for low-order polynomials were worked out numerically by Smith (1918) and theoretically by Guest (1958). In the same year Hoel (1958) found the $D$-optimal designs. The results of Hoel and Guest were the same and prompted Kiefer and Wolfowitz to prove Theorem 5.5.2 for general regression functions. Since then numerous "Equivalence Theorems" have been proven. Theorems 5.4.7, 5.4.8, and 5.4.9 are quite general and include many others as special cases.

The result in Theorem 5.5.2 is actually quite simple. If $|M(\xi)| > 0$ and we let

$$\phi(\xi) = \log |M(\xi)|,$$

then the $D$-optimal design maximizes $\phi(\xi)$. Since $\phi$ is concave in $\xi$, a local maximum will be a global maximum. Then the design $\xi^*$ will be the $D$-optimal design if and only if

$$g(\alpha) = \phi((1 - \alpha)\xi^* + \alpha\xi)$$

has a nonpositive derivative at $\alpha = 0$ for all $\xi$ which is equivalent to

$$\text{trace}[M^{-1}(\xi^*)(M(\xi) - M(\xi^*))] \leq 0 \qquad \text{for all } \xi.$$

This in turn can be seen to hold if it holds for all "one-point" designs $\xi_x$ $(x \in \mathcal{X})$ for which $M(\xi_x) = f(x)f^T(x)$. In this case $\xi^*$ is $D$-optimal if and only if

$$f^T(x) M^{-1}(\xi^*) f(x) \leq k \qquad \text{for all } x \in \mathcal{X}.$$

Although Theorem 5.5.2 shows that (in the approximate design theory) the $G$- and $D$-optimal design coincide, there is a difference in the statistical interpretation of the two criteria. A $D$-optimal design minimizes the volume of the ellipsoid of concentration for the unknown parameter vector $\theta$, while a $G$-optimal design minimizes the worst possible variance of the estimator of the response function $f^T(x)\theta$ over the design space $\mathcal{X}$. Thus there is another measure of efficiency that is used to compare competing designs with $|M(\xi)| \neq 0$, namely

$$\text{eff}^G(\xi) := \frac{k}{\max_{x \in \mathcal{X}} f^T(x) M^{-1}(\xi) f(x)}.$$

The quantity $\text{eff}^G(\xi)$ is called the *G-efficiency* of the design $\xi$.

In the remaining part of this section we will consider the univariate polynomial regression model of degree $m$,

$$Y = \sum_{j=0}^{m} \theta_j x^j + \varepsilon, \tag{5.5.5}$$

where the explanatory variable varies in the design space $\mathcal{X} = [a, b](-\infty < a < b < \infty)$. In order to apply the preceding Theorem 5.5.1 in the univariate polynomial regression, define

$$g(x) = f_m((b - a)x + a) = A f_m(x), \tag{5.5.6}$$

where $f_m(x) = (1, x, \ldots, x^m)^T$ and $A$ in a lower triangular matrix with diagonal elements $1, b - a, \ldots, (b - a)^m$. In this case the moment matrix of a design $\xi_{a,b}$ in the model (5.5.5) is obtained as

$$M_m(\xi) = \int_a^b f_m(x) f_m^T(x) d\xi_{a,b}(x) = \int_0^1 g(x) g^T(x) d\xi(x), \tag{5.5.7}$$

where $g$ is a given by (5.5.6) and $\xi$ is the linear transformation of $\xi_{a,b}$ on the interval $[0, 1]$. By Theorem 5.5.1 the design maximizing $|\int_0^1 g(x) g^T(x) d\xi(x)|$ is the same as the $D$-optimal design for the univariate polynomial regression (5.5.2) on the interval $[0, 1]$. Consequently the $D$-optimal design $\xi_{a,b}^D$ for the linear model (5.5.5) on the interval $[a, b]$ can be obtained from the $D$-optimal design $\xi_{0,1}^D$ on the interval $[0, 1]$ by a linear transformation:

$$\xi_{a,b}^D(\{x\}) = \xi_{0,1}^D\left(\left\{\frac{x - a}{b - a}\right\}\right).$$

This means that the $D$-optimal design remains "essentially" unchanged if the experimenter uses different units of measurements. This property is also reflected in terms of canonical moments. Let $\xi$ denote the linear transformation of the measure $\xi_{a,b}$ on the interval $[0,1]$; then a straightforward calculation shows that

$$|M_m(\xi_{a,b})| = \left|\int_0^1 f_m((b-a)x+a)f_m((b-a)x+a)^T d\xi(x)\right|$$

$$= (b-a)^{m(m+1)}\left|\int_0^1 f_m(x)f_m^T(x)d\xi(x)\right|$$

$$= (b-a)^{m(m+1)}\underline{H}_{2m}(\xi)$$

$$= (b-a)^{m(m+1)}\prod_{j=1}^m (\zeta_{2j-1}\zeta_{2j})^{m-j+1}. \tag{5.5.8}$$

Here $\underline{H}_{2m}(\xi) = |c_{i+j}|_{i,j=0}^m$ denotes the Hankel determinant of the moments of $\xi$ defined in (1.4.3), $\zeta_1 = p_1$, $\zeta_j = q_{j-1}p_j$ $(j \geq 2)$, and $p_1, p_2, \ldots$ denote the canonical moments of $\xi$ that coincide with the canonical moments of $\xi_{a,b}$, by Theorem 1.3.2. Now the canonical moments of the design maximizing (5.5.8) are independent of $[a,b]$, and by Theorem 1.3.2, D-optimal designs for univariate polynomial regression on different intervals are related by a simple linear transformation.

**Theorem 5.5.3** *The D-optimal design $\xi_m^D$ for the full parameter $\theta$ in the univariate polynomial regression model of degree $m$ on the interval $[-1,1]$ has equal masses at the $m+1$ zeros of the polynomial*

$$(x^2 - 1)P_m'(x),$$

*where $P_m'$ denotes the derivative of the mth Legendre polynomial defined in (2.1.20). The D-optimal design on the interval $[a,b]$ is obtained from $\xi_m^D$ by the linear transformation*

$$\xi_{a,b}(\{x\}) = \xi\left(\left\{\frac{2x-b-a}{b-a}\right\}\right). \tag{5.5.9}$$

*Proof*  The second part follows from the discussion in the previous paragraph. For the first part we apply (5.5.8) with $b = -a = 1$ and obtain

$$|M_m(\xi)| = 2^{m(m+1)}\left(\prod_{j=1}^{m-1} q_{2j}^{m-j}p_{2j}^{m-j+1}\right)\left(\prod_{j=1}^m (q_{2j-1}p_{2j-1})^{m-j+1}\right)p_{2m},$$

where $p_1, \ldots, p_{2m}$ denote the first $2m$ canonical moments of the measure $\xi$. This expression is maximized in terms of canonical moments for

$$p_{2j-1} = \frac{1}{2}, \quad j = 1, \ldots, m,$$

$$p_{2j} = \frac{m-j+1}{2(m-j)+1}, \quad j = 1, \ldots, m-1, \tag{5.5.10}$$

$$p_{2m} = 1.$$

By Corollary 4.3.3 ($z = 1$, $n = m$) it follows that the $D$-optimal design on the interval $[-1, 1]$ must have equal masses at the zeros of the polynomial

$$(x^2 - 1)C_{m-1}^{(3/2)}(x),$$

where $C_{m-1}^{(3/2)}(x)$ is the $(m-1)$th ultraspherical polynomial orthogonal with respect to the measure $(1 - x^2)dx$. The assertion now follows from (2.1.20) and (2.1.24).      ∎

The criterion value of the $D$-optimal design $\xi^D$ for the parameter vector $\theta$ can be directly obtained from (5.5.10) and (5.5.8)

$$\phi_0(M_m(\xi_m^D)) = |M_m(\xi_m^D)|^{1/(m+1)} = \left(\frac{b-a}{2}\right)^m \tag{5.5.11}$$

$$\times \left[\left(\frac{m}{2m-1}\right)^m \prod_{j=2}^{m}\left(\frac{(m-j+1)^2}{(2(m-j)+1)(2(m-j)+3)}\right)^{m-j+1}\right]^{1/(m+1)}$$

which can be used to calculate the $D$-efficiency of a given design $\xi$ in the polynomial regression of degree $m$

$$\text{eff}_m^D(\xi) = \left(\frac{|M_m(\xi)|}{|M_m(\xi_m^D)|}\right)^{1/(m+1)}. \tag{5.5.12}$$

As a further application of Corollary 4.3.3, we consider the design problem for the estimation of the "highest coefficient" $\theta_m$ in the univariate polynomial regression (5.5.5), which was already discussed for the quadratic case $m = 2$ in Example 5.4.5. Following the discussion of Section 5.4, we put $K = e_m = (0, \ldots, 0, 1)^T \in \mathbb{R}^{m+1}$. The range inclusion $\xi \in \mathcal{A}(e_m)$ is satisfied if and only if $|M(\xi)| \neq 0$, and the information "matrix" for the parameter $K^T\theta = \theta_m$ is given by

$$C_{e_m}(M(\xi)) = (e_m^T M_m^{-1}(\xi)e_m)^{-1} = \frac{|M_m(\xi)|}{|M_{m-1}(\xi)|}, \tag{5.5.13}$$

where

$$M_{m-1}(\xi) = \int_a^b f_{m-1}(x) f_{m-1}^T(x) d\xi(x) = (c_{i+j})_{i,j=0}^{m-1}$$

denotes the information matrix in a polynomial regression of degree $m - 1$. A design maximizing $|M_m(\xi)|/|M_{m-1}(\xi)|$ is called $D_1$-*optimal* where the index reflects the fact that the design is optimal for the estimation of the "highest coefficient" $\theta_m$.

**Theorem 5.5.4**   *The $D_1$-optimal design $\xi_m^{D_1}$ in the univariate polynomial regression of degree $m$ on the interval $[-1, 1]$ has equal masses $1/m$ at the zeros of the Chebyshev polynomial of the second kind $U_{m-1}(x)$ and mass $1/2m$ at the points $-1$ and $1$. The $D_1$-optimal design on the interval $[a, b]$ is obtained by the transformation (5.5.9).*

*Proof*   From (5.5.13) and (5.5.8) it follows that the canonical moments of the $D_1$-optimal design are independent of the interval $[a, b]$, which proves the second part of the Theorem. For the first part we maximize

$$C_{e_m}(M(\xi)) = \frac{|M_m(\xi)|}{|M_{m-1}(\xi)|} = 2^{2m} \prod_{j=1}^m \zeta_{2j-1} \zeta_{2j}$$

in terms of the canonical moments of $\xi$ and obtain

$$p_j = \tfrac{1}{2}, \qquad j = 1, \ldots, 2m - 1$$

$$\tag{5.5.14}$$

$$p_{2m} = 1.$$

The assertion now follows from Corollary 4.3.3 $(z = 0, n = m)$, observing (2.1.19).   ∎

The value of the criterion for the $D_1$-optimal design on the interval $[a, b]$ is given by

$$C_{e_m}(M(\xi_m^{D_1})) = (b - a)^{2m} (\tfrac{1}{2})^{4m-2}, \tag{5.5.15}$$

and (5.5.13) and (5.5.8) yield for the $D_1$-*efficiency* of a design $\xi$ on the interval $[a, b]$,

$$\mathrm{eff}_m^{D_1}(\xi) = \frac{C_{e_m}(M(\xi))}{C_{e_m}(M(\xi_m^{D_1}))} = 2^{4m-2} \prod_{j=1}^m \zeta_{2j-1} \zeta_{2j}, \tag{5.5.16}$$

where $\zeta_1 = p_1$, $\zeta_j = q_{j-1}p_j$ $(j > 1)$ and $p_1, p_2, \ldots$ denote the canonical moments of the design $\xi$ on the interval $[a, b]$. We will use this representation of the efficiency in the construction of optimal designs for model discrimination in Chapter 6.

## 5.6 $D_s$-OPTIMAL DESIGNS FOR UNIVARIATE POLYNOMIAL REGRESSION

In Section 5.5 we discussed the problem of optimum experimental design for estimating the "highest" coefficient in a univariate polynomial regression. In many cases the experimenter is not only interested in the estimation of $\theta_m$ but also in the $s$ parameters $\theta_{m-s+1}, \ldots, \theta_m$ corresponding to the $s$ "highest" powers $x^{m-s+1}, \ldots, x^m$ in the polynomial model. Define

$$K_s = \begin{bmatrix} 0 \\ I_s \end{bmatrix} \in \mathbb{R}^{(m+1) \times s},$$

where $I_s$ is the $s \times s$ identity matrix and $0$ is a $(m + 1 - s) \times s$ matrix with all elements zero. Then $K_s^T \boldsymbol{\theta} = (\theta_{m+1-s}, \ldots, \theta_m)^T$. Similarly decompose the information matrix of a design $\xi$

$$M_m(\xi) = \int_a^b \boldsymbol{f}_m(x)\boldsymbol{f}_m^T(x)d\xi(x) = \begin{pmatrix} M_{11}(\xi) & M_{12}(\xi) \\ M_{21}(\xi) & M_{22}(\xi) \end{pmatrix},$$

where $M_{22}(\xi) \in \mathbb{R}^{s \times s}$, $M_{11}(\xi) \in \mathbb{R}^{(m+1-s) \times (m+1-s)}$ $M_{21}(\xi) = M_{12}(\xi)^T \in \mathbb{R}^{s \times (m+1-s)}$. A design satisfies the range inclusion $\xi \in \mathcal{A}(K_s)$ if and only if $|M(\xi)| \neq 0$ and standard results on partitioned matrices show that the information matrix for the parameter subsystem $(\theta_{m+1-s}, \ldots, \theta_m)^T$ is given by the $s \times s$ matrix

$$C_{K_s}(M(\xi)) = M_{22 \cdot 1}(\xi) := M_{22}(\xi) - M_{21}(\xi)M_{11}^{-1}(\xi)M_{12}(\xi).$$

The $\phi_0$-optimal design for the "highest $s$ coefficients" $K_s^T \boldsymbol{\theta}$ is called the $D_s$-*optimal design,* and it maximizes

$$\phi_0(C_{K_s}(M(\xi))) = |M_{22 \cdot 1}(\xi)|^{1/s} = \left(\frac{|M_m(\xi)|}{|M_{11}(\xi)|}\right)^{1/s}, \tag{5.6.1}$$

where the last equality follows from a standard result on partitioned matrices. Theorem 5.4.9 shows that a design $\xi$ is $D_s$-optimal if and only if

$$d_s(x, \xi) = (\boldsymbol{g}_2(x) - \boldsymbol{h}(x))^T M_{22 \cdot 1}^{-1}(\xi)(\boldsymbol{g}_2(x) - \boldsymbol{h}(x)) \leq s \tag{5.6.2}$$

for all $x \in [a, b]$. Here the vector $\boldsymbol{h}$ is defined by $\boldsymbol{h}(x) = M_{21}(\xi)M_{11}^{-1}(\xi)\boldsymbol{g}_1(x)$ and $\boldsymbol{g}_1(x) = (1, \ldots, x^{m-s})^T$, $\boldsymbol{g}_2(x) = (x^{m+1-s}, \ldots, x^m)^T$ denote the corresponding partition of the vector of regression functions $\boldsymbol{f}_m(x) = (1, \ldots, x^m)^T$.

**Theorem 5.6.1**  *The $D_s$-optimal design $\xi_m^{D_s}$ for the univariate polynomial regression model (5.5.5) on the interval $[-1, 1]$ is supported at the zeros of the polynomial*

$$(x^2 - 1)\left( P'_s(x)U_{m-s}(x) - P'_{s-1}(x)U_{m-s-1}(x) \right),$$

*where $P'_\ell(x)$ denotes the derivative of the $\ell$th Legendre polynomial defined in (2.1.20) and $U_\ell(x)$ is the $\ell$th Chebyshev polynomial of the second kind defined in (2.1.6). The masses at the support points $-1 = x_0 < x_1 < \ldots < x_{m-1} < x_m = 1$ are given by*

$$\xi_m^{D_s}(\{x_j\}) = \frac{2}{2m + 1 + U_{2(m-s)}(x_j)}, \qquad j = 0, \ldots, m. \qquad (5.6.3)$$

*The $D_s$-optimal design on the interval $[a, b]$ is obtained by the linear transformation (5.5.9).*

*Proof*  The second part of the assertion follows from (5.5.8) and (5.6.1), observing that $M_{11}(\xi) = M_{m-s}(\xi) = (c_{i+j})_{ij=0}^{m-s}$. On the interval $[-1, 1]$ we obtain

$$\{\phi_0(C_{K_s}(M(\xi)))\}^s = \frac{|M_m(\xi)|}{|M_{m-s}(\xi)|}$$

$$= 2^{s(2m+1-s)} \prod_{j=1}^{m-s} (\zeta_{2j-1}\zeta_{2j})^s \prod_{j=m-s+1}^{m} (\zeta_{2j-1}\zeta_{2j})^{m-j+1},$$

and a maximization with respect to the canonical moments of $\xi$ yields

$$p_{2j-1} = \frac{1}{2}, \qquad j = 1, \ldots, m,$$

$$p_{2j} = \frac{1}{2}, \qquad j = 1, \ldots, m - s, \qquad (5.6.4)$$

$$p_{2j} = \frac{m - j + 1}{2(m - j) + 1}, \qquad j = m - s + 1, \ldots, m - 1,$$

$$p_{2m} = 1.$$

Thus the assertion follows from Theorem 4.4.3 ($n = m, r = m - s$). The proof of this theorem did not include a proof of the representation (5.6.3) of the weights of the $D_s$-optimal design. The reason for this gap is that the representation (5.6.3) relies heavily on the equivalence theorem (5.6.2) that was not available in Chapter 4. We are now in a position to complete the proof of Theorem 4.4.3, which is essentially equivalent to the assertion of the present theorem. To this end observe that the left-hand side of (5.6.2) can be rewritten in the form

$$d_s(x, \xi) = f_m^T(x)M_m^{-1}(\xi)f_m(x) - f_{m-s}^T(x)M_{m-s}^{-1}(\xi)f_{m-s}(x), \qquad (5.6.5)$$

where $f_{m-s}(x) = g_1(x) = (1, x, \ldots, x^{m-s})^T$, $M_{m-s}(\xi) = M_{11}(\xi)$. Equation (5.6.5) follows by writing

$$d(x, \xi) = f_m^T(x) M_m^{-1}(\xi) f_m(x) = (C f_m(x))^T (C M_m(\xi) C^T)^{-1}(C f(x)),$$

where

$$C = \begin{pmatrix} I_{m-s+1} & 0 \\ -M_{21} M_{11}^{-1} & I_s \end{pmatrix}.$$

For $j = 0, \ldots, m$ let

$$L_j(x) = \prod_{i=0, i \neq j}^{m} \frac{x - x_i}{x_j - x_i} \tag{5.6.6}$$

denote the $j$th *Lagrange interpolation polynomial* with nodes $x_0 < x_1 < \ldots < x_m$, and define $L(x) = (L_0(x), \ldots, L_m(x))^T = L f_m(x)$, where $L$ is an appropriate $(m + 1) \times (m + 1)$ matrix. Note that $L_j(x_i) = \delta_{ij}$, where $\delta_{ij}$ denotes Kronecker's symbol. If $e_i$ denotes the $(i + 1)$th unit vector in $\mathbb{R}^{m+1}$, this implies that $L(x_i) L^T(x_i) = e_i e_i^T$. Consequently any design $\xi$ with support $x_0, \ldots, x_m$ and masses $\xi(\{x_j\})$ satisfies

$$f_m^T(x) M^{-1}(\xi) f_m(x) = f_m^T(x) \left[ \int L^{-1} L(x) L^T(x) (L^{-1})^T d\xi(x) \right]^{-1} f_m(x)$$

$$= L^T(x) \left[ \sum_{i=0}^{m} \xi(\{x_i\}) L(x_i) L^T(x_i) \right]^{-1} L(x)$$

$$= \sum_{j=0}^{m} \frac{L_j^2(x)}{\xi(\{x_j\})}. \tag{5.6.7}$$

For the second term in (5.6.5) note that the canonical moments of the measure $\xi_m^{D_s}$ up to the order $2m - 2s + 1$ coincide with the canonical moments of the arc-sine distribution of Example 1.7.2. From Corollary 2.3.6 it therefore follows that the orthonormal polynomials with respect to the measure $d\xi_m^{D_s}(x)$ coincide with the corresponding polynomials for the arc-sine distribution whenever the degree is less or equal $m - s$. The Chebyshev polynomials $T_j(x)$ are orthogonal with respect to the arc-sine measure, and we obtain from (2.1.12),

$$\int_{-1}^{1} T_j(x) T_k(x) d\xi_m^{D_s}(x) = 0, \qquad j \neq k, j, k \leq m - s,$$

$$\tag{5.6.8}$$

$$\int_{-1}^{1} T_j^2(x) d\xi_m^{D_s}(x) = \tfrac{1}{2}, \qquad 1 \leq j \leq m - s.$$

Now let $T(x) = (T_0(x), \ldots, T_{m-s}(x))^T = Tf_{m-s}(x)$, where $T$ is an appropriate $(m - s + 1) \times (m - s + 1)$ matrix. Then the second term in (5.6.5) is given by

$$
f_{m-s}^T(x) M_{m-s}^{-1}(\xi_m^{D_s}) f_{m-s}(x)
$$

$$
= f_{m-s}^T(x) \left[ T^{-1} \int_{-1}^{1} T(x) T^T(x) d\xi_m^D(x) (T^{-1})^T \right]^{-1} f_{m-s}(x)
$$

$$
= T^T(x) \, \mathrm{diag}\,(1, 2, \ldots, 2) T(x) = 1 + 2 \sum_{j=1}^{m-s} T_j^2(x)
$$

$$
= m - s + \tfrac{1}{2} + \tfrac{1}{2} U_{2(m-s)}(x) \tag{5.6.9}
$$

Here the equality in the second line follows from the orthogonality relations (5.6.8), while the third line is a consequence the identity

$$
\sum_{j=1}^{r} \cos^2 j\theta = \frac{1}{2} \left( r + \frac{\cos\,(r+1)\theta \sin r\theta}{\sin \theta} \right) = \frac{1}{2} \left( r - \frac{1}{2} + \frac{\sin\,(2r+1)\theta}{2 \sin \theta} \right)
$$

and the trigonometric representations (2.1.5) and (2.1.6) for the Chebyshev polynomials of the first and second kind. Observing (5.6.5), (5.6.7), and (5.6.9), the equivalence condition (5.6.2) of Theorem 5.4.9 reduces to

$$
d_s(x, \xi_m^{D_s}) = \sum_{j=0}^{m} \frac{L_j^2(x)}{\xi_m^{D_s}(\{x_j\})} - (m - s) - \frac{1}{2} - \frac{1}{2} U_{2m-2s}(x) \leq s \tag{5.6.10}
$$

for all $x \in [-1, 1]$. Now by Theorem 5.4.9 there is equality in (5.6.10) for every support point of the $D_s$-optimal design $\xi_m^{D_s}$, that is,

$$
\frac{1}{\xi_m^{D_s}(\{x_j\})} - \frac{1}{2}(2m + 1 + U_{2m-2s}(x_j)) = 0, \qquad j = 0, \ldots, m.
$$

This proves the representation of the weights in Theorem 4.4.3 ($r = m - s$) and therefore completes the proof of Theorem 5.6.1.                                    ∎

## 5.7   $\phi_p$-OPTIMAL DESIGNS FOR THE HIGHEST TWO COEFFICIENTS

In the previous sections the canonical moments of the optimal designs do not depend on the design space $[a, b]$, which reflected the nice "invariance property" that optimal designs on different intervals are related by linear transformations. This property depends on the particular structure of the determinant criterion $\phi_0(C_K(M(\xi)))$ which was illustrated in Theorem 5.5.1. In general, $\phi_p$-optimal designs do not satisfy invariance properties of this type, and we will illustrate these difficulties in the present section.

Consider the univariate polynomial regression (5.5.5) on the interval $[-b, b]$ $(b > 0)$, and assume that the experimenter is interested into the highest two coefficients $(\theta_{m-1}, \theta_m)^T = K_2^T \theta$, where

$$K_2 = \begin{bmatrix} 0 \\ I_2 \end{bmatrix} \in \mathbb{R}^{(m+1) \times 2}$$

is defined in the same way as in the previous section. A design satisfies the range inclusion $\xi \in \mathcal{A}(K_2)$ if and only if $|M(\xi)| \neq 0$, and a $\phi_p$-optimal design for the parameter subsystem $K_2^T \theta = (\theta_{m-1}, \theta_m)^T$ maximizes the function

$$\phi_p(C_{K_2}(M(\xi))) = (\tfrac{1}{2} \text{ trace } (C_{K_2}(M(\xi)))^p)^{1/p}$$

$$= \{\tfrac{1}{2}[\lambda_1^p(C_{K_2}(M(\xi))) + \lambda_2^p(C_{K_2}(M(\xi)))]\}^{1/p},$$

where $\lambda_1(C_{K_2}(M(\xi)))$ and $\lambda_2(C_{K_2}(M(\xi)))$ denote the eigenvalues of the matrix $C_{K_2}(M(\xi)) \in \mathbb{R}^{2 \times 2}$. If $\bar{\xi}$ denotes the reflection of $\xi$ at the origin, i.e. $\bar{\xi}(\{x\}) = \xi(\{-x\})$, then it follows that

$$M_m(\bar{\xi}) = D M_m(\xi) D,$$

where $D = \text{diag } (1, -1, 1, -1, \ldots) \in \mathbb{R}^{(m+1) \times (m+1)}$. It is now easy to see that the matrices $C_{K_2}(M(\xi))$ and $C_{K_2}(M(\bar{\xi}))$ have the same eigenvalues and that consequently

$$\phi_p(C_{K_2}(M_m(\bar{\xi}))) = \phi_p(C_{K_2}(M_m(\xi))).$$

By Theorem 5.4.4 the function $\xi \longrightarrow \phi_p(C_{K_2}(M(\xi)))$ is concave on $\{\xi \,||\, M(\xi)| \neq 0\}$. This implies that

$$\phi_p(C_{K_2}(M_m(\tfrac{1}{2}(\xi + \bar{\xi})))) \geq \phi_p(C_{K_2}(M_m(\xi))),$$

and there exists a symmetric $\phi_p$-optimal design for $(\theta_{m-1}, \theta_m)^T$. The following result was established by Gaffke (1987) for the interval $[-1, 1]$ by using characterizations of design optimality and admissibility.

**Theorem 5.7.1** *Let $-\infty < p < 1$, $\xi_m^p$ denote the design that is supported at the zeros $-b = x_0 < x_1 < \ldots < x_{m-1} < x_m = b$ of the polynomial*

$$(x^2 - b^2)\left\{ U_{m-1}\left(\frac{x}{b}\right) + \beta(b) U_{m-3}\left(\frac{x}{b}\right) \right\}$$

*with masses*

$$\xi_m^p(\mp b) = \frac{1 + \beta(b)}{2[m + (m - 2)\beta(b)]},$$

$$\xi_m^p(x_j) = \left[ m - 1 - \frac{(1 + \beta(b))U_{m-2}(\frac{x_j}{b})}{U_m(\frac{x_j}{b}) + \beta(b)U_{m-2}(\frac{x_j}{b})} \right]^{-1}, \qquad j = 1, \ldots, m - 1,$$

*where $\beta(b)$ is the unique solution in $[0, 1)$ of the equation*

$$\left( \frac{1 - \beta}{2} \right)^{1-p} - b^{2p}\beta = 0 \tag{5.7.1}$$

*and $U_\ell(x)$ denotes the $\ell$th Chebyshev polynomial of the second kind. The design $\xi_m^p$ is $\phi_p$-optimal for the highest two coefficients in a polynomial regression on the interval $[-b, b]$*

*Proof* By the previous paragraph we can restrict ourselves to symmetric designs $\xi$ on the interval $[-b, b]$. Let $\underline{R}_j(x)$ denote the $j$th monic orthogonal polynomial with respect to the symmetric measure $d\xi(x)$, and define $\underline{R}(x) = (\underline{R}_0(x), \ldots, \underline{R}_m(x))^T$. By Corollary 2.3.6 it follows that

$$\underline{R}(x) = Q\boldsymbol{f}_m(x), \tag{5.7.2}$$

where $Q = (q_{ij})_{i,j=0}^m$ is a lower triangular matrix with diagonal elements $q_{jj} = 1$ $(j = 0, \ldots, m)$ and

$$q_{ij} = 0 \qquad \text{if } i + j \text{ is odd.}$$

Therefore we obtain $QK_2 = K_2$ and from (5.7.2) for a symmetric design $\xi$

$$\{C_{K_2}(M(\xi))\}^{-1} = K_2^T \left[ Q^{-1} \int_{-b}^b \underline{R}(x)\underline{R}^T(x)d\xi(x)(Q^{-1})^T \right]^{-1} K_2$$

$$= (QK_2)^T \text{ diag } (w_0^{-1}, \ldots, w_m^{-1})QK_2 = \begin{bmatrix} w_{m-1}^{-1} & 0 \\ 0 & w_m^{-1} \end{bmatrix}, \tag{5.7.3}$$

where $w_j = \int_{-b}^b \underline{R}_j^2(x)d\xi(x)$ denotes the $L_2$-norm of the $j$th monic orthogonal polynomial with respect to the measure $d\xi(x)$. By Remark 2.3.7 the $L_2$-norm is obtained as

$$w_j = \int_{-b}^b \underline{R}_j^2(x)d\xi(x) = (2b)^{2j} \prod_{i=1}^j \zeta_{2i-1}\zeta_{2i} = b^{2j} \prod_{i=1}^j q_{2i-2}p_{2i}, \tag{5.7.4}$$

where $p_1, p_2, \ldots$ denote the canonical moments of $\xi$, and the last equality follows from the symmetry of $\xi$ and Corollary 1.3.4. Combining (5.7.3) and (5.7.4) yields that the determination of a $\phi_p$-optimal design for $(\theta_{m-1}, \theta_m)^T$ reduces to the problem of maximizing the function

$$2^{1/p} \phi_p(C_K(M(\xi))) = (w_{m-1}^p + w_m^p)^{1/p}$$

$$= \left( b^{2m-2} \prod_{j=1}^{m-1} q_{2j-2} p_{2j} \right) \left\{ 1 + b^{2p} q_{2m-2}^p p_{2m}^p \right\}^{1/p}.$$

Straightforward algebra gives for the canonical moments of the optimal design

$$p_{2j-1} = \frac{1}{2}, \qquad j = 1, \ldots, m,$$

$$p_{2j} = \frac{1}{2}, \qquad j = 1, \ldots, m-2,$$

$$p_{2m-2} = \frac{1 + \beta(b)}{2}, \qquad p_{2m} = 1,$$

where $\beta(b)$ is the unique solution of (5.7.1) in the interval $[0, 1)$. The assertion of Theorem 5.7.1 now follows from Theorem 4.4.4 (in the case $b = 1$) and from Theorem 1.3.2 for arbitrary $b$. ∎

Note that the $\phi_p$-optimal design on the interval $[-b', b']$ cannot be obtained from the optimal design on $[-b, b]$ because for $b' \neq b$ the solutions $\beta(b)$ and $\beta(b')$ of (5.7.1) are different whenever $p \neq 0$. In the case $p = 0$ we obtain $\beta(b) = \frac{1}{3}$ independently of $b$, which reflects the particular role of the determinant criteria.

**Theorem 5.7.2** *If $b^2 \leq 2$, let $\xi_m^{-\infty}$ denote the design that puts equal mass $1/m$ at the zeros of the $(m-1)$th Chebyshev polynomial of the second kind $U_{m-1}(x/b)$ and mass $1/2m$ at the points $-b$ and $b$. If $b^2 > 2$, let $\xi_m^{-\infty}$ be given by Theorem 5.7.1 where $\beta(b) = 1 - 2/b^2$. Then $\xi_m^{-\infty}$ is E-optimal for estimating the parameters $(\theta_{m-1}, \theta_m)^T$ in the polynomial regression model on the interval $[-b, b]$*

*Proof* By the discussion in the proof of Theorem 5.7.1, we can restrict ourselves to symmetric designs that maximize

$$\phi_{-\infty}(C_K(M(\xi))) = b^{2m-2} \min \left\{ \prod_{j=1}^{m-1} q_{2j-2} p_{2j}, b^2 \prod_{j=1}^{m} q_{2j-2} p_{2j} \right\}.$$

If we maximize $b^2 \prod_{j=1}^{m} q_{2j-2} p_{2j}$ subject to the restriction

$$\prod_{j=1}^{m-1} q_{2j-2} p_{2j} \geq b^2 \prod_{j=1}^{m} q_{2j-2} p_{2j}, \tag{5.7.5}$$

we obtain $p_{2j} = \frac{1}{2}, j = 1, \ldots, m - 1, p_{2m} = 1$ whenever $b^2 \leq 2$. In the other case $b^2 > 2$ the maximum is attained at a point with equality in (5.7.5), that is, $p_{2m-2} = 1 - 1/b^2, p_{2j} = \frac{1}{2} (j = 1, \ldots, m - 2), p_{2m} = 1$. The assertion now follows from Corollary 4.3.3 ($b^2 \leq 2$) and Theorem 4.4.4 ($b^2 > 2$). ∎

## 5.8 DESIGNS FOR MULTIVARIATE POLYNOMIAL REGRESSION

In this section we will concentrate on the multivariate polynomial regression model introduced in Example 5.2.4,

$$Y = \alpha_0 + \sum_{i=1}^{q} \alpha_i t_i + \sum_{1 \leq i_1 \leq i_2 \leq q} \alpha_{i_1, i_2} t_{i_1} t_{i_2} + \cdots \qquad (5.8.1)$$

$$+ \sum_{1 \leq i_1 \leq \ldots \leq i_m \leq q} \alpha_{i_1, \ldots, i_m} \prod_{j=1}^{m} t_{i_j} + \varepsilon,$$

where $t = (t_1, \ldots, t_n)^T$ denotes the explanatory variable. In the following we will call the model (5.8.1) the *complete q-way mth degree polynomial model* and assume that the design space is the $q$-dimensional cube $\mathcal{X} = [-1, 1]^q$. The regression functions in this model are the $N_{q,m} = \binom{m+q}{m}$ functions of the form (5.2.18). As a simple example consider the case $m = 2$ and $q = 2$, for which the deterministic part in (5.8.1) reduces to the 2-way quadratic polynomial with six parameters:

$$\alpha_0 + \alpha_1 t_1 + \alpha_2 t_2 + \alpha_{1,2} t_1 t_2 + \alpha_{1,1} t_1^2 + \alpha_{2,2} t_2^2. \qquad (5.8.2)$$

In many cases the experimenter has a more precise knowledge about the underlying regression and does not want to use the complete model (5.8.1) but a model with some missing terms. For example, it could be clear by physical considerations that there is no *interaction* between the variables $t_1$ and $t_2$ and no quadratic term of $t_1$ in the 2-way quadratic polynomial (5.8.2). In this case the polynomial regression in two variables is *incomplete* and (5.8.2) becomes a model with four parameters:

$$\alpha_0 + \alpha_1 t_1 + \alpha_2 t_2 + \alpha_{2,2} t_2^2. \qquad (5.8.3)$$

In general, an *incomplete q-way mth degree polynomial regression* can be defined as follows: Let for $m_1, \ldots, m_q \in \{0, \ldots, m\}$ with $\sum_{j=1}^{q} m_j \leq m$ ,

$$\tau_{m_1, \ldots, m_q} = \begin{cases} 1 & \text{if } \prod_{j=1}^{q} t_j^{m_j} \text{ appears in the multivariate} \\ & \text{polynomial regression (5.8.1),} \\ 0 & \text{else,} \end{cases}$$

denote $N_{q,m}$ given numbers with values 0 or 1, and define a linear model by

$$Y = \sum_{\substack{m_1,\ldots,m_q \in \{0,\ldots,m\} \\ \sum_{j=1}^{q} m_j \leq m}} \tau_{m_1,\ldots,m_q} \left( \alpha_{m_1,\ldots,m_q} \prod_{j=1}^{q} t_j^{m_j} \right) + \varepsilon . \qquad (5.8.4)$$

The multivariate polynomial regression (5.8.4) has

$$N_{q,m,\tau} := \sum_{\substack{m_1,\ldots,m_q \in \{0,\ldots,m\} \\ \sum_{j=1}^{q} m_j \leq m}} \tau_{m_1,\ldots,m_q} \qquad (5.8.5)$$

parameters, and the indicator functions $\tau_{m_1,\ldots,m_q}$ specify the monomials defined by (5.2.18) that appear in the incomplete $q$-way $m$th degree polynomial regression model (5.8.4). The quadratic regression example (5.8.3) is obtained for the choice $m = 2$ and $q = 2$ and

$$\tau_{0,0} = \tau_{1,0} = \tau_{0,1} = \tau_{0,2} = 1, \quad \tau_{1,1} = \tau_{2,0} = 0 . \qquad (5.8.6)$$

As a further example consider the case $m = 2$, $q = 3$, and

$$\tau_{0,0,0} = \tau_{1,0,0} = \tau_{0,1,0} = \tau_{0,0,1} = \tau_{1,1,0} = \tau_{2,0,0} = 1 ,$$
$$\tau_{1,0,1} = \tau_{0,1,1} = \tau_{0,2,0} = \tau_{0,0,2} = 0 , \qquad (5.8.7)$$

then an incomplete 3-way quadratic polynomial becomes

$$\alpha_0 + \alpha_1 t_1 + \alpha_2 t_2 + \alpha_3 t_3 + \alpha_4 t_1 t_2 + \alpha_5 t_1^2. \qquad (5.8.8)$$

The vector of regression functions $\boldsymbol{f}(t)$ in the model (5.8.4) consists of the $N_{q,m,\tau}$ monomials $\prod_{j=1}^{q} t_j^{m_j}$ satisfying $\sum_{j=1}^{q} m_j \leq m$ and $\tau_{m_1,\ldots,m_q} = 1$. The moment matrix of a design $\eta$ on the design space $\mathcal{X}$ is given by

$$M(\eta) = \int_{[-1,1]^q} \boldsymbol{f}(t)\boldsymbol{f}(t)^T \, d\xi(t) . \qquad (5.8.9)$$

and optimal designs maximize information functionals of this matrix.

In general, the determination of optimal designs for this model is an extremely hard problem and has only been solved numerically in very special cases (see Farrell, Kiefer, and Walbran, 1967; Lim and Studden, 1988). It turns out in the following discussion that very efficient optimal designs can be generated by the following procedure: We replace the class $\Xi$ of all designs on $\mathcal{X}$ by the class of all *product designs* on the cube $[-1,1]^q$,

$$\Xi_q = \{\eta \in \Xi | \eta = \xi_1 \times \ldots \times \xi_q\} \qquad (5.8.10)$$

(here $\xi_j$ denotes a design on the interval $[-1, 1]$) and then maximize an optimality criterion within the class $\Xi_q$. A design maximizing a criterion within the class of all product designs $\Xi_q$ is called an *optimal product design* We will illustrate this procedure for the *D*-optimality criterion for the full parameter vector and for the $D_s$- optimality criterion. In many incomplete polynomial regression models it turns out that the optimal product designs with respect to these criteria can be conveniently described in terms of the canonical moments of their factor measures.

We will start our investigations with the *D*-optimality criterion and consider a product design $\eta = \xi_1 \times \ldots \times \xi_q \in \Xi_q$. In what follows $p_{1(j)}, p_{2(j)}, p_{3(j)}, \ldots$ denote the canonical moments of the $j$ factor $\xi_j$ $(j = 1, \ldots, q)$ and $q_{i(j)} = 1 - p_{i(j)}$ $(i \geq 1, j = 1, \ldots, q)$. A concavity argument similar to that used in Section 5.7 yields that the factors $\xi_j$ of the *D*- (and $D_s$-)optimal product design must be symmetric, which is equivalent to $p_{2i-1(j)} = \frac{1}{2}$, by Corollary 1.3.4. An application of Corollary 2.3.6 shows that the monic orthogonal polynomials $\{W_{i(j)}(t_j)\}_{i \geq 0}$ with respect to the measure $d\xi_j(t_j)$ satisfy the recursion $(q_{0(j)} = 1)$,

$$W_{i+1(j)}(t_j) = t_j W_{i(j)}(t_j) - q_{2i-2(j)} p_{2i(j)} W_{i-1(j)}(t_j), \qquad i \geq 0, \qquad (5.8.11)$$

with initial conditions $W_{-1(j)}(t_j) = 0$, $W_{0(j)}(t_j) = 1$.

Let $f(t)$ denote the vector of regression functions in the model (5.8.4). We define a corresponding vector $g_\eta(t)$ of products of monic orthogonal polynomials as follows: If the $l$th component of $f(t)$ is $\prod_{j=1}^{q} t_j^{m_j}$ $(m_1, \ldots, m_q \in \{0, \ldots, m\}, \sum_{j=1}^{q} m_j \leq m, \tau_{m_1, \ldots, m_q} = 1)$, then the $l$th component of $g_\eta$ is defined as $\prod_{j=1}^{q} W_{m_j(j)}(t_j)$ (note that $g_\eta$ depends on $\eta = \xi_1 \times \cdots \times \xi_q$). To determine *D*-optimal product designs in the incomplete model (5.8.4) explicitly, we make the following basic assumption:

**Assumption 5.8.1** *There exists a symmetric product design $\eta$ with nonsingular information matrix (5.8.9), and there exists a permutation matrix $Q$ such that*

$$Q g_\eta(t) = A_\eta Q f(t), \qquad (5.8.12)$$

*where $A_\eta$ is a lower triangular matrix with determinant 1.*

It follows readily from the recursive relation (5.8.11) that the existence of a symmetric product measure $\eta'$ (with nonsingular information matrix) and a permutation matrix $Q$ satisfying (5.8.12) implies that (5.8.12) is satisfied for all symmetric product designs $\eta \in \Xi_q$ with $\det M(\eta) > 0$ (with the same matrix $Q$). A necessary and sufficient condition for Assumption 5.8.1 is that if $\tau_{m_1, \ldots, m_q} = 1$, then $\tau_{m'_1, \ldots, m'_q} = 1$ for every set $(m'_1, \ldots, m'_q)$ such that $m'_j \leq m_j$ and $m_j$ and $m'_j$ have the same parity for $j = 1, \ldots, q$. Thus Assumption 5.8.1 is obviously fulfilled for the complete $q$-way $m$th degree polynomial.

Throughout this section we define

$$m_j^* := \max\{m_j \mid \tau_{m_1, \ldots, m_q} = 1\} \qquad (5.8.13)$$

as the largest exponent of the variable $t_j$ in the incomplete regression model (5.8.7) ($j = 1, \ldots, q$) and

$$\tau(i, j, k) := \sum_{\substack{\sum_{l=1}^{q} m_l \leq k \\ m_j = i}} \tau_{m_1, \ldots, m_q} \qquad (5.8.14)$$

as the number of terms in the model (5.8.7) that are of degree smaller than or equal to $k$ and that contain the factor factor $t_j^i$ ($i = 1, \ldots, k, j = 1, \ldots, q$).

**EXAMPLE 5.8.2**   Consider the model (5.8.8), and define

$$f(t) = (1, t_1, t_2, t_3, t_1 t_2, t_1^2)^T.$$

Then, by the recursion (5.8.11), it follows that

$$g_\eta(t) = (1, t_1, t_2, t_3, t_1 t_2, t_1^2 - p_{2(1)})^T \qquad (\eta = \xi_1 \times \xi_2 \times \xi_3).$$

A suitable choice for the permutation matrix $Q$ is $Q = I_6$, and the lower triangular matrix $A_\eta$ in (5.8.12) is given by

$$A_\eta = \begin{bmatrix} 1 & 0 & 0 & 0 & 0 & 0 \\ 0 & 1 & 0 & 0 & 0 & 0 \\ 0 & 0 & 1 & 0 & 0 & 0 \\ 0 & 0 & 0 & 1 & 0 & 0 \\ 0 & 0 & 0 & 0 & 1 & 0 \\ -p_{2(1)} & 0 & 0 & 0 & 0 & 1 \end{bmatrix}.$$

The largest exponents of the variables $t_1$, $t_2$ and $t_3$ are $m_1^* = 2$, $m_2^* = m_3^* = 1$, respectively, and the quantities $\tau(i, j, 2)$ in (5.8.14) are given by

$$\tau(1, 1, 2) = 2, \quad \tau(1, 2, 2) = 2, \quad \tau(1, 3, 2) = 1,$$

$$\tau(2, 1, 2) = 1, \quad \tau(2, 2, 2) = 0, \quad \tau(2, 3, 2) = 0.$$

**Theorem 5.8.3**   *The D-optimal product design for an incomplete polynomial regression (5.8.4) satisfying Assumption 5.8.1 is given by $\eta^D = \xi_1^* \times \cdots \times \xi_q^*$. Here, for $j = 1, \ldots, q$, $\xi_j^*$ is a probability measure on the interval $[-1, 1]$ which is* *uniquely determined by its canonical moments*

$$p_{2l-1(j)} = \frac{1}{2}, \qquad l = 1, \ldots, m_j^*,$$  (5.8.15)

$$p_{2l(j)} = \frac{\sum_{i=l}^{m} \tau(i, j, m)}{\sum_{i=l}^{m} \tau(i, j, m) + \sum_{i=l+1}^{m} \tau(i, j, m)}, \qquad l = 1, \ldots, m_j^* - 1,$$

$$p_{2m_j^*(j)} = 1,$$

*where $m_j^*$ and $\tau(i, j, m)$ are defined in (5.8.13) and (5.8.14), respectively.*

*Proof* The discussion following (5.8.10) shows that the optimal product design must have symmetric factors which implies that $p_{2i-1(j)} = \frac{1}{2}$ whenever these are defined. By Assumption 5.8.1 the information matrix (5.8.9) of a symmetric product design $\eta = \xi_1 \times \cdots \times \xi_q$ is

$$M(\eta) = \int_{[-1,1]^d} Q^T A_\eta^{-1} Q \, g_\eta(t) \, g_\eta^T(t) \, Q^T \, (A_\eta^T)^{-1} Q \, d\eta(t),$$

and the determinant of this matrix is given by

$$\det(M(\eta)) = \det\left( \int_{[-1,1]^d} g_\eta(t) \, g_\eta^T(t) \, d\eta(t) \right)$$

$$= \prod_{\sum_{l=1}^{q} m_l \le m} \left( \prod_{j=1}^{q} \left[ \int_{-1}^{1} W_{m_j(j)}^2(t_j) \, d\xi_j(t_j) \right] \right)^{\tau_{m_1, \ldots, m_q}}$$  (5.8.16)

$$= \prod_{j=1}^{q} \prod_{i=1}^{m} \left[ \int_{-1}^{1} W_{i(j)}^2(t_j) \, d\xi_j(t_j) \right]^{\tau(i,j,m)},$$

with $\tau(i, j, m)$ defined in (5.8.14). The second line in (5.8.16) follows from the first by using the orthogonality relation for the polynomials $W_{i(j)}(t_j)$ which implies that the matrix

$$\int_{[-1,1]^d} g_\eta(t) \, g_\eta^T(t) \, d\eta(t)$$

is diagonal with elements given by the integrals of the squares of the coefficients of $g_\eta(t)$. The maximization of $\det(M(\eta))$ with respect to the product design $\eta = \xi_1 \times \cdots \times \xi_q$ can now be carried out as a maximization of the factors

$$A_j(\xi_j) = \prod_{i=1}^{m} \left[ \int_{-1}^{1} W_{i(j)}^2(t_j) \, d\xi_j(t_j) \right]^{\tau(i,j,m)}$$

with respect to the (univariate) measure $\xi_j$. From Remark 2.3.7 we can calculate the $L_2$-norm of $W_{i(j)}(t_j)$ and obtain

$$A_j(\xi_j) = \prod_{i=1}^{m} \left[ 2^{2i} \prod_{l=1}^{i} q_{2l-2(j)} p_{2l-1(j)} q_{2l-1(j)} p_{2l(j)} \right]^{\tau(i,j,m)}$$

$$= \prod_{l=1}^{m} [q_{2l-2(j)} p_{2l(j)}]^{\sum_{i=l}^{m} \tau(i,j,m)},$$

where $p_{i(j)}$ denotes the $i$th canonical moments of $\xi_j$ and $p_{2i-1(j)} = \frac{1}{2}$ (by symmetry of $\xi_j$). A maximization of $A_j(\xi_j)$ in terms of the even canonical moments $p_{2l(j)}$ yields (5.8.15). Note that $\sum_{i=l}^{m} \tau(i,j,m) = 0$ whenever

$$l > m_j^* = \max\{m_j | \tau_{m_1,\ldots,\,m_q} = 1\}.$$

The uniqueness of the $D$-optimal product design follows because $p_{2m_j(j)} = 1$ implies that the corresponding vector of moments is a boundary point of the moment space for which the representation $\xi_j^*$ is unique. ∎

Theorem 5.8.3 provides a complete solution of the $D$-optimal design problem in the class of product designs for all incomplete models satisfying Assumption 5.8.1. The optimal design is uniquely characterized in terms of the canonical moments of its factors. The support points and the weights of the $j$th factor $\xi_j^*$ with canonical moments (5.8.15) can be identified by the methods explained in Section 3.6. Analytical results as in the univariate case are only possible in very special cases which are discussed below.

**EXAMPLE 5.8.4** Consider the model (5.8.8). By Theorem 5.8.3 and (5.8.7) the canonical moments of $\xi_1^*$, $\xi_2^*$, $\xi_3^*$ for the $D$-optimal product design $\eta^* = \xi_1^* \times \xi_2^* \times \xi_3^*$ are given by

$$p_{1(1)} = p_{3(1)} = \tfrac{1}{2}, \quad p_{2(1)} = \tfrac{3}{4}, \quad p_{4(1)} = 1,$$

and

$$p_{1(j)} = \tfrac{1}{2}, \quad p_{2(j)} = 1 \quad (j = 2, 3).$$

The corresponding measures of the factors of the optimal product design can be found by Corollary 4.2.2 resulting in

$$\xi_1^* = \begin{pmatrix} -1 & 0 & 1 \\ \frac{3}{8} & \frac{1}{4} & \frac{3}{8} \end{pmatrix}, \quad \xi_2^* = \xi_3^* = \begin{pmatrix} -1 & 1 \\ \frac{1}{2} & \frac{1}{2} \end{pmatrix}.$$

In the following we specialize Theorem 5.8.3 to the case of the complete $q$-way $m$th degree polynomial regression (5.8.1) where the situation is more transparent. In this case the $D$-optimal product design can be found analytically, thus generalizing Theorem 5.5.3.

**Corollary 5.8.5** *The D-optimal product design for the complete q-way mth degree polynomial regression model* (5.8.1) *on the q-dimensional cube* $[-1, 1]^q$ *is given by* $\eta_D^{(q)} = \xi \times \ldots \times \xi$, *where the design* $\xi$ *on the interval* $[-1, 1]$ *has equal masses* $1/(q + m)$ *at the* $m - 1$ *zeros of the* $(m - 1)$th *ultraspherical polynomial* $C_{m-1}^{(q/2+1)}(x)$ *and masses* $(q + 1)/(2(q + m))$ *at the boundary points* $-1$ *and* 1.
    *The D-optimal product design on the cube* $[a_1, b_1] \times [a_2, b_2] \times \ldots \times [a_q, b_q]$ *is obtained by transforming the design* $\xi$ *linearly to the corresponding intervals* $[a_j, b_j]$ $(j = 1, \ldots, q)$.

*Proof*  In the complete model we have $\tau_{m_1, \ldots, m_q} = 1$ for all $m_1, \ldots, m_q \in \{0, \ldots, m\}$ with $\sum_{j=1}^{q} m_j \leq m$ and

$$\tau(i, j, m) = \sum_{\sum_{k \neq j} m_k \leq m - i} 1 = \binom{m - i + q - 1}{q - 1}.$$

Theorem 5.8.3 gives the canonical moments of the factors of the $D$-optimal product design as

$$p_{2l(j)} = \frac{\sum_{i=l}^{m} \binom{m-i+q-1}{q-1}}{\sum_{i=l}^{m} \binom{m-i+q-1}{q-1} + \sum_{i=l+1}^{m} \binom{m-i+q-1}{q-1}} = \frac{\binom{m-l+q}{q}}{\binom{m-l+q-1}{q} + \binom{m-l+q}{q}}$$

$$= \frac{q + m - l}{q + 2(m - l)}, \qquad l = 1, \ldots, m,$$

independently of $j = 1, \ldots, q$. Here the second equality is a consequence of the identity

$$\sum_{j=0}^{i} \binom{q - 1 + j}{q - 1} = \binom{q + i}{q}, \tag{5.8.17}$$

which follows by a simple induction. The assertion of the Corollary can now be obtained by an application of Corollary 4.3.3. ∎

**EXAMPLE 5.8.6**  Consider a $q$-way quadratic polynomial. The $D$-optimal product design is $\eta_D^{(q)} = \xi \times \ldots \times \xi$ where each factor $\xi$ has masses proportional to $(q + 1) : 2 : (q + 1)$ at the points $-1$, 0, and 1. The following gives the $D$-efficiencies (see Definition 5.4.6, $p = 0$, $K = I$) of the optimal product design with respect to the $D$-optimal design for various values of $q$. The results indicate that optimal product designs perform very efficiently:

| $q$ | 1 | 2 | 3 | 4 | 5 |
|---|---|---|---|---|---|
| $D$-efficiency | 1.0000 | 0.8855 | 0.9950 | 0.9954 | 0.9960 |

As a second example consider a 3-way cubic model on the interval $[-1, 1]$. The $D$-optimal product design is given by $\eta_D^{(3)} = \xi \times \xi \times \xi$, where $\xi$ has masses proportional to $4 : 2 : 2 : 4$ at the points $-1$, $x_1$, $x_2$, $1$, and $x_1$, $x_2$ are the zeros of the polynomial $C_2^{(5/2)}(x)$. By Example 2.1.5 we have

$$C_2^{(5/2)}(x) = \frac{5}{2}[7x^2 - 1],$$

which shows that $\xi$ is given by

$$\xi = \begin{pmatrix} -1 & \dfrac{-1}{\sqrt{7}} & \dfrac{1}{\sqrt{7}} & 1 \\ \dfrac{1}{3} & \dfrac{1}{6} & \dfrac{1}{6} & \dfrac{1}{3} \end{pmatrix}.$$

The final result in this section is concerned with the $D_s$-optimality criterion for the incomplete $q$-way $m$th degree polynomial regression (5.8.4). This might be useful if the main interest of the experimenter are the parameters corresponding to the "highest" powers in the model, which are the monomials of the form

$$\prod_{j=1}^{q} t_j^{m_j} \quad \text{with} \quad \tau_{m_1, \ldots, m_q} = 1 \quad \text{and} \quad \sum_{j=1}^{q} m_j \in \{m - s + 1, \ldots, m\}. \quad (5.8.18)$$

As in Section 5.6 we consider a partition of the information matrix

$$M(\eta) = \begin{pmatrix} M_{11}(\eta) & M_{12}(\eta) \\ M_{21}(\eta) & M_{22}(\eta) \end{pmatrix},$$

where $M_{11}(\eta)$ denotes the information matrix of the design $\eta$ in the incomplete model of degree $n = m - s < m$. A $D_s$-optimal design maximizes

$$|M_{22}(\eta) - M_{21}(\eta)M_{11}^{-1}(\eta)M_{12}(\eta)| = \frac{|M(\eta)|}{|M_{11}(\eta)|},$$

which is proportional to the determinant of the inverse covariance matrix of the Gauss-Markov estimator for the $s = m - n$ "highest" coefficients $[n < m]$ corresponding to the monomials (5.8.18). The following result specifies the canonical moments of the $D_s$-optimal product design. Here we use the definition $\sum_{i=k}^{l} \beta_i = 0$ whenever $l < k$.

**Theorem 5.8.7** *A $D_s$-optimal product design for the "highest $s = m - n$ $(n < m)$ coefficients" specified by (5.8.18) in an incomplete polynomial regression (5.8.4) on the cube $[-1, 1]^q$ satisfying Assumption 5.8.1 is given by $\eta^{D_s} = \hat{\xi}_1 \times \cdots \times \hat{\xi}_q$. Here, for $j = 1, \ldots, q$, the jth factor $\hat{\xi}_j$ of the $D_s$-optimal product design $\eta^{D_s}$ has canonical moments*

$$p_{2l-1(j)} = \frac{1}{2}, \qquad l = 1, \ldots, m_j^*, \qquad (5.8.19)$$

$$p_{2l(j)} = \frac{\sum_{i=l}^{m} \tau(i, j, m) - \sum_{i=l}^{n} \tau(i, j, n)}{\sum_{i=l}^{m} \tau(i, j, m) - \sum_{i=l}^{n} \tau(i, j, n) + \sum_{i=l+1}^{m} \tau(i, j, m) - \sum_{i=l+1}^{n} \tau(i, j, n)},$$

$$l = 1, \ldots, \min\{n, m_j^*\}$$

$$p_{2l(j)} = \frac{\sum_{i=l}^{m} \tau(i, j, m)}{\sum_{i=l}^{m} \tau(i, j, m) + \sum_{i=l+1}^{m} \tau(i, j, m)}, \qquad l = \min\{n, m_j^*\} + 1, \ldots, m_j^*,$$

*with the convention that this sequence terminates at $p_{2l_0(j)}$ whenever $p_{2l_0(j)}$ is 0 or 1 $(1 \le l_0 \le m_j^*)$ and $\frac{0}{0}$ is defined as 0.*

*Proof* The symmetry (which implies that $p_{2l-1(j)} = \frac{1}{2}$) is obtained by standard arguments using the concavity of the optimality criterion. A similar reasoning as in the proof of Theorem 5.8.3 yields that the $D_s$-optimal product design is given by $\eta^{D_s} = \hat{\xi}_1 \times \cdots \times \hat{\xi}_q$, where $\hat{\xi}_j$ maximizes the factor

$$B_j(\xi_j) = \frac{\prod_{i=1}^{m} \left( \int_{-1}^{1} W_{i(j)}^2(t_j) \, d\xi_j(t_j) \right)^{\tau(i,j,m)}}{\prod_{i=1}^{n} \left( \int_{-1}^{1} W_{i(j)}^2(t_j) \, d\xi_j(t_j) \right)^{\tau(i,j,n)}}.$$

Expressing $B_j(\xi_j)$ in terms of canonical moments of the measure $\xi_j$, we obtain from Remark 2.3.7 and straightforward algebra that

$$B_j(\xi_j) = \prod_{l=1}^{n} \left( q_{2l-2(j)} p_{2l(j)} \right)^{\sigma(l,j,m) - \sigma(l,j,n)} \prod_{l=n+1}^{m} \left( q_{2l-2(j)} p_{2l(j)} \right)^{\sigma(l,j,m)} \qquad (5.8.20)$$

with $\sigma(l, j, m) = \sum_{i=l}^{m} \tau(i, j, m)$. If $n < m_j^*$, it is easy to see that $\sigma(l, j, m) - \sigma(l, j, n) > 0$ $(l = 1, \ldots, n)$ and that $\sigma(l, j, m) > 0$ $(l = n+1, \ldots, m_j^*)$. Thus the assertion follows directly by maximizing (5.8.20). The remaining case $n \ge m_j^*$ is more delicate. In this case $B_j(\xi_j)$

reduces to

$$B_j(\xi_j) = \prod_{l=1}^{m_j^*} \left(q_{2l-2(j)} p_{2l(j)}\right)^{n_l}, \tag{5.8.21}$$

where

$$n_l = \sigma(l, j, m) - \sigma(l, j, n) = \sum_{i=l}^{m_j^*} \tau(i, j, m) - \tau(i, j, n)$$

$$= \sum_{i=l}^{m_j^*} \sum_{\substack{n+1 \le \sum_{k=1}^{q} m_k \le m \\ m_j = i}} \tau_{m_1, \ldots, m_q}. \tag{5.8.22}$$

In the last line we used $\tau(i, j, r) = 0$ for $i > m_j^*$ and $r \in \{n, m\}$ [which is immediate from the definition of $m_j^*$ in (5.8.13)]. From (5.8.22) we have $n_1 \ge n_2 \ge \cdots \ge n_{m_j^*} \ge 0$, and defining $l_0 + 1 = \min\{l \mid 1 \le l \le m_j^*, n_j = 0\}$, we obtain from (5.8.21)

$$B_j(\xi_j) = \prod_{l=1}^{l_0} \left(q_{2l-2(j)} p_{2l(j)}\right)^{n_l}.$$

If $l_0 = 0$, then $\hat{\xi}_j$ can be chosen arbitrarily. We put $p_{2(j)} = 0$ which corresponds to the assertion of the theorem and obtain the factor design $\hat{\xi}_j$ with mass 1 at the point 0. If $l_0 \ge 1$, we obtain $p_{2l_0(j)} = 1$ and

$$p_{2l(j)} = \frac{n_l}{n_l + n_{l+1}}, \qquad 1 \le l \le l_0,$$

which is (5.8.22) in the case $n \ge m_j^*$.   ∎

**EXAMPLE 5.8.8**   Consider the model (5.8.8). If the experimenter is interested into the estimation of the parameters corresponding to the quadratic terms $t_1 t_2$ and $t_1^2$, a $D_1$-optimal design may be appropriate. By Theorem 5.8.7 (with $m = 2$, $n = 1$, $m_1^* = 2$, $m_2^* = 1$, $m_3^* = 1$) and (5.8.7), it follows that the $D_1$-optimal product design $\eta^{D_1} = \hat{\xi}_1 \times \hat{\xi}_2 \times \hat{\xi}_3$, where $\hat{\xi}_2, \hat{\xi}_3$ have canonical moments

$$p_{1(1)} = p_{3(1)} = \tfrac{1}{2}, \quad p_{2(1)} = \tfrac{2}{3}, \quad p_{4(1)} = 1,$$

$$p_{1(2)} = \tfrac{1}{2}, \quad p_{2(2)} = 1,$$

$$p_{1(3)} = \tfrac{1}{2}, \quad p_{2(3)} = 0.$$

The corresponding designs $\hat{\xi}_j$ ( $j = 1, 2, 3$) can be obtained from Corollary 4.2.2 and the $D_1$-optimal product design has equal masses at the points $(-1, -1, 0)$, $(-1, 0, 0)$, $(-1, 1, 0)$, $(1, -1, 0)$, $(1, 0, 0)$, $(1, 1, 0)$. Note that this design has a singular information matrix for estimating all parameters in model (5.8.8) and that other $D_1$-optimal product designs can be obtained by using arbitrary symmetric measures for the design $\hat{\xi}_3$—see the proof of Theorem 5.8.7.

**Corollary 5.8.9** *The $D_1$-optimal product design for the complete q-way mth degree polynomial regression model (5.8.1) on the q-dimensional cube $[-1, 1]^q$ is given by $\eta^{D_1} = \xi \times \ldots \times \xi$, where the design $\xi$ on the interval $[-1, 1]$ has equal masses $1/(q + m - 1)$ at the $m - 1$ zeros of the $(m - 1)$th ultraspherical polynomial $C_{m-1}^{((q+1)/2)}(x)$ and masses $q/(2(q + m - 1))$ at the boundary points $-1$ and $1$.*

*The $D_1$-optimal product design on the cube $[a_1, b_1] \times [a_2, b_2] \times \ldots \times [a_q, b_q]$ is obtained by transforming the design $\xi$ linearly to the corresponding intervals $[a_j, b_j]$ ($j = 1, \ldots, q$).*

*Proof* The result follows from Theorem 5.8.7 ($n = m - 1$, $m_j^* = m$, $j = 1, \ldots, q$), (5.8.17) and an application of Corollary 4.3.3. ∎

# CHAPTER 6

# Discrimination and Model Robust Designs

## 6.1 INTRODUCTION

So far it has been assumed that the linear model (5.2.1) is known by the experimenter. In many applications precise knowledge about the form of the regression function is not available, and the analysis of the data is performed in two steps. In the first step the data is used to identify an appropriate regression model, and the second step might consist of performing some statistical analysis in the determined model. For example, if a cubic regression model is assumed by the experimenter, the results of the experiments will typically be used to test whether a quadratic model would be more appropriate. In this case "good" designs have to address at least three different tasks: (1) the problem of testing the hypothesis $H_0 : \theta_3 = 0$ for the "highest" coefficient in the cubic model, (2) the problem of estimating the parameters in the full cubic polynomial if the test rejects the hypothesis $H_0$, (3) the problem of estimating the parameters in the reduced quadratic regression model if the test does not reject the hypothesis $H_0$.

An optimal design for one task may be exceptionally inefficient for another. To find a design that performs well for a number of different tasks, a common approach is to include all aspects of interest in one design criterion. The resulting criteria are usually called *composite design criteria* or *mixtures of optimality criteria*. This chapter illustrates the application of canonical moments in this field for the one-dimensional polynomial regression. Thereby we concentrate ourselves on two specific situations. In Sections 6.2 and 6.3 it is assumed that the main interest of the experimenter is the identification of the degree of the underlying polynomial regression, and an optimal design for this task has to be constructed. Optimal designs for this problem are called *optimal discrimination designs*. Second, in Section 6.4 we consider the case where the main interest of the experimenter is the estimation of a response function which is known to belong to the class of polynomials up to degree $m$. In this case "good" designs should allow the efficient estimation of all parameters in any of the polynomial models up to degree $m$ and are called *model robust designs*.

**170**

## 6.2 DISCRIMINATION DESIGNS WITH GEOMETRIC MEANS

Throughout this chapter we consider the univariate polynomial regression model

$$Y = \sum_{i=0}^{m} \theta_{mi} x^i + \varepsilon, \tag{6.2.1}$$

where $\boldsymbol{\theta}_m = (\theta_{m0}, \ldots, \theta_{mm})^T$ is the vector of unknown parameters, the vector of regression functions is $\boldsymbol{f}_m(x) = (1, x, \ldots, x^m)^T$ and the design space is assumed to be the interval $[a, b]$. If the experimenter wants to check if the degree of the polynomial regression is $m$ or $m - 1$, he could use the $D_1$-optimal design given in Theorem 5.5.4. From the discussion in Example 5.2.5 such a design maximizes the power of the $F$-test for the hypothesis $H_0 : \theta_{mm} = 0$ in the model (6.2.1) and is therefore appropriate for detecting any departures from the polynomial regression of degree $m - 1$. However, in most applications the degree of the regression (6.2.1) is not known to be $m - 1$ or $m$ before the experiments are carried out. Thus, if the "first" test does not reject the hypothesis $H_0 : \theta_{mm} = 0$, then the experimenter usually performs a "second" test in the "reduced" model of degree $m - 1$

$$Y = \sum_{i=0}^{m-1} \theta_{m-1,i} x^i + \varepsilon,$$

for the hypothesis $H_0 : \theta_{m-1,m-1} = 0$ in order to check if the degree of the regression is less or equal than $m - 2$. An optimal design for this test should maximize the power of a different test in a different model, and it is not clear that the $D_1$-optimal design for the polynomial regression of degree $m - 1$ is an efficient choice for testing hypotheses about the "highest" coefficients in polynomial models of lower degree.

**EXAMPLE 6.2.1** Consider the polynomial regression (6.2.1) of degree $m = 4$. The $D_1$-optimal design for this model is obtained from Theorem 5.5.4 as

$$\xi_4^{D_1} = \begin{pmatrix} -1 & -\frac{1}{\sqrt{2}} & 0 & \frac{1}{\sqrt{2}} & 1 \\ \frac{1}{8} & \frac{1}{4} & \frac{1}{4} & \frac{1}{4} & \frac{1}{8} \end{pmatrix}.$$

The $D_1$-efficiencies of this design for testing the hypothesis that the "highest" coefficient in a cubic, quadratic, linear model vanishes are listed in Table 6.2.1. For the calculation we use (5.5.16) and the fact that $\xi_4^{D_1}$ has canonical moments $p_i = \frac{1}{2}$ $(i \le 7)$, $p_8 = 1$. For example, we observe that $\xi_4^{D_1}$ is only 50% efficient for testing the hypothesis $H_0 : \theta_{22} = 0$ in a quadratic regression. The table also

**Table 6.2.1** $D_1$-efficiencies of the designs $\xi_4^{D_1}, \xi_{0,\beta_U}, \xi_{0,\tilde{\beta}_U}, \xi_{-\infty,\beta_U}, \xi_{-\infty,\tilde{\beta}_U}$ in a polynomial regression of degree $m = 1, 2, 3, 4$

| m | $\xi_4^{D_1}$ | $\xi_{0,\beta_U}$ | $\xi_{0,\tilde{\beta}_U}$ | $\xi_{-\infty,\beta_U}$ | $\xi_{-\infty,\tilde{\beta}_U}$ |
|---|---|---|---|---|---|
| 4 | 1.0 | 0.8359 | 0.8533 | 0.625 | 0.6667 |
| 3 | 0.5 | 0.6269 | 0.64 | 0.625 | 0.6667 |
| 2 | 0.5 | 0.5878 | 0.6 | 0.625 | 0.6667 |
| 1 | 0.5 | 0.5714 | 0.5 | 0.625 | 0.5 |

contains the corresponding efficiencies of the following four designs

$$\xi_{0,\beta_U} = \begin{pmatrix} -1 & -\sqrt{\frac{3}{7}} & 0 & \sqrt{\frac{3}{7}} & 1 \\ \frac{1}{5} & \frac{1}{5} & \frac{1}{5} & \frac{1}{5} & \frac{1}{5} \end{pmatrix}, \tag{6.2.2}$$

$$\xi_{0,\tilde{\beta}_U} = \begin{pmatrix} -1 & -\sqrt{\frac{2}{5}} & 0 & \sqrt{\frac{2}{5}} & 1 \\ \frac{1}{6} & \frac{5}{24} & \frac{1}{4} & \frac{5}{24} & \frac{1}{6} \end{pmatrix}, \tag{6.2.3}$$

$$\xi_{-\infty,\beta_U} = \begin{pmatrix} -1 & -\sqrt{\frac{3}{8}} & 0 & \sqrt{\frac{3}{8}} & 1 \\ \frac{1}{4} & \frac{1}{6} & \frac{1}{6} & \frac{1}{6} & \frac{1}{4} \end{pmatrix}, \tag{6.2.4}$$

$$\xi_{-\infty,\tilde{\beta}_U} = \begin{pmatrix} -1 & -\sqrt{\frac{1}{3}} & 0 & \sqrt{\frac{1}{3}} & 1 \\ \frac{3}{16} & \frac{3}{16} & \frac{1}{4} & \frac{3}{16} & \frac{3}{16} \end{pmatrix}. \tag{6.2.5}$$

These designs perform better in the models of lower degree and will be explained later.

The general problem of identifying the degree of a polynomial regression can be formulated in the following way: Consider the model (6.2.1). Based on a sample of observations the experimenter wants to identify the appropriate deterministic model

$$h_l(x) = \sum_{i=0}^{l} \theta_{li} x^i = \boldsymbol{\theta}_l^T \boldsymbol{f}_l(x), \qquad l = 1, \ldots, m,$$

where $\boldsymbol{f}_l(x) = (1, x, \ldots, x^l)^T$ denotes the vector of monomials up to the order $l$ and $\boldsymbol{\theta}_l = (\theta_{l0}, \ldots, \theta_{ll})^T \in \mathbb{R}^{l+1}$ is the vector of unknown parameters in the polynomial model of degree $l = 1, \ldots, m$. Let

$$\mathcal{F}_m = \{h_l(x) | l = 1, \ldots, m\} \tag{6.2.6}$$

denote the class of all possible (deterministic) polynomial regression models up to degree $m$. Anderson (1962) studied the following decision rule for this problem. For a given set of levels $(\alpha_1, \ldots, \alpha_m)$ the procedure chooses the largest integer in $\{1, \ldots, m\}$ for which the $F$-test in the model $h_j(x)$ rejects the hypothesis $H_0: \theta_{jj} = 0$ at the levels $\alpha_j$ $(j = 1, \ldots, m)$. It is well known that Anderson's method has several optimality properties (see Anderson, 1962, or Spruill, 1990). In the following $Y_1, \ldots, Y_n$ denote $n$ independent normally distributed observations with common variance $\sigma^2 > 0$ and mean given by one of the models in (6.2.6), that is,

$$Y = X_l \theta_l + \varepsilon \qquad \text{for some } l = 0, \ldots, m,$$

where $Y = (Y_1, \ldots, Y_n)^T$, $\varepsilon \sim \mathcal{N}(0, \sigma^2 I_n)$, $\theta_l = (\theta_{l0}, \ldots, \theta_{ll})^T$ and $X_l = (x_i^j)_{i=1,\ldots,n}^{j=0,\ldots,l}$ is the design matrix in the regression of degree $l$ $(l = 0, \ldots, m)$. For Anderson's procedure the probability of choosing too many functions is independent of the matrix $X_l$ provided that $X_l$ has full rank (see Anderson, 1962). If the model is $h_l(x) = h_{l-1}(x) + \theta_{ll} x^l$ it follows from Example 5.2.5 that the $F$-distribution of the test statistic for the hypothesis $H_0: \theta_{ll} = 0$ has the non-centrality parameter

$$\delta_l^2 = \frac{\theta_{ll}^2}{\sigma^2} (e_l^T (X_l^T X_l)^{-1} e_l)^{-1}, \tag{6.2.7}$$

where $e_l = (0, \ldots, 0, 1)^T \in \mathbb{R}^{l+1}$ denotes the $(l+1)$th unit vector. Consequently the probability of deciding in favor of $h_{l-1}(x)$ (when the model is in fact $h_l(x)$) is a decreasing function of $\delta_l^2$ and a "good" choice of a design for model discrimination should make the quantities $\delta_1^2, \ldots, \delta_m^2$ as large as possible.

   In the approximate design theory the quantities corresponding to the non-centrality parameters $\delta_j^2$ in (6.2.7) are the ratios of the determinants

$$\delta_l^2(\xi) = \frac{n}{\sigma^2} \theta_{ll}^2 (e_l^T M_l^{-1}(\xi) e_l)^{-1} = \frac{n}{\sigma^2} \theta_{ll}^2 \frac{|M_l(\xi)|}{|M_{l-1}(\xi)|},$$

where

$$M_l(\xi) = \int_a^b f_l(x) f_l(x)^T d\xi(x) = (c_{i+j})_{i,j=0}^l \tag{6.2.8}$$

denotes the moment matrix of the design $\xi$ in the polynomial regression of degree $l$. Consequently a "good" design for the discrimination between the polynomial models in the class $\mathcal{F}_m$ should make the $D_1$-efficiencies

$$\text{eff}_l^{D_1}(\xi) = \frac{\delta_l^2(\xi)}{\sup_\eta \delta_l^2(\eta)} \tag{6.2.9}$$

$$= \frac{|M_l(\xi)|/|M_{l-1}(\xi)|}{\sup_\eta |M_l(\eta)|/|M_{l-1}(\eta)|} = \frac{2^{4l-2}}{(b-a)^{2l}} \frac{|M_l(\xi)|}{|M_{l-1}(\xi)|}$$

as large as possible. As expected, a simultaneous maximization of the efficiencies

$$\text{eff}_1^{D_1}(\xi), \ldots, \text{eff}_m^{D_1}(\xi)$$

is impossible, and we have to restrict ourselves to the maximization of real valued functions of these quantities.

**Definition 6.2.2**   *Let $\beta_1, \ldots, \beta_m$ denote nonnegative numbers with sum $1$, $\beta_m > 0$ and $\xi_{0,\beta}$ denote a design with nonsingular moment matrix $M_m(\xi_{0,\beta})$. The design $\xi_{0,\beta}$ is called a $\Psi_0$-optimal discriminating design for the class $\mathcal{F}_m$ with respect to the the prior $\beta = (\beta_1, \ldots, \beta_m)$ if and only if $\xi_{0,\beta}$ maximizes the weighted geometric mean*

$$\Psi_0^\beta(\xi) = \prod_{l=1}^m \left(\text{eff}_l^{D_1}(\xi)\right)^{\beta_l} = \prod_{l=1}^m \left(\frac{2^{4l-2}}{(b-a)^{2l}} \frac{|M_l(\xi)|}{|M_{l-1}(\xi)|}\right)^{\beta_l}. \tag{6.2.10}$$

Note that the a $\Psi_0$-optimal discriminating design guarantees the estimability of all parameters in the polynomial regression of degree $m$ and that the best design does not depend on the constant

$$\prod_{l=1}^m \left(\frac{2^{2(2l-1)}}{(b-a)^{2l}}\right)^{\beta_l}$$

which reflects a particular property of the geometric mean criterion. The vector $\beta = (\beta_1, \ldots, \beta_m)$ defines a prior on the set of polynomial models up to degree $\mathcal{F}_m$. The weight $\beta_l$ reflects the experimenters belief about the adequacy of the model of degree $l$ or the importance of an error of the second kind in the test of the hypothesis $H_0 : \theta_{ll} = 0$. For example, if $\beta = (0, \ldots, 0, 1)$ the optimality criterion (6.2.10) gives the $D_1$-criterion and the corresponding design could be used for a test of whether the degree of the polynomial is $m - 1$ or $m$. Similarly, if it is clear from physical considerations that there is at least a linear trend in the data, one could use priors of the form $\tilde{\beta} = (0, \beta_2, \ldots, \beta_m)$ in the optimality criterion (6.2.10). We will show with the following theorem that the designs $\xi_{0,\beta_U}$ and $\xi_{0,\tilde{\beta}_U}$ in Example 6.2.1 are in fact the $\Psi_0$-optimal discriminating designs for the class $\mathcal{F}_4$ with respect to the "uniform" priors $\beta_U = (\frac{1}{4}, \frac{1}{4}, \frac{1}{4}, \frac{1}{4})$ and $\tilde{\beta}_U = (0, \frac{1}{3}, \frac{1}{3}, \frac{1}{3})$, respectively.

**Theorem 6.2.3** *The $\Psi_0$-optimal discriminating design for the class $\mathcal{F}_m$ with respect to the prior $\beta = (\beta_1, \ldots, \beta_m)$ $(\beta_m > 0)$ is uniquely determined by its canonical moments*

$$p_{2i} = \frac{\sigma_i}{\sigma_i + \sigma_{i+1}}, \quad i = 1, \ldots, m-1, \quad p_{2m} = 1,$$

$$p_{2i-1} = \frac{1}{2}, \quad i = 1, \ldots, m,$$

*where $\sigma_i = \sum_{l=i}^{m} \beta_l, i = 1, \ldots, m$.*

*Proof* By definition of the criterion $\Psi_0^\beta$ we have to maximize the function in (6.2.10), which reduces by (5.5.16) to

$$\Psi_0^\beta(\xi) = C \prod_{l=1}^{m} \prod_{j=1}^{l} (q_{2j-2} p_{2j-1} q_{2j-1} p_{2j})^{\beta_l}$$

$$= C \prod_{j=1}^{m} \prod_{l=j}^{m} (q_{2j-2} p_{2j-1} q_{2j-1} p_{2j})^{\beta_l}$$

$$= C \prod_{j=1}^{m} (q_{2j-1} p_{2j-1})^{\sigma_j} \prod_{j=1}^{m-1} q_{2j}^{\sigma_{j+1}} p_{2j}^{\sigma_j} p_{2m}^{\sigma_m},$$

where $p_1, p_2, \ldots$ denote the canonical moments of the design $\xi$ $(q_0 = 1)$ and the constant $C$ does not depend on the design $\xi$. The assertion now follows by a straightforward maximization of this function in terms of the canonical moments. ∎

**EXAMPLE 6.2.4** Consider the uniform prior $\beta_U = (\frac{1}{4}, \frac{1}{4}, \frac{1}{4}, \frac{1}{4})$ for the class $\mathcal{F}_4$ of polynomial regression models up to degree 4. The quantities $\sigma_j$ are given by

$$\sigma_1 = 1, \quad \sigma_2 = \frac{3}{4}, \quad \sigma_3 = \frac{1}{2}, \quad \sigma_4 = \frac{1}{4}.$$

By Theorem 6.2.3 the canonical moments of the $\Psi_0$-optimal discriminating design for the class $\mathcal{F}_4$ with respect to the uniform prior $\beta_U$ are $p_{2i-1} = \frac{1}{2}$, $(i = 1, \ldots, 4)$, and

$$p_2 = \frac{4}{7}, \quad p_4 = \frac{3}{5}, \quad p_6 = \frac{2}{3}, \quad p_8 = 1.$$

The corresponding design $\xi_{0,\tilde{\beta}_U}$ is again obtained by an application of Corollary 4.2.2 and is given in formula (6.2.2) in Example 6.2.1.

If the experimenter is sure that there is at least a linear term in his regression model, he could use a different prior that does not put any weight on the linear model. For example, if $\tilde{\beta}_U = (0, \frac{1}{3}, \frac{1}{3}, \frac{1}{3})$, the quantities $\sigma_j = \sum_{l=j}^{m} \beta_l$ are

$$\sigma_1 = 1, \quad \sigma_2 = 1, \quad \sigma_3 = \tfrac{2}{3}, \quad \sigma_4 = \tfrac{1}{3}.$$

The canonical moments of the $\Psi_0$-optimal discriminating design for the class $\mathcal{F}_4$ with respect to the "uniform" prior $\tilde{\beta}_U$ are $p_{2i-1} = \frac{1}{2}$, $(i = 1, \ldots, 4)$ and

$$p_2 = \tfrac{1}{2}, \quad p_4 = \tfrac{3}{5}, \quad p_6 = \tfrac{2}{3}, \quad p_8 = 1.$$

The corresponding design $\xi_{0,\tilde{\beta}_U}$ is again obtained by an application of Corollary 4.2.2 and is given in formula (6.2.3) in Example 6.2.1.

Note that for the uniform prior $\beta_U = (1/m, \ldots, 1/m)$ Definition 6.2.2 gives the $D$-optimality criterion

$$\Psi_0^{\beta_U}(\xi) = \frac{2^{2m}}{(b-a)^{m+1}} |M_m(\xi)|^{1/m}.$$

Thus the $D$-optimal design for polynomial regression of degree $m$ is also the $\Psi_0$-optimal discriminating design with respect to the uniform prior for the class $\mathcal{F}_m$ and is explicitly given in Theorem 5.5.3. In general, for an arbitrary prior $\beta$, the $\Psi_0$-optimal discriminating design with respect to $\beta$ cannot be described analytically. However, the following theorem specifies an important subclass of priors for which explicit solutions of the design problem are available:

**Theorem 6.2.5**   *Let for $l = 1, \ldots, m$ and $z \geq 0$,*

$$\beta_l(z) = z \frac{\Gamma(m - l + z)\Gamma(m)}{\Gamma(m - l + 1)\Gamma(m + z)}, \tag{6.2.11}$$

*$\beta(z) = (\beta_1(z), \ldots, \beta_m(z))$. On the interval $[-1, 1]$ the $\Psi_0$-optimal discriminating design $\xi_{0,\beta(z)}$ for the class $\mathcal{F}_m$ with respect to the prior $\beta(z)$ puts masses $(z + 1)/(2(m + z))$ at the points $-1, 1$ and masses $1/(m + z)$ at the $m - 1$ zeros of the $(m - 1)$th ultraspherical polynomial $C_{m-1}^{(z/2+1)}(x)$. On the interval $[a, b]$, the $\Psi_0$-optimal discriminating design for the class $\mathcal{F}_m$ with respect to the prior $\beta(z)$ can be obtained from $\xi_{0,\beta(z)}$ by the linear transformation (5.5.9).*

*Proof*   The proof is performed in two steps. We first show that (6.2.11) defines a prior on the set $\{1, \ldots, m\}$ and then apply Theorem 6.2.3 and Corollary 4.3.3 in order to prove the assertion. For the first step we remark that the case $z = 0$ in

(6.2.11) is understood as the limit

$$\beta(0) = \lim_{z \to 0} \beta(z) = (0, \ldots, 0, 1)$$

and that the relation

$$\sum_{l=i}^{m} \frac{\Gamma(q + l + 1 - i)}{\Gamma(l + 1 - i)} \frac{\Gamma(m + z - l)}{\Gamma(m + 1 - l)} \qquad (6.2.12)$$

$$= \frac{\Gamma(q + 1)}{\Gamma(z + q + 1)} \frac{\Gamma(z)}{\Gamma(m + 1 - i)} \Gamma(m + z + q - i + 1).$$

holds for all $m \in \mathbb{N}$, $q \in \mathbb{N}_0$ and $z \in \mathbb{R} \setminus \{0, -1, -2, \ldots\}$. In order to show (6.2.12), define $f_{q,i}(z)$ as the left-hand side. It then follows from the functional equation of the Gamma function ($q \geq 1, i \geq 1$) that

$$f_{q,i-1}(z) = f_{q,i}(z) + q\, f_{q-1,i-1}(z).$$

If we note that $f_{q,m}(z) = \Gamma(q + 1)\Gamma(z)$ and $f_{q,m+1}(z) = 0$ for all $q \in \mathbb{N}_0, m \in \mathbb{N}$, the identity (6.2.12) is then obtained by induction.

Note that (6.2.11) is a special case of the normalizing condition for the measure of orthogonality corresponding to the Hahn polynomials in Example 2.3.9 ($\alpha = q$, $\beta = z - 1$, $N = m - i$).

The fact that the vector $\beta(z)$ in (6.2.11) defines a prior on the set $\{1, \ldots, m\}$ is now obtained by putting $q = 0$ and $i = 1$ in (6.2.12) and by treating the case $z = 0$ separately. We also use (6.2.12) to calculate

$$\sigma_i(z) = \sum_{l=i}^{m} \beta_l(z) = \frac{\Gamma(m)}{\Gamma(m + z)} \frac{\Gamma(m + z + 1 - i)}{\Gamma(m + 1 - i)}, \qquad i = 1, \ldots, m.$$

By Theorem 6.2.3 the canonical moments of the $\Psi_0$-optimal discriminating design with respect to the prior $\beta(z)$ are obtained as

$$p_{2i-1} = \tfrac{1}{2}, \qquad i = 1, \ldots, m,$$

$p_{2m} = 1$ and

$$p_{2i} = \frac{\sigma_i(z)}{\sigma_i(z) + \sigma_{i+1}(z)} = \frac{m - i + z}{2(m - i) + z}, \qquad i = 1, \ldots, m - 1.$$

The assertion now follows from Corollary 4.3.3. ∎

Note that the case $z = 0$ in (6.2.11) gives the $D_1$-optimality criterion and that $z = 1$ gives the uniform prior with equal weight for all models in $\mathcal{F}_m$. For increasing $z$ the prior $\beta(z)$ in (6.2.11) puts more weight on the models of lower degree. For example, if $z = 2$, the weights are proportional to

$$m : m - 1 : \ldots : 2 : 1,$$

while $z = 3$ gives weights proportional to

$$(m + 1)m : m(m - 1) : \ldots : 6 : 2.$$

Theorem 6.2.3 defines a very interesting and useful mapping, say, $\mathcal{H}_m$, from the set of all prior distributions

$$\mathcal{B}_{\overline{m}}^{\geq} := \left\{ (\beta = (\beta_1, \ldots, \beta_m) | \beta_l \geq 0, \beta_m > 0, \sum_{l=1}^{m} \beta_l = 1 \right\} \tag{6.2.13}$$

into the set of all symmetric probability measures on a given interval, say, $[a, b]$, with $m + 1$ support points including the boundary points $a$ and $b$ and with even canonical moments greater or equal than $\frac{1}{2}$

$$\mathcal{P}_{\overline{m}}^{\geq} = \{ \xi \in \Xi | p_{2i-1} = \tfrac{1}{2}, 1 > p_{2i} \geq \tfrac{1}{2}, p_{2m} = 1 \}. \tag{6.2.14}$$

This mapping can be extended in an obvious way to the set

$$\mathcal{B}_m = \left\{ (\beta = (\beta_1, \ldots, \beta_m) | \sigma_i = \sum_{l=i}^{m} \beta_l > 0 (i \leq m), \sum_{l=1}^{m} \beta_l = 1 \right\} \tag{6.2.15}$$

with image

$$\mathcal{P}_m^* = \{ \xi \in \Xi | p_{2i-1} = \tfrac{1}{2}, 0 < p_{2i} < 1, p_{2m} = 1 \}. \tag{6.2.16}$$

In the following we will show that this map is one to one:

**Theorem 6.2.6** *The mapping $\mathcal{H}_m$ defined by Theorem 6.2.3 from the set $\mathcal{B}_m$ in (6.2.15) onto the set $\mathcal{P}_m^*$ in (6.2.16) is one to one, $\mathcal{H}_m(\mathcal{B}_{\overline{m}}^{\geq}) = \mathcal{P}_{\overline{m}}^{\geq}$, and for $\xi \in \mathcal{P}_m^*$ the inverse map of $\mathcal{H}_m$ is given by $\mathcal{H}_m^{-1}(\xi) = (\beta_1, \ldots, \beta_m)$, where*

$$\beta_l = \prod_{j=1}^{l-1} \frac{q_{2j}}{p_{2j}} \left( 1 - \frac{q_{2l}}{p_{2l}} \right), \qquad l = 1, \ldots, m. \tag{6.2.17}$$

*Proof* The proof is performed by an explicit calculation of the inverse map. The probability measure $\xi \in \mathcal{P}_m^*$ is uniquely determined by its canonical moments $p_i \in (0, 1)$ $(i \leq 2m - 1)$, $p_{2m} = 1$. Since $\mathcal{H}_m$ maps the set $\mathcal{B}_m$ in $\mathcal{P}_m^*$, we have to determine a unique vector $(\beta_1, \ldots, \beta_m)$ for which $\xi$ maximizes the function $\Psi_0^\beta$ in (6.2.10). The canonical moments of a probability measure $\xi$ maximizing

$\Psi_0^\beta$ are given in Theorem 6.2.3, and we obtain the equations

$$\frac{\sigma_i}{\sigma_i + \sigma_{i+1}} = p_{2i}, \qquad i = 1, \ldots, m-1,$$

or equivalently

$$\sigma_{i+1} = \frac{q_{2i}}{p_{2i}} \sigma_i, \qquad i = 1, \ldots, m-1.$$

Observing that $\sigma_1 = \sum_{l=1}^{m} \beta_l = 1$, we have

$$\sigma_{i+1} = \prod_{j=1}^{i} \frac{q_{2j}}{p_{2j}}, \qquad i = 1, \ldots, m-1,$$

and the definition of $\sigma_l$ in Theorem 6.2.3 shows (note that $p_{2m} = 1$) that

$$\beta_l = \sigma_l - \sigma_{l+1} = \prod_{j=1}^{l-1} \frac{q_{2j}}{p_{2j}} \left(1 - \frac{q_{2l}}{p_{2l}}\right), \qquad l = 1, \ldots, m. \tag{6.2.18}$$

This proves that the given probability measure $\xi$ maximizes $\Psi_0^\beta$ if and only if the prior $\beta$ is given by (6.2.18) (note that $\sigma_i > 0$). Therefore the inverse map of $\mathcal{H}_m$ exists and is defined by (6.2.18). The assertion $\mathcal{H}_m(\mathcal{B}_m^{\geq}) = \mathcal{P}_m^{\geq}$ is immediate from these considerations. ∎

Roughly speaking, Theorem 6.2.6 shows that every symmetric design $\xi$ is a $\Psi_0$-optimal discriminating design with respect to a "prior" that can be calculated from the canonical moments of $\xi$. However, some care is necessary with this interpretation. In general, not every design $\xi \in \mathcal{P}_m^*$ yields a prior $\beta = \mathcal{H}_m^{-1}(\xi) \in \mathcal{B}_m^{\geq}$ because some components of $\beta$ could become negative and negative weights cannot be interpreted as the experimenters belief about the adequacy of different models. Thus a design $\xi$ whose inverse image $\mathcal{H}_m^{-1}(\xi)$ has negative components is expected to be inefficient for model discrimination. Positive weights $\beta_l$ are obtained in (6.2.17) if and only if all canonical moments of even order are greater or equal than $\frac{1}{2}$.

**EXAMPLE 6.2.7** Consider the class $\mathcal{F}_5$ of polynomial models up to degree 5 on the interval $[-1, 1]$ and the distribution $\xi_B$ with masses proportional to $1 : 5 : 10 : 10 : 5 : 1$ at the points $-1, -\frac{3}{5}, -\frac{1}{5}, \frac{1}{5}, \frac{3}{5}$, and $1$, respectively. This is the Binomial distribution with parameters $p = \frac{1}{2}$ and $n = 5$ transformed to the interval $[-1, 1]$. By Example 2.3.8 it follows that the canonical moments of even order of $\xi_B$ are given by $p_{2i} = i/5$ ($i = 1, \ldots, 5$), while the canonical moments of odd order are $\frac{1}{2}$. By Theorem 6.2.6 the design $\xi_B$ maximizes the function $\Psi_0^\beta$ in (6.2.10), where

$$\beta = \mathcal{H}_5^{-1}(\xi_B) = (-3, -2, 2, 3, 1),$$

which does not define a prior on the class of polynomials $\mathcal{F}_5$. This indicates that the Binomial distribution $\xi_B$ should not be used as a discriminating design for polynomial regression. In fact, observing (5.5.16), it follows that the $D_1$-efficiencies of the design $\xi_B$ in the polynomial regression models of degree 1, . . ., 5 are only 0.2, 0.256, 0.3686, 0.4719, and 0.3775, respectively.

## 6.3   DISCRIMINATION DESIGNS BASED ON $\Psi_p$-MEANS

The $\Psi_0$-optimal discriminating designs maximize a weighted product of the ratios of consecutive Hankel determinants $|M_l(\xi)|/|M_{l-1}(\xi)|$. Instead of the geometric mean one could also consider a sum of the efficiencies or, more generally, a weighted $p$-mean.

**Definition 6.3.1**   *Let $-\infty \leq p < 1$ and $\beta_1, \ldots, \beta_m$ denote nonnegative numbers with sum equal one and $\beta_m > 0$, and let $\xi_{p,\beta}$ denote a design with nonsingular moment matrix $M_m(\xi)$. The design $\xi_{p,\beta}$ is called a $\Psi_p$-optimal discriminating design for the class $\mathcal{F}_m$ with respect to the the prior $\beta = (\beta_1, \ldots, \beta_m)$ if and only if $\xi_{p,\beta}$ maximizes the weighted p-mean of $D_1$-efficiencies*

$$\Psi_p^\beta(\xi) = \left[\sum_{l=1}^m \beta_l \left(\text{eff}_l^{D_1}(\xi)\right)^p\right]^{1/p} \tag{6.3.1}$$

$$= \left[\sum_{l=1}^m \beta_l \left(\frac{2^{4l-2}}{(b-a)^{2l}} \frac{|M_l(\xi)|}{|M_{l-1}(\xi)|}\right)^p\right]^{1/p}.$$

The geometric mean criterion of Definition 6.2.2 appears as the limit case

$$\Psi_0^\beta(\xi) = \lim_{p \to 0} \Psi_p^\beta(\xi), \tag{6.3.2}$$

by applying the l'Hospital rule to $\log \Psi_p^\beta$. Another case of interest is the maximin criterion which is obtained when $p$ tends to $-\infty$, namely

$$\Psi_{-\infty}^\beta(\xi) = \lim_{p \to -\infty} \Psi_p^\beta(\xi) = \min\{\text{eff}_l^{D_1}(\xi)|\beta_l > 0, l = 1, \ldots, m\} \tag{6.3.3}$$

Note that the definition of the $\Psi_{-\infty}$-optimal discriminating design with respect to the prior $\beta$ does not depend on the specific size of the weights $\beta_l$. It only depends on $\beta$ through the positive components of the prior. Although these criteria can easily be expressed in terms of canonical moments of the design $\xi$, the resulting maximization problems for these criteria are more complicated than in the geometric case ($p = 0$). For a complete description of the canonical moments of the $\Psi_p$-optimal discriminating designs, some equivalence theorems for these optimality criteria are needed.

**Theorem 6.3.2**

1. *Let* $-\infty < p < 1$, *then* $\xi_{p,\beta}$ *is the* $\Psi_p$-*optimal discriminating design for the class* $\mathcal{F}_m$ *with respect to the prior* $\beta = (\beta_1, \ldots, \beta_m)$ *if and only if* $|M_m(\xi_{p,\beta})| \neq 0$ *and the inequality*

$$\sum_{l=1}^{m} \beta_l (\text{eff}_l^{D_1}(\xi_{p,\beta}))^p \frac{(e_l^T M_l^{-1}(\xi_{p,\beta}) f_l(x))^2}{e_l^T M_l^{-1}(\xi_{p,\beta}) e_l} \leq \sum_{l=1}^{m} \beta_l (\text{eff}_l^{D_1}(\xi_{p,\beta}))^p \quad (6.3.4)$$

*holds for all* $x \in [a, b]$. *Moreover there is equality in* (6.3.4) *for all support points of the* $\Psi_p$-*optimal discriminating design with respect to the prior* $\beta$.

2. *Let* $p = -\infty$ *and define for a design* $\xi \in \Xi$ *with* $|M_m(\xi)| \neq 0$

$$\mathcal{N}^{D_1}(\xi) = \left\{ j \in \{1, \ldots, m\} \,\middle|\, \beta_j > 0, \text{eff}_j^{D_1}(\xi) = \Psi_{-\infty}^{\beta}(\xi) \right\}. \quad (6.3.5)$$

*A design* $\xi_{-\infty, \beta}$ *is a* $\Psi_{-\infty}$-*optimal discriminating design with respect to the prior* $\beta = (\beta_1, \ldots, \beta_m)$ *if and only if* $|M_m(\xi_{-\infty, \beta})| \neq 0$ *and for any* $l \in \mathcal{N}^{D_1}(\xi_{-\infty, \beta})$ *there exist a nonnegative number* $\alpha_l$ *such that*

$$\sum_{l \in \mathcal{N}^{D_1}(\xi_{-\infty, \beta})} \alpha_l = 1 \quad (6.3.6)$$

*and such that the inequality*

$$\sum_{l \in \mathcal{N}^{D_1}(\xi_{-\infty, \beta})} \alpha_l \frac{(e_l^T M_l^{-1}(\xi_{-\infty, \beta}) f_l(x))^2}{e_l^T M_l^{-1}(\xi_{-\infty, \beta}) e_l} \leq 1 \quad (6.3.7)$$

*holds for all* $x \in [a, b]$. *Moreover there is equality in* (6.3.7) *for all support points of the* $\Psi_{-\infty}$-*optimal discriminating design with respect to the prior* $\beta$.

*Proof* The proof will be performed by an application of the Equivalence Theorems 5.4.7 and 5.4.8 in Section 5.4. We assume that $\beta_l > 0$ for all $l = 1, \ldots, m$ and remark that the general case $\beta_l \geq 0$ can be obtained by the same reasoning, where the corresponding components with $\beta_l = 0$ are omitted. We also consider only the case $p \neq 0$; a proof for the statement for $p = 0$ is given in Lemma 7.3.2 and the subsequent discussion. Since $\beta_m > 0$, we have $|M_m(\xi_{p,\beta})| \neq 0$. Define

$$\gamma_l = [\beta_l^{1/p} e_l^T M_l^{-1}(\xi_l^{D_1}) e_l)]^{1/2} = \left[ \beta_l^{1/p} \frac{2^{4l-2}}{(b-a)^{2l}} \right]^{1/2}, \quad (6.3.8)$$

where $\xi_l^{D_1}$ denotes the $D_1$-optimal design for the polynomial of degree $l$, $M_l(\xi)$ is the moment matrix of $\xi$ in the polynomial regression of degree $l$, and the last equality follows from (5.5.15). Let $\xi$ denote a design with nonsingular information matrix $M_m(\xi)$, and define block diagonal matrices

$$\widetilde{M}(\xi) = \begin{pmatrix} M_1(\xi) & & \\ & \ddots & \\ & & M_m(\xi) \end{pmatrix} \in \mathbb{R}^{m(m+3)/2 \times m(m+3)/2}, \qquad (6.3.9)$$

$$K = \begin{pmatrix} \gamma_1^{-1} e_1 & & \\ & \ddots & \\ & & \gamma_m^{-1} e_m \end{pmatrix} \in \mathbb{R}^{m(m+3)/2 \times m},$$

and a diagonal matrix

$$C_K(\widetilde{M}(\xi)) = (K^T \widetilde{M}(\xi)^{-1} K)^{-1}$$

$$= \begin{pmatrix} \gamma_1^2 (e_1^T M_1^{-1}(\xi) e_1)^{-1} & & \\ & \ddots & \\ & & \gamma_m^2 (e_m^T M_m^{-1}(\xi) e_m)^{-1} \end{pmatrix}$$

$$= \begin{pmatrix} \beta_1^{1/p} \mathrm{eff}_1^{D_1}(\xi) & & \\ & \ddots & \\ & & \beta_m^{1/p} \mathrm{eff}_m^{D_1}(\xi) \end{pmatrix} \in \mathbb{R}^{m \times m},$$

where all other entries in these matrices are zero. If $\phi_p$ denotes the matrix mean of Definition 5.3.1, then the $\Psi_p$-optimalty criterion in Definition 6.3.1 can be written as

$$\Psi_p^\beta(\xi) = \left[ \sum_{l=1}^m \beta_l \left( \mathrm{eff}_l^{D_1}(\xi) \right)^p \right]^{1/p} \qquad (6.3.10)$$

$$= [\mathrm{trace}\, C_K(\widetilde{M}(\xi))^p]^{1/p} = m^{1/p} \phi_p(C_K(\widetilde{M}(\xi))).$$

Here the case $p = -\infty$ is understood as the corresponding limit as explained in (6.3.2) and (6.3.3). Therefore the determination of the $\Psi_p$-optimal discriminating design reduces to the maximization of the function $\phi_p \circ C_K$ in the set $\widetilde{\mathcal{M}}$ of all "generalized" moment matrices of the form (6.3.9) for which $M_m(\xi)$ is nonsingular. In the following discussion we have to distinguish two cases:

First, $-\infty < p < 1$. By Theorem 5.4.7, $\widetilde{M} = \widetilde{M}(\xi_{p,\beta})$ maximizes $\phi_p \circ C_K$ if and only if the inequality

$$\text{trace } \widetilde{A}\widetilde{M}^{-1}KC_K(\widetilde{M})^{p+1}K^T\widetilde{M}^{-1} \leq \text{trace } C_K(\widetilde{M})^p \qquad (6.3.11)$$

holds for all $\widetilde{A} \in \widetilde{\mathcal{M}}$ and there is equality in (6.3.11) for $A = \widetilde{M}(\xi_{p,\beta})$. Since any matrix $\widetilde{A} \in \widetilde{\mathcal{M}}$ can be represented as a convex combination of matrices of the form

$$\widetilde{F}(x) = \begin{pmatrix} f_1(x)f_1^T(x) & \\ & \ddots & \\ & & f_m(x)f_m^T(x) \end{pmatrix} \in \mathbb{R}^{m(m+3)/2 \times m(m+3)/2},$$

it follows that the inequality (6.3.11) holds for all $\widetilde{A} \in \widetilde{\mathcal{M}}$ if and only if it holds for all matrices $\widetilde{F}(x)$ with $x \in [a,b]$. Therefore we obtain

$$\text{trace } \widetilde{F}(x)\widetilde{M}^{-1}KC_K(\widetilde{M})^{p+1}K^T\widetilde{M}^{-1}$$

$$= \sum_{l=1}^{m} \beta_l(\text{eff}\,_l^{D_1}(\xi_{p,\beta}))^p \frac{(e_l^T M_l^{-1}(\xi_{p,\beta})f_l(x))^2}{e_l^T M_l^{-1}(\xi_{p,\beta})e_l}$$

$$\leq \text{trace } C_K(\widetilde{M})^p$$

$$= \sum_{l=1}^{m} \beta_l(\text{eff}\,_l^{D_1}(\xi_{p,\beta}))^p$$

for all $x \in [a,b]$, which gives (6.3.4). Moreover, if $\xi_{p,\beta}$ is the $\Psi_p$-optimal discriminating design with respect to the prior $\beta$, there is equality in (6.3.11) for $\widetilde{A} = \widetilde{M}(\xi_{p,\beta})$, and consequently there must be equality in (6.3.4) for all support points of $\xi_{p,\beta}$.

For the second part of the assertion we note that the criterion in (6.3.3) only depends on $\beta$ through the positive components of the prior. Therefore we assume in the above discussion $\beta_l = 1/m$, $l = 1, \ldots, m$ and obtain by Theorem 5.4.8 that $\xi_{-\infty,\beta}$ is $\Psi_{-\infty}$-optimal with respect to the prior $\beta$ if and only if the corresponding moment matrix $\widetilde{M} = \widetilde{M}(\xi_{-\infty,\beta})$ satisfies

$$\text{trace } \widetilde{A}\widetilde{M}^{-1}KC_K(\widetilde{M})EC_K(\widetilde{M})K^T\widetilde{M}^{-1} \leq \lambda_{\min}(C_K(\widetilde{M})) = \Psi_{-\infty}^{\beta}(\xi_{-\infty,\beta})$$

for all $\widetilde{A} \in \widetilde{\mathcal{M}}$. Here $E$ is some $m \times m$ matrix in the set $\mathcal{E}$ defined in Theorem 5.4.8. It is easy to see that the left-hand side depends only on the diagonal elements of the matrix $E$. Therefore it is sufficient to replace the set $\mathcal{E}$ by the convex hull of the set of diagonal matrices

$$\left\{\text{diag}(\alpha_1, \ldots, \alpha_m) | \alpha_l = 0 \text{ if } l \notin \mathcal{N}^{D_1}(\xi_{-\infty,\beta}), \alpha_l \geq 0, \sum_{l=1}^{m} \alpha_l = 1\right\}$$

The assertion now follows by the same reasoning as in part a) observing that $\beta_1 = \ldots = \beta_m = 1/m$.                                                                      ∎

**Corollary 6.3.3** *For* $-\infty < p < 1$, $\xi_{p,\beta}$ *is the* $\Psi_p$-*optimal discriminating design for the class* $\mathcal{F}_m$ *with respect to the prior* $\beta$ *if and only if* $\xi_{p,\beta}$ *is the* $\Psi_0$-*optimal discriminating design for the class* $\mathcal{F}_m$ *with respect to the prior* $\beta' = (\beta'_1, \ldots, \beta'_m)$, *where*

$$\beta'_l = \frac{\beta_l (\mathrm{eff}_l^{D_1}(\xi_{p,\beta}))^p}{\sum_{j=1}^m \beta_j (\mathrm{eff}_j^{D_1}(\xi_{p,\beta}))^p}, \qquad l = 1, \ldots, m.$$

*Let* $\mathcal{N}^{D_1}(\xi_{-\infty,\beta})$ *be defined in* (6.3.5). *Then* $\xi_{-\infty,\beta}$ *is the* $\Psi_{-\infty}$-*optimal discriminating design for the class* $\mathcal{F}_m$ *with respect to the prior* $\beta$ *if and only if there exists a nonnegative vector* $\alpha = (\alpha_1, \ldots, \alpha_m)$ *with* $\alpha_l = 0$ *for all* $l \notin \mathcal{N}^{D_1}(\xi_{-\infty,\beta})$ *such that* $\xi_{-\infty,\beta}$ *is the* $\Psi_0$-*optimal discriminating design for the class* $\mathcal{F}_m$ *with respect to the prior* $\alpha$.

**Theorem 6.3.4** *Let* $-\infty < p < 1$, *then the* $\Psi_p$-*optimal discriminating design for the class* $\mathcal{F}_m$ *with respect to the prior* $\beta$ *is uniquely determined by its canonical moments* $p_1, \ldots, p_{2m-1}, p_{2m}$ *where*

$$p_{2l-1} = \tfrac{1}{2}, \qquad l = 1, \ldots, m, \tag{6.3.12}$$

$p_{2m} = 1$, *and* $(p_2, \ldots, p_{2m-2})$ *is the unique solution of the system of* $m - 1$ *equations*

$$\beta_m 2^{2p(m-l)} \prod_{i=l+1}^{m-1} (p_{2i}^{p+1} q_{2i}^{p-1})(1 - p_{2l})^{p-1}(2p_{2l} - 1) = \beta_l \tag{6.3.13}$$

$(l = 1, \ldots, m - 1)$.

*Proof*   Let $p_1, p_2, \ldots$ denote the canonical moments of a design $\xi$ with $|M(\xi)| > 0$. From (5.5.16) we have $(q_0 = 1)$,

$$\mathrm{eff}_l^{D_1}(\xi) = 2^{4l-2} \prod_{j=1}^l (q_{2j-2} p_{2j-1} q_{2j-1} p_{2j}),$$

and by Definition 6.3.1 the canonical moments of the optimal discriminating designs are independent of the design space $[a, b]$. Moreover the canonical moments of odd order of $\xi_{p,\beta}$ are given by $\tfrac{1}{2}$ and $p_{2m} = 1$. This follows because $\Psi_p^\beta$ is an increasing function of $p_{2m}$ and $p_{2i-1}(1 - p_{2i-1})$ $(i = 1, \ldots, m)$. Therefore by Theorem 1.2.5 and Corollary 1.3.4 the $\Psi_p$-optimal discriminating design with respect to the prior $\beta$ is unique and symmetric. For such a design the $D_1$-

efficiency in the polynomial of degree $l$ is given by

$$\mathrm{eff}_l^{D_1}(\xi) = 2^{2l-2} \prod_{j=1}^{l}(q_{2j-2}p_{2j}).$$

In the case $p = 0$ the optimality criterion (6.3.1) reduces to the geometric mean considered in Section 6.2, and it is straightforward to show that the canonical moments of the $\Psi_0$-optimal discriminating design with respect to the prior $\beta$ in Theorem 6.2.3 are the unique solution of (6.3.12) and (6.3.13) for $p = 0$. Second, if the $\Psi_p$-optimal discriminating $\xi_{p,\beta}$ design with respect to the prior $\beta$ has canonical moments $(\frac{1}{2}, p_2, \frac{1}{2}, \ldots, \frac{1}{2}, p_{2m-2}, \frac{1}{2}, 1)$, then it follows from Theorem 6.2.6 that $\xi_{p,\beta}$ is $\Psi_0$-optimal with respect to a prior $\beta^* = (\beta_1^*, \ldots, \beta_m^*)$ if and only if

$$\beta_l^* = \left(1 - \frac{q_{2l}}{p_{2l}}\right) \prod_{j=1}^{l-1} \frac{q_{2j}}{p_{2j}}, \qquad l = 1, \ldots, m.$$

By Corollary 6.3.3 we obtain that $\xi_{p,\beta}$ is $\Psi_p$-optimal with respect to the prior $\tilde{\beta} = (\tilde{\beta}_1, \ldots, \tilde{\beta}_m)$ if and only if

$$
\tilde{\beta}_l = \beta_l^* \frac{(\mathrm{eff}_l^{D_1}(\xi_{p,\beta}))^{-p}}{\displaystyle\sum_{j=1}^{m} \beta_j^*(\mathrm{eff}_j^{D_1}(\xi_{p,\beta}))^{-p}}
$$

$$
= \frac{2^{-2p(l-1)}\left(\prod_{i=1}^{l} q_{2i-2}p_{2i}\right)^{-p}\prod_{i=1}^{l-1}\frac{q_{2i}}{p_{2i}}\left(1 - \frac{q_{2l}}{p_{2l}}\right)}{\displaystyle\sum_{j=1}^{m} 2^{-2(j-1)p}\left(\prod_{i=1}^{j} q_{2i-2}p_{2i}\right)^{-p}\prod_{i=1}^{j-1}\frac{q_{2i}}{p_{2i}}\left(1 - \frac{q_{2j}}{p_{2j}}\right)}
$$

$(l = 1, \ldots, m)$. Since the map $\mathcal{H}_m$ from the set of all "priors" $\mathcal{B}_m$ in (6.2.15) onto the set $\mathcal{P}_m^*$ in (6.2.16) is one to one, it now follows from Corollary 6.3.3 that $\tilde{\beta} = \beta$. Solving these equations with respect to the canonical moments yields $p_{2l} = \frac{1}{2}$ if $\beta_l = 0$, and

$$
\frac{\beta_m}{\beta_l} = \frac{\tilde{\beta}_m}{\tilde{\beta}_l} = \frac{2^{-2p(m-l)}\prod_{i=l}^{m-1}\frac{q_{2i}}{p_{2i}}}{\prod_{i=l+1}^{m}(q_{2i-2}p_{2i})^p\left(1 - \frac{q_{2l}}{p_{2l}}\right)}
$$

whenever $\beta_l > 0$ $(l = 1, \ldots, m - 1)$. Thus for a given prior $\beta = (\beta_1, \ldots, \beta_m)$ the canonical moments of even order of the $\Psi_p$- optimal discriminating design with respect to $\beta$ have to satisfy the equations in (6.3.13) while the canonical moments of odd order are $\frac{1}{2}$ which is (6.3.12). The assertion of the theorem now follows by observing that the equation $\alpha(2x - 1) = (1 - x)^{1-p}$ has a unique solution in the interval $(0, 1)$ whenever $\alpha > 0$ and $p < 1$. ∎

**EXAMPLE 6.3.5**  If all components of the prior $\beta$ are positive, it can be shown that (6.3.13) is equivalent to

$$\beta_{l+1}(2p_{2l} - 1)\, p_{2l+2}^{1+p}2^{2p} = \beta_l(1 - p_{2l})^{1-p}(2p_{2l+2} - 1)$$

For example, let $m = 3, p = -1$, then the system of equations in (6.3.13) reduces to two quadratic equations for $p_2, p_4$ ($p_6 = 1$). The optimal designs are obtained from Corollary 4.2.2 for the interval $[-1, 1]$ and by a linear transformation for arbitrary intervals $[a, b]$. For example, if $\tilde{\beta}_U = (0, \frac{1}{2}, \frac{1}{2})$, then the even canonical moments of $\xi_{-1,\tilde{\beta}_U}$ are $p_2 = \frac{1}{2}$, $p_4 = (5 - \sqrt{5})/4$, $p_6 = 1$. The corresponding design on $[-1, 1]$ has masses 0.2043, 0.2957, 0.2957, and 0.2043 at the points $-1, -0.3931, 0.3931$, and 1, respectively. If the experimenter wants to discriminate between a linear and quadratic regression with some possibility of testing for a cubic trend he could use a prior with more weight for the quadratic model, such as $\beta^* = (0, \frac{4}{5}, \frac{1}{5})$. The $\Psi_{-1}-$ optimal discriminating design with respect to the prior $\beta^*$ has even canonical moments $p_2 = \frac{1}{2}$, $p_4 = (17 - \sqrt{17})/16$, $p_6 = 1$ and masses 0.2230, 0.2770, 0.2770, and 0.2230 at the points $-1, -0.3124, 0.3124$, and 1, respectively.

**Theorem 6.3.6**  Let $p = -\infty$, then the $\Psi_{-\infty}$-optimal discriminating design for the class $\mathcal{F}_m$ with respect to the prior $\beta$ is uniquely determined by its canonical moments $p_1, \ldots, p_{2m-1}, p_{2m} = 1$ where $p_{2i-1} = \frac{1}{2}$ $(i = 1, \ldots, m)$ and $p_{2m-2}, \ldots, p_2$ are defined recursively by

$$P_{2(m-j)} = \max\left\{ \kappa_{m-j}\left(1 - \left(\frac{1}{2}\right)^{2j} \prod_{i=m-j+1}^{m-1} (q_{2i}p_{2i})^{-1}\right), \frac{1}{2} \right\}, \qquad (6.3.14)$$

where

$$\kappa_j = \begin{cases} 1 & \text{if } \beta_j > 0, \\ 0 & \text{if } \beta_j = 0. \end{cases}$$

*Proof*  Consider the design $\xi_{-\infty,\beta}$ with canonical moments defined by (6.3.14), and let

$$\gamma_j := \kappa_{m-j}\left(1 - \left(\frac{1}{2}\right)^{2j} \prod_{i=m-j+1}^{m-1} (q_{2i}p_{2i})^{-1}\right) \qquad (j = 1, \ldots, m - 1).$$

If $\gamma_j \geq \frac{1}{2}$, then it is easy to see [observing (6.3.14)] that

$$\text{eff}^{D_1}_{m-j}(\xi_{-\infty,\beta}) = \text{eff}^{D_1}_m(\xi_{-\infty,\beta}).$$

On the other hand, if $\gamma_j < \frac{1}{2}$, then $p_{2(m-j)} = q_{2(m-j)} = \frac{1}{2}$, by (6.3.14), and it follows that $(1 - \gamma_j)^{-1} < 2$, which implies that

$$\frac{\operatorname{eff}_m^{D_1}(\xi_{-\infty,\beta})}{\operatorname{eff}_{m-j}^{D_1}(\xi_{-\infty,\beta})} = \frac{q_{2(m-j)}}{1 - \gamma_j} < 1,$$

whenever $\kappa_{m-j} = 1$. Consequently the set $\mathcal{N}^{D_1}(\xi_{-\infty,\beta})$ defined by formula (6.3.5) in Theorem 6.3.2 is given by $(\gamma_0 = 1)$,

$$\mathcal{N}^{D_1}(\xi_{-\infty,\beta}) = \{j \in \{1, \ldots, m\} \mid \gamma_{m-j} \geq \tfrac{1}{2}\}.$$

By Theorem 6.2.6 the design $\xi_{-\infty,\beta}$ maximizes the geometric weighted mean

$$\prod_{l=1}^{m} \left(\frac{|M_l(\xi)|}{|M_{l-1}(\xi)|}\right)^{\alpha_l},$$

where the weights $\alpha_l$ are defined in (6.2.17). If $l \notin \mathcal{N}^{D_1}(\xi_{-\infty,\beta})$, we have $\gamma_{m-l} < \frac{1}{2}$, $p_{2l} = \frac{1}{2}$, and this implies $\alpha_l = 0$. On the other hand, if $\gamma_{m-l} \geq \frac{1}{2}$, it follows from $p_{2l} \geq \frac{1}{2}$ that $\alpha_l \geq 0$. Therefore $\xi_{-\infty,\beta}$ is $\Psi_0$-optimal with respect to the prior $\alpha = (\alpha_1, \ldots, \alpha_m)$ and the assertion follows by an application of Corollary 6.3.3. ∎

**EXAMPLE 6.3.7** Consider the class $\mathcal{F}_4$ of polynomial regression models up to degree 4 with uniform prior $\beta_U = (\frac{1}{4}, \frac{1}{4}, \frac{1}{4}, \frac{1}{4})$. The odd canonical moments of the $\Psi_{-\infty}$-optimal discriminating design with respect to the prior $\beta_U$ are given by $p_{2i-1} = \frac{1}{2}$ $(i = 1, \ldots, 4)$, while the even canonical moments can be calculated recursively by (6.3.14),

$$p_8 = 1, \quad p_6 = \tfrac{3}{4}, \quad p_4 = \tfrac{2}{3}, \quad p_2 = \tfrac{5}{8}.$$

On the interval $[-1, 1]$, the corresponding design is obtained from Corollary 4.2.2 and given by (6.2.4). If we use the prior $\tilde{\beta}_U = (0, \frac{1}{3}, \frac{1}{3}, \frac{1}{3})$, then the even canonical moments of the $\Psi_{-\infty}$-optimal discriminating design with respect to $\tilde{\beta}_U$ are

$$p_8 = 1, \quad p_6 = \tfrac{3}{4}, \quad p_4 = \tfrac{2}{3}, \quad p_2 = \tfrac{1}{2},$$

and the corresponding design is given in (6.2.5) in Example 6.2.1. Note that $\Psi_{-\infty}$-optimal discriminating designs with respect to these "uniform priors" produce equal $D_1$-efficiencies in all models with positive weight (see Table 6.2.1).

**Theorem 6.3.8** *If $[a, b] = [-1, 1]$, then the $\Psi_{-\infty}$-optimal discriminating design $\xi_{-\infty,\beta}$ for the class $\mathcal{F}_m$ with respect to a prior with positive components puts masses*

$3/(2(m+2))$ *at the points* $-1$ *and* $1$ *and masses* $1/(m+2)$ *at the* $m-1$ *zeros of the derivative of the mth Chebyshev polynomial of the second kind* $U'_m(x)$. *The* $\Psi_{-\infty}$- *optimal design on the interval* $[a,b]$ *is obtained from* $\xi_{-\infty,\beta}$ *by the linear transformation* (5.5.9).

*Proof* By Theorem 6.3.6 the canonical moments of the $\Psi_{-\infty}$-optimal discriminating design with respect to the uniform prior $\beta_U$ are given by $p_{2i-1} = \frac{1}{2}$ $(i = 1, \ldots, m)$ and by

$$p_{2i} = \frac{m-i+2}{2(m-i)+2}, \qquad i = 1, \ldots, m,$$

where the last identity follows from (6.3.14) and an induction argument. The assertion is now an immediate consequence of Corollary 4.3.3 $(n = m, z = 2)$, (2.1.24), and (2.1.19) which show that

$$2C^{(2)}_{m-1}(x) = \frac{d}{dx} C^{(1)}_m(x) = U'_m(x) \qquad \blacksquare$$

## 6.4 MODEL ROBUST DESIGNS

As was pointed out in Section 6.1, a mixture of optimality criteria (or a composite optimality criterion) includes different aspects in one optimality criterion. In Sections 6.2 and 6.3 we considered a situation where the main interest of the experiment is the identification of a model that is assumed to belong to the class $\mathcal{F}_m$ of polynomials up to degree $m$. In this section we concentrate on the efficient estimation of the parameters in a polynomial regression with a degree only known to be less or equal than $m$. To this end let $\xi_l^D$ denote the $D$-optimal design on the interval $[a,b]$ for the polynomial regression (6.2.1) of degree $l$, and define for $l = 1, \ldots, m$

$$\text{eff}_l^D(\xi) = \left( \frac{|M_l(\xi)|}{|M_l(\xi_l^D)|} \right)^{1/(l+1)} \tag{6.4.1}$$

as the $D$-efficiency of the design $\xi$ in the polynomial regression model of degree $l$ (see Definition 5.4.6, $p = 0$, $K = I_{l+1}$). The value of the determinant in the denominator of (6.4.1) is obtained from (5.5.11) as

$$|M_l(\xi_l^D)| = \left( \frac{b-a}{2} \right)^{l(l+1)} \left( \frac{l}{2l-1} \right)^l \prod_{j=2}^{l} \left( \frac{(l-j+1)^2}{(2(l-j)+1)(2(l-j)+3)} \right)^{l+1-j}. \tag{6.4.2}$$

**Definition 6.4.1**   *Let* $-\infty \le p < 1$ *and* $\beta_1, \ldots, \beta_m$ *denote nonnegative numbers with sum equal one and* $\beta_m > 0$, *and let* $\xi_{p,\beta}$ *denote a design with nonsingular moment matrix* $M_m(\xi_{p,\beta})$. *The design* $\xi_{p,\beta}$ *is called* $\Phi_p$-*optimal for the class* $\mathcal{F}_m$ *with respect to the the prior* $\beta = (\beta_1, \ldots, \beta_m)$ *if and only if* $\xi_{p,\beta}$ *maximizes the weighted p-mean of D-efficiencies*

$$\Phi_p^\beta(\xi) = \left[ \sum_{l=1}^m \beta_l \left( \mathrm{eff}_l^D(\xi) \right)^p \right]^{1/p} = \left[ \sum_{l=1}^m \beta_l \left( \frac{|M_l(\xi)|}{|M_l(\xi_l^D)|} \right)^{p/(l+1)} \right]^{1/p}. \tag{6.4.3}$$

Again the geometric mean criterion is

$$\Phi_0^\beta(\xi) = \lim_{p \to 0} \Phi_p^\beta(\xi) = \prod_{l=1}^m \left( \frac{|M_l(\xi)|}{|M_l(\xi_l^D)|} \right)^{\beta_l/(l+1)}, \tag{6.4.4}$$

and the limit $p \to -\infty$ gives the maximin criterion

$$\Phi_{-\infty}^\beta(\xi) = \lim_{p \to -\infty} \Phi_p^\beta(\xi) = \min\{\mathrm{eff}_l^D(\xi) | \beta_l > 0, l = 1, \ldots, m\}. \tag{6.4.5}$$

The calculation of the $\Phi_p$-optimal designs for the class $\mathcal{F}_m$ with respect to the prior $\beta$ is performed by similar arguments as presented in Sections 6.2 and 6.3 for the model discrimination designs, and we present the main results without giving detailed proofs. The interested reader is referred to the work of Dette and Studden (1995).

**Theorem 6.4.2**   *Let* $p \in (-\infty, 1)$, *then the* $\Phi_p$-*optimal design for the class* $\mathcal{F}_m$ *with respect to the prior* $\beta$ *is uniquely determined by its canonical moments* $(\frac{1}{2}, p_2, \frac{1}{2}, \ldots, p_{2m-2}, \frac{1}{2}, 1)$, *where* $(p_2, \ldots, p_{2m-2})$ *is the unique solution of the system of equations*

$$\frac{\beta_{l+1}}{l+2} \left( 1 - 2\frac{q_{2l}}{p_{2l}} + \frac{q_{2l}q_{2l+2}}{p_{2l}p_{2l+2}} \right) \left( \prod_{j=1}^{l+1} (q_{2j-2}p_{2j})^j \right)^{p/[(l+1)(l+2)]} \tag{6.4.6}$$

$$= \frac{\beta_l}{l+1} \frac{q_{2l}}{p_{2l}} \left( 1 - 2\frac{q_{2l+2}}{p_{2l+2}} + \frac{q_{2l+2}q_{2l+4}}{p_{2l+2}p_{2l+4}} \right) C_l^p (l = 1, \ldots, m-1),$$

*which satisfies*

$$1 - 2\frac{q_{2l}}{p_{2l}} + \frac{q_{2l}q_{2l+2}}{p_{2l}p_{2l+2}} \ge 0, \qquad l = 1, \ldots, m. \tag{6.4.7}$$

*Here*

$$C_l = \frac{|M_{l+1}(\xi_{l+1}^D)|^{1/(l+2)}}{|M_l(\xi_l^D)|^{1/(l+1)}} \tag{6.4.8}$$

$$= \left[\frac{l^{l^2}(l+1)^{(l+1)^2}(2l-1)^l}{(2l+1)^{(l+1)(2l+1)}} \prod_{j=2}^{l}\left\{\frac{(l+1-j)^2}{(2(l-j)+1)(2(l-j)+3)}\right\}^{-(l+1-j)}\right]^{1/[(l+1)(l+2)]}$$

$(l = 1, \ldots, m-1)$, *and the lth equation in* (6.4.6) *has to be replaced by the equation*

$$1 - 2\frac{q_{2l}}{p_{2l}} + \frac{q_{2l}q_{2l+2}}{p_{2l}p_{2l+2}} = 0 \tag{6.4.9}$$

*whenever* $\beta_l = 0$ $(l = 1, \ldots, m-1)$.

**Theorem 6.4.3**  *The $\Phi_0$-optimal design for the class $\mathcal{F}_m$ with respect to the prior $\beta = (\beta_1, \ldots, \beta_m)$ is uniquely determined by its canonical moment sequence*

$$p_{2i-1} = \frac{1}{2}, \qquad i = 1, \ldots, m,$$

$$p_{2i} = \frac{\sigma_i}{\sigma_i + \sigma_{i+1}}, \qquad i = 1, \ldots, m-1,$$

$$p_{2m} = 1,$$

*where the numbers $\sigma_i$ are defined by*

$$\sigma_i = \sum_{l=i}^{m}\frac{l+1-i}{l+1}\beta_l, \qquad i = 1, \ldots, m.$$

**Theorem 6.4.4.**  *Let for $l = 1, \ldots m$ and $z \geq 0$*

$$\beta_l(z) = \frac{l+1}{m+2z-1}\frac{\Gamma(m-l+z-1)\Gamma(z+1)\Gamma(m)}{\Gamma(m-l+1)\Gamma(z-1)\Gamma(m+z-1)}, \tag{6.4.10}$$

$\beta(z) = (\beta_1(z), \ldots, \beta_m(z))$. *If $[a, b] = [-1, 1]$, then the $\Phi_0$-optimal design for the class $\mathcal{F}_m$ with respect to the prior $\beta(z)$ puts masses $(z+1)/(2(m+z))$ at the points $-1, 1$ and masses $1/(m+z)$ at the $m-1$ zeros of the $(m-1)$th ultraspherical polynomial $C_{m-1}^{(z/2+1)}(x)$. On the interval $[a, b]$, the $\Phi_0$-optimal discriminating design for the class $\mathcal{F}_m$ with respect to the prior $\beta(z)$ can be obtained by the linear transformation* (5.5.9).

We will now concentrate on the maximin criterion (6.4.5) for which the solution of the design problem is more transparent. The following discussion will make frequent use of the quantities

$$
a_l = \begin{cases} \dfrac{(l+1)^{l+1}(2l-1)^{2l-1}}{(l-1)^{l-1}(2l+1)^{2l+1}} & \text{if } l = 2, \\[4mm] \dfrac{2^4}{3^6} & \text{if } l = 1. \end{cases} \tag{6.4.11}
$$

**Theorem 6.4.5**  *The $\Phi_{-\infty}$-optimal design $\xi_{-\infty}$ for the class $\mathcal{F}_m$ with respect to a prior $\beta$ with positive components is uniquely determined by its canonical moments $(\frac{1}{2}, p_2, \frac{1}{2}, \ldots, \frac{1}{2}, p_{2m-2}, \frac{1}{2}, 1)$, where the canonical moments (of even order) $p_2, \ldots, p_{2m-2}$ are given by the continued fractions*

$$
p_{2l} = 1 - \frac{a_l}{|1} - \frac{a_{l+1}}{|1} - \cdots - \frac{a_{m-1}}{|1} \qquad (l = 2, \ldots, m-1) \tag{6.4.12}
$$

*and $p_2$ is the largest root in the interval $[0, 1]$ of the equation*

$$
p_2(1 - p_2)^2 = \frac{16}{729 p_4^2}. \tag{6.4.13}
$$

**EXAMPLE 6.4.6**  Consider the class $\mathcal{F}_3$ of polynomials up to degree 3 and a given prior $\beta^U = (\frac{1}{3}, \frac{1}{3}, \frac{1}{3})$ for the linear, quadratic and cubic regression. Here Theorem 6.4.2 gives two equations for $(p_2, p_4)$,

$$
\left(1 - 2\frac{q_2}{p_2} + \frac{q_2 q_4}{p_2 p_4}\right)(p_2(q_2 p_4)^2)^{p/6} = \frac{3}{2}\frac{q_2}{p_2}\left(1 - 2\frac{q_4}{p_4}\right) \cdot C_1^p,
$$

$$
\left(1 - \frac{2q_4}{p_4}\right)\{p_2(q_2 p_4)^2 q_4^3\}^{p/12} = \frac{4}{3}\frac{q_4}{p_4} \cdot C_2^p, \tag{6.4.14}
$$

and $p_2, p_4$ have to satisfy (6.4.7). The optimal design puts masses $\alpha, \frac{1}{2} - \alpha, \frac{1}{2} - \alpha$, and $\alpha$ at the points $-1, -t, t$, and 1, where $t = p_2 q_4$ and $\alpha = p_2 p_4/(2(q_2 + p_2 p_4))$ (see Corollary 4.2.2). The solution of (6.4.14) was determined using the Newton-Raphson algorithm, which gives the $\Phi_p$-optimal design in the case $-\infty < p < 1$. The remaining case $p = -\infty$ can be obtained from Theorem 6.4.5. The corresponding designs and $D$-efficiencies are given in Table 6.4.1, which contains in its first row the $D$-optimal design for the cubic polynomial. Note that $\Phi_p$-optimal designs are relatively robust with respect to different values of the parameter $p$ in the optimality criterion.

**Table 6.4.1**   $D$- and $\Phi_{p,\beta}$-optimal designs for the class $\mathcal{F}_3$ of polynomials up to degree 3 with respect to a uniform prior

| $p$ | $\xi(\pm 1)$ | $\xi(\pm t)$ | $t$ | eff$_1^D$ | eff$_2^D$ | eff$_3^D$ |
|---|---|---|---|---|---|---|
| $\xi_3^D$ | 0.25 | 0.25 | 0.44721 | 0.7746 | 0.8653 | 1.0 |
| $\xi_{0,\beta^u}$ | 0.31944 | 0.18056 | 0.40105 | 0.8348 | 0.9143 | 0.9542 |
| $\xi_{p,\beta^u}$ | 0.32345 | 0.17655 | 0.40059 | 0.8388 | 0.9134 | 0.9494 |
| $\xi_{-2,\beta^u}$ | 0.32703 | 0.17297 | 0.40047 | 0.8423 | 0.9141 | 0.9448 |
| $\xi_{-3,\beta^u}$ | 0.33021 | 0.16979 | 0.40059 | 0.8455 | 0.9137 | 0.9407 |
| $\xi_{-\infty,\beta^u}$ | 0.36634 | 0.13366 | 0.42695 | 0.8840 | 0.8840 | 0.8840 |

Note: First two columns: weights; third column: interior positive support point.

**Table 6.4.2**   $D$- and $\Phi_{p,\tilde{\beta}}$-optimal designs for the class $\mathcal{F}_3$ of polynomials up to degree 3 with respect to the prior $\tilde{\beta} = \left(\frac{3}{16}, \frac{12}{16}, \frac{1}{16}\right)$

| $p$ | $\xi(\pm 1)$ | $\xi(\pm t)$ | $t$ | eff$_1^D$ | eff$_2^D$ | eff$_3^D$ |
|---|---|---|---|---|---|---|
| $\xi_3^D$ | 0.25 | 0.25 | 0.44721 | 0.7746 | 0.8653 | 1.0 |
| $\xi_{0,\beta^u}$ | 0.34167 | 0.15833 | 0.19124 | 0.8336 | 0.9833 | 0.7327 |
| $\xi_{-1,\beta^u}$ | 0.34178 | 0.15822 | 0.21194 | 0.8353 | 0.9758 | 0.7645 |
| $\xi_{-2,\beta^u}$ | 0.34228 | 0.15772 | 0.22807 | 0.8372 | 0.9719 | 0.7864 |
| $\xi_{-3,\beta^u}$ | 0.34304 | 0.15696 | 0.24122 | 0.8392 | 0.9684 | 0.8025 |
| $\xi_{-\infty,\beta^u}$ | 0.36634 | 0.13366 | 0.42695 | 0.8840 | 0.8840 | 0.8840 |

Note: First two columns: weights; third column: interior positive support point.

Obviously this property of the $\Phi_p$-optimal designs also depends on the given prior $\beta$. As an example of a stronger dependence of the design $\xi_{p,\beta}$ on the parameter $p$, we consider the case $m = 3$ (linear, quadratic, or cubic regression) and the prior $\tilde{\beta}_1 = \frac{3}{16}, \tilde{\beta}_2 = \frac{12}{16}, \tilde{\beta}_3 = \frac{1}{16}$ (more weight on the linear and quadratic model). The results are listed in Table 6.4.2.

**Theorem 6.4.7**   Let $\beta = (0, \ldots, 0, \beta_k, \ldots, \beta_m)$ $(2 \leq k \leq m)$. The $\Phi_{-\infty}$-optimal design for the class $\mathcal{F}_m$ with respect to the prior $\beta$ is uniquely determined by its canonical moments $p_2, \ldots, p_{2m}$, where $p_{2k+2}, \ldots, p_{2m}$ are given by the continued fractions

$$p_{2l} = 1 - \frac{a_l}{|1} - \cdots - \frac{a_{m-2}}{|1} - \frac{a_{m-1}}{|1}, \qquad l = k+1, \ldots, m$$

[*with $a_l$ defined in (6.4.11)*], *and* $p_{2k}, \ldots, p_2$ *are the unique solution of the system*

$$p_{2l} = \frac{3p_{2l+2} - 1}{4p_{2l+2} - 1}, \qquad l = k - 1, \ldots, 1,$$

$$\frac{C_k^{(k+1)(k+2)}}{(p_{2k+2})^{k+1}} = \prod_{j=1}^{k} (q_{2j-2} p_{2j})^j (q_{2k})^{k+1}$$

[*with $C_k$ defined in (6.4.8)*].

# CHAPTER 7

# Applications in Approximation Theory

## 7.1 INTRODUCTION

Chapters 5 and 6 illustrated several applications of the theory of canonical moments in the field of optimal approximate design for polynomial regression which is closely related to the theory of orthogonal polynomials, an important field in approximation theory. In the present chapter we discuss some applications of canonical moments and the theory of optimal design in this area. Section 7.2 demonstrates how canonical moments can be used for deriving the limit distribution of the zeros of the Jacobi polynomials $P_n^{(\alpha_n, \beta_n)}(x)$ when the degree and the parameters tend to infinity. The zeros of these polynomials are used for numerical quadrature and (as illustrated in Chapters 5 and 6) in the theory of optimal design. As a second application we derive some identities for sums of squares of orthogonal polynomials. These generalize the well-known trigonometric identity and have applications in mathematical physics. Finally Section 7.4 considers a generalization of the classical Chebyshev approximation problem

$$\min_{a_0, \ldots, a_{m-1}} \sup_{x \in [-1, 1]} \left| x^m - \sum_{j=0}^{m-1} a_j x^j \right| \qquad (7.1.1)$$

which investigates the best approximation of $x^m$ (with respect to the sup-norm on the interval $[-1, 1]$) by polynomials of degree $m - 1$. While the results of Section 7.2 are based directly on the representation of the canonical moments of the uniform distribution on the zeros of Jacobi polynomials in Theorem 4.3.1, our approach in Sections 7.3 and 7.4 combines the general theory of equivalence and duality for approximate designs in Section 5.4 with the concept of canonical moments and orthogonal polynomials.

## 7.2 ASYMPTOTIC ZERO DISTRIBUTION OF ORTHOGONAL POLYNOMIALS

A very simple and important set of orthogonal polynomials are the Chebyshev polynomials of the first kind

$$T_n(x) = \cos(n \arccos x) \qquad (x \in [-1, 1]), \tag{7.2.1}$$

which are defined in Example 2.1.1 and are orthogonal with respect to the arc-sine distribution with density

$$\frac{1}{\pi\sqrt{1 - x^2}}, \qquad x \in (-1, 1).$$

The recurrence relation for these polynomials is given in (2.1.7) and the $n$ zeros of $T_n(x)$ can be calculated explicitly as

$$x_{k,n} = \cos\left(\frac{2k - 1}{2n} \pi\right), \qquad k = 1, \dots, n. \tag{7.2.2}$$

For $y \in [-1, 1]$ let

$$N_n(y) := \frac{1}{n} \#\{x \le y \mid T_n(x) = 0\} \tag{7.2.3}$$

denote the relative proportion of zeros of $T_n(x)$ that are less than or equal to $y$. Obviously the function $N_n(y)$ defines a distribution function on the interval $[-1, 1]$ and the asymptotic behavior of this function is of some interest. From

$$\lim_{n\to\infty} N_n(y) = \lim_{n\to\infty} \sum_{x_{k,n} \le y} \frac{1}{n} = \lim_{n\to\infty} \frac{1}{n} \sum_k I\left\{k \ge \frac{n}{\pi} \arccos y + \frac{1}{2}\right\}$$

$$= 1 - \frac{1}{\pi} \arccos y = \frac{1}{\pi} \int_{-1}^{y} \frac{dx}{\sqrt{1 - x^2}}, \tag{7.2.4}$$

it follows that the distribution function $N_n$ converges to the distribution function of the arc-sine measure. In other words, the probability measure with equal mass at the zeros of the Chebyshev polynomials of the first kind converges weakly to the arc-sine measure.

Note that for a general system of orthogonal polynomials the derivation in (7.2.4) is not possible because it requires precise bounds on the zeros of the $n$th orthogonal polynomial. However, there is a simple alternative derivation of (7.2.4) that is based on canonical moments and can be transferred to a broader class of orthogonal polynomials. Let $\xi_n$ denote the probability measure with equal mass at the zeros of $T_n(x)$, i.e.

$$N_n(z) = \int_{-1}^{z} d\xi_n(t).$$

By (2.1.17), the $n$th Chebyshev polynomial $T_n(x)$ and the Jacobi polynomials $P_n^{(-1/2,-1/2)}(x)$ differ only by a constant factor. In other words, both polynomials have the same zeros. By part (2) of Theorem 4.3.1 the measure $\xi_n$ has canonical moments $p_j^{(n)} = \frac{1}{2}$, $(j = 1, \ldots, 2n-1)$, $p_{2n}^{(n)} = 0$. This implies that

$$\lim_{n \to \infty} p_j^{(n)} = p_j = \tfrac{1}{2} \qquad j \geq 1. \tag{7.2.5}$$

The mapping between the ordinary and canonical moments is one to one and continuous, and it follows that the moments of $\xi_n$ converge to the moments of the distribution corresponding to the sequence of canonical moments $(\frac{1}{2}, \frac{1}{2}, \frac{1}{2}, \ldots)$. Example 3.6.8 shows that the probability measure corresponding to this sequence is the arc-sine measure which is determined by its moments. Now well-known methods of probability theory (Feller, 1966, p. 263) show that $\xi_n$ converges weakly to the arc-sine distribution, because the moments are converging.

Surprisingly the limit relation (7.2.4) is not only valid for the Chebyshev polynomials of the first kind but for a very large class of orthogonal polynomials on the interval $[-1, 1]$. General results regarding the asymptotic zero distribution of orthogonal polynomials can be obtained from logarithmic potential theory. Roughly speaking, the asymptotic zero distribution of a system of orthogonal polynomials on a compact set is given by the equilibrium distribution of the logarithmic energy. The arc-sine measure is actually the equilibrium measure of the logarithmic energy on the interval $[-1, 1]$. Consequently it appears as the limit distribution of the zeros of many systems of orthogonal polynomials on the interval $[-1, 1]$. For more details of this approach, we refer to the work of Stahl and Totik (1992).

In what follows we will use the canonical moment approach in order to investigate the asymptotic zero distribution of the generalized Jacobi polynomials $P_n^{(\alpha_n,\beta_n)}(x)$ where $\alpha_n, \beta_n > -1$. The term generalized is used in order to emphasize that the sequence

$$\left\{ P_n^{(\alpha_n,\beta_n)}(x) \right\}_{n=0,1,2,\ldots}$$

does not necessarily define a sequence of orthogonal polynomials because the parameters $\alpha_n, \beta_n$ depend on the degree $n$. However, $P_n^{(\alpha_n,\beta_n)}(x)$ is the $n$th orthogonal polynomial with respect to the (fixed) measure $(1-x)^{\alpha_n}(1+x)^{\beta_n}dx$, so by Lemma 2.1.3 all of the zeros of $P_n^{(\alpha_n,\beta_n)}(x)$ are real, simple, and located in the interval $(-1, 1)$. In what follows let

$$N_n^{(\alpha_n,\beta_n)}(y) = \frac{1}{n}\#\left\{x \le y \mid P_n^{(\alpha_n,\beta_n)}(x) = 0\right\} \tag{7.2.6}$$

denote the relative proportion of zeros of $P_n^{(\alpha_n,\beta_n)}(x)$ that are less than or equal to $y$.

**Theorem 7.2.1** *If* $\lim_{n\to\infty} \frac{\alpha_n}{n} = a$ *and* $\lim_{n\to\infty} \frac{\beta_n}{n} = b$ $(a, b \ge 0)$, *then*

$$\lim_{n\to\infty} N_n^{(\alpha_n,\beta_n)}(y) = \frac{2+a+b}{2\pi} \int_{r_1}^{y} \frac{\sqrt{(r_2 - x)(x - r_1)}}{1 - x^2} dx, \qquad r_1 \le y \le r_2,$$

*where*

$$r_{1,2} := \frac{b^2 - a^2 \pm 4\sqrt{(a+1)(b+1)(a+b+1)}}{(2+a+b)^2}.$$

*Proof* Let $\xi_n^{(\alpha_n,\beta_n)}$ denote the uniform distribution on the zeros of the $n$th generalized Jacobi polynomial $P_n^{(\alpha_n,\beta_n)}(x)$. By part 2 of Theorem 4.3.1 the canonical moments of $\xi^{(\alpha_n,\beta_n)}$ are given by

$$p_{2i}^{(n)} = \frac{n - i}{2(n - i) + \alpha_n + \beta_n + 1}, \qquad i = 1, \ldots, n,$$

$$\tag{7.2.7}$$

$$p_{2i-1}^{(n)} = \frac{\beta_n + n - i + 1}{2(n - i + 1) + \alpha_n + \beta_n}, \qquad i = 1, \ldots, n,$$

and this implies for all $i \in \mathbb{N}$ that

$$h = \lim_{n\to\infty} p_{2i}^{(n)} = \frac{1}{2 + a + b} = p_{2i},$$

$$\tag{7.2.8}$$

$$g = \lim_{n\to\infty} p_{2i-1}^{(n)} = \frac{b+1}{2 + a + b} = p_{2i-1}.$$

The result now follows from Theorem 4.5.2, Theorem 1.3.5, and Theorem 1.3.2. More precisely, Theorem 4.5.2 $(g + h \le 1, g \ge h)$ shows that the measure $\tilde{\xi}^{(s)}$ corresponding to the sequence

$$\tfrac{1}{2}, g, \tfrac{1}{2}, h, \tfrac{1}{2}, g, \tfrac{1}{2}, \ldots$$

on the interval $[-1, 1]$ is absolute continuous with density

$$\frac{1}{2\pi h} \frac{\sqrt{4\mu - (x^2 - \eta)^2}}{|x|(1 - x^2)} I\{|x^2 - \eta| < 2\sqrt{\mu}\},$$

where $\eta = g(1 - h) + h(1 - g)$, $\mu = g(1 - g)h(1 - h)$. By Theorem 1.3.5 the measure $\tilde{\xi}$ corresponding to the sequence (7.2.8) on the interval $[0, 1]$ is related to $\tilde{\xi}^{(s)}$ by $\tilde{\xi}([0, x]) = \tilde{\xi}^{(s)}([-\sqrt{x}, \sqrt{x}])$. Therefore the density of $\tilde{\xi}$ is given by

$$h(x) = \frac{1}{2h\pi} \frac{\sqrt{4\mu - (x - \eta)^2}}{x(1 - x)} I\{|x - \eta| < 2\sqrt{\mu}\}. \qquad (7.2.9)$$

Since all zeros of $P_n^{(\alpha_n, \beta_n)}(x)$ are located in the interval $(-1, 1)$, the limit distribution $\xi$ satisfies supp$(\xi) \subset (-1, 1)$. By Theorem 1.3.2, $\xi$ is induced through $\tilde{\xi}$ by the linear transformation $y = 2x - 1$, and the assertion follows by transforming the density in (7.2.9) onto $[-1, 1]$,

$$\frac{d\xi}{dx} = \frac{1}{2} h\left(\frac{y + 1}{2}\right) = \frac{1}{2h\pi} \frac{\sqrt{16\mu - (y + 1 - 2\eta)^2}}{(1 - y^2)} I\{|y + 1 - 2\eta| < 4\sqrt{\mu}\}$$

and observing that $2\eta - 1 \pm 4\sqrt{\mu} = r_{1,2}$.                                    ∎

Note that for the special choice $\lim_{n\to\infty}(\alpha_n/n) = \lim_{n\to\infty}(\beta_n/n) = 0$, Theorem 7.2.1 gives the arc-sine distribution as limit, which shows that the asymptotic distribution of the Jacobi polynomials $P_n^{(\alpha,\beta)}(x)$ is still the arc-sine measure independent of the parameters $\alpha, \beta > -1$.

The following results provide the limit distribution of the zeros of the polynomial $P_n^{(\alpha_n, \beta_n)}(x)$ if the parameters $\alpha_n$ and $\beta_n$ tend to infinity with a larger order than $n$. The proofs are similar to the proof of Theorem 7.2.1 and are left to the reader.

**Theorem 7.2.2**   *Let* $\lim_{n\to\infty} \alpha_n/n = \infty$, $\lim_{n\to\infty} \beta_n/n = \infty$ *and* $\lim_{n\to\infty} \alpha_n/\beta_n = c > 0$; *then*

$$\lim_{n\to\infty} N_n^{(\alpha_n, \beta_n)}\left(\sqrt{\frac{n}{\alpha_n}} y - \frac{\alpha_n - \beta_n}{\alpha_n + \beta_n}\right) = \frac{2}{\pi\sigma^2} \int_{-\sigma}^{y} \sqrt{\sigma^2 - x^2} dx, \qquad |y| \leq \sigma,$$

*where* $\sigma = 4c/(1 + c)^{3/2}$.

**Theorem 7.2.3**   *Let* $\lim_{n\to\infty} \alpha_n/n = \infty$ *and* $\lim_{n\to\infty} \beta_n/n = b \geq 0$; *then*

$$\lim_{n\to\infty} N_n^{(\alpha_n, \beta_n)}\left(\frac{n}{\alpha_n} y - 1\right) = \frac{1}{4\pi} \int_{s_1}^{y} \frac{\sqrt{(s_2 - x)(x - s_1)}}{x} dx, \qquad s_1 \leq y \leq s_2,$$

*where* $s_{1,2} = 2(2 + b) \pm 4\sqrt{1 + b}$.

**Theorem 7.2.4** *Let* $\lim_{n\to\infty} \alpha_n/n = \infty$, $\lim_{n\to\infty} \beta_n/n = \infty$ *and* $\lim_{n\to\infty} \alpha_n/\beta_n = \infty$; *then*

$$\lim_{n\to\infty} N_n^{(\alpha_n,\beta_n)} \left( \frac{4\sqrt{n\beta_n}}{\alpha_n} y - \frac{\alpha_n - \beta_n}{2n + \alpha_n + \beta_n} \right) = \frac{2}{\pi} \int_{-1}^{y} \sqrt{1 - x^2}\, dx$$

*for all* $-1 \leq y \leq 1$.

## 7.3  IDENTITIES FOR ORTHOGONAL POLYNOMIALS

Our main interest in this section are some generalizations of the trigonometric identity

$$\sin^2 m\theta + \cos^2 m\theta = 1, \tag{7.3.1}$$

which actually is an identity for squares of orthogonal polynomials. To see this, we use the trigonometric representation for the Chebyshev polynomials of the first and second kind defined in (2.1.5) and (2.1.6) and rewrite (7.3.1) as

$$(1 - x^2)U_{m-1}^2(x) + T_m^2(x) = 1 \qquad \forall x \in \mathbb{R}. \tag{7.3.2}$$

Note that the Chebyshev polynomials of the first kind $T_n(x)$ are orthogonal with respect to the arc-sine distribution $\mu^T$ with density

$$w^T(x) = \frac{1}{\pi}(1 - x^2)^{-1/2},$$

while the polynomials of the second kind are orthogonal with respect to a measure $\mu^U$ with density

$$w^U(x) = (1 - x^2)w^T(x) = \frac{1}{\pi}\sqrt{1 - x^2}.$$

To illustrate the main ideas of this section, we will present a proof of the trigonometric identity (7.3.1) [or equivalently (7.3.2)] that is based on the approximate theory of optimal design.

Let $\Xi$ denote the set of all probability measures on the interval $[-1, 1]$, $\boldsymbol{f}_m(x) = (1, x, \ldots, x^m)^T$ the vector of monomials up to the order $m$, and for $\xi \in \Xi$,

$$M_m(\xi) = \int_{-1}^{1} \boldsymbol{f}_m(x)\boldsymbol{f}_m^T(x)d\xi(x) \tag{7.3.3}$$

the moment or Hankel matrix of order $2m$. Consider the $D_1$-optimal design problem in a univariate polynomial regression of degree $m$ on the interval $[-1, 1]$, which is the problem of maximizing

$$(e_m^T M_m^{-1}(\xi) e_m)^{-1} = \frac{|M_m(\xi)|}{|M_{m-1}(\xi)|} \tag{7.3.4}$$

(see the end of Section 5.5) over the set of all probability measures on the interval $[-1, 1]$. Here $e_j = (0, \ldots, 0, 1)^T$ is the $(j + 1)$th unit vector in $\mathbb{R}^{j+1}$ $(j = 0, \ldots, m)$. The $D_1$-optimal design $\xi_m^{D_1}$ is given in Theorem 5.5.4. The general equivalence Theorem 5.4.9 ($p = -1$, $K = e_m$) shows that a design $\xi$ maximizes (7.3.4) if and only if the inequality

$$(e_m^T M_m^{-1}(\xi) f_m(x))^2 \le e_m^T M_m^{-1}(\xi) e_m \tag{7.3.5}$$

holds for all $x \in [-1, 1]$ and there is equality in (7.3.5) for all support points of the $D_1$-optimal design. Obviously the function $\hat{P}_m(x) = e_m^T M_m^{-1}(\xi) f_m(x)$ defines a polynomial of degree $m$, and the identity

$$\int_{-1}^1 \hat{P}_m(x) f_m^T(x) d\xi(x) = e_m^T M_m^{-1}(\xi) \int_{-1}^1 f_m(x) f_m^T(x) d\xi(x) = e_m^T$$

shows that $\hat{P}_m(x)$ is the $m$th orthogonal polynomial with respect to the measure $d\xi(x)$. Since this orthogonality does not depend on special properties of the $D_1$-optimal design, we summarize this result in the following lemma:

**Lemma 7.3.1** *Let $\xi$ denote a measure on the interval $[-1, 1]$ such that $|M_m(\xi)| \ne 0$. The polynomials*

$$P_l(x, \xi) = \left(e_l^T M_l^{-1}(\xi) e_l\right)^{-1/2} e_l^T M_l^{-1}(\xi) f_l(x), \qquad l = 0, \ldots, m, \tag{7.3.6}$$

*are orthonormal with respect to the measure $d\xi(x)$.*

Observing the characterization of the $D_1$-optimal design $\xi_m^{D_1}$ by the inequality (7.3.5) and Lemma 7.3.1, it follows that the $m$th orthonormal polynomial $P_m(x, \xi_m^{D_1})$ with respect to the measure $d\xi_m^{D_1}(x)$ satisfies

$$P_m^2(x, \xi_m^{D_1}) \le 1 \qquad \forall x \in [-1, 1]. \tag{7.3.7}$$

It is shown in the proof of Theorem 5.5.4 that the canonical moments of the $D_1$-optimal design are given by $p_j = \frac{1}{2}$ $j = 1, \ldots, 2m - 1$ and $p_{2m} = 1$ [see (5.5.14)] and that they coincide with the first $2m - 1$ canonical moments of the arc-sine distribution. By Corollary 2.3.5 this means that $P_m(x, \xi_m^{D_1})$ is proportional to the $m$th orthogonal polynomial $T_m(x)$ with respect to the arc-sine distribution. Discussing equality in (7.3.5) [or equivalently in (7.3.7)] shows that $P_m(1, \xi_m^{D_1}) = 1 = T_m(1)$ and $P_m(x, \xi_m^{D_1})$ coincides with the $m$th Chebyshev polynomial of the first kind. The inequality (7.3.7) can now be rewritten as a well-known property of the Chebyshev polynomial of the first kind, that is,

$$T_m^2(x) \leq 1 \qquad \forall x \in [-1, 1]. \tag{7.3.8}$$

Moreover by Theorem 5.4.9 there is equality in (7.3.8) for all $m + 1$ support points of the $D_1$-optimal design, which are given by the zeros of the polynomial $(1 - x^2) U_{m-1}(x)$, by Theorem 5.5.4. Therefore the polynomials $T_m^2(x) - 1$ and $(x^2 - 1) U_{m-1}^2(x)$ are of degree $2m$, nonpositive on $[-1, 1]$, and vanish at the $m + 1$ support points of the $D_1$-optimal design. A simple counting argument shows that they must be proportional. Since both polynomials have leading coefficient $2^{2m-2}$, they are identical, which proves the trigonometric identity (7.3.1).

Although this proof appears to be complicated, it provides the possibility of further extensions of the "trigonometric identity" to more general systems of orthogonal polynomials on the interval $[-1, 1]$. In order to present a transparent representation, we restrict ourselves to the symmetric case. For systems of polynomials orthogonal with respect to a nonsymmetric measure the reader is referred to Dette (1993b). A careful inspection shows that the derivation of (7.3.7) is based on three main arguments:

- The optimal design problem (here $D_1$-optimality).
- The equivalence theorem, more precisely an inequality of the form (7.3.5).
- The explicit solution of the design problem in Theorem 5.5.4.

It turns out that the appropriate generalization of the $D_1$-optimality criterion is the geometric mean of $D_1$-efficiencies $\Psi_0^\beta$ defined in Definition 6.2.2,

$$\Psi_0^\beta(\xi) = \prod_{l=1}^m \left( \frac{|M_l(\xi)|}{|M_{l-1}(\xi)|} \right)^{\beta_l} = \prod_{l=1}^m (e_l^T M_l^{-1}(\xi) e_l)^{-\beta_l}. \tag{7.3.9}$$

Note that the constant in the definition (6.2.10) has no impact on the optimization problem and is omitted for the sake of simplicity. The prior $\beta = (\beta_1, \ldots, \beta_m)^T$ varies in the set $\mathcal{B}_m$ defined by (6.2.15); that is, $(\beta_1, \ldots, \beta_m)$ are real weights such that

$$\sigma_i = \sum_{l=i}^m \beta_l > 0, \quad 1 \leq i \leq m, \quad \sigma_m = \sum_{l=1}^m \beta_l = 1. \tag{7.3.10}$$

The probability measure which maximizes (7.3.9) can be characterized by the Equivalence Theorem 6.3.2 ($p = 0$). However, this characterization is only valid for positive weights because its derivation requires the concavity of the function $\Psi_0^\beta$. For more general weights satisfying (7.3.10) the equivalent condition (6.3.4) is only necessary.

**Lemma 7.3.2** *If* $(\beta_1, \ldots, \beta_m) \in \mathcal{B}_m$ *and* $\xi^*$ *maximizes the function* $\Psi_0^\beta(\xi)$ *over the class of all probability measures on the interval* $[-1, 1]$, *then for all* $x \in [-1, 1]$,

$$\sum_{l=1}^{m} \beta_l \frac{(e_l^T M_l^{-1}(\xi^*) f_l(x))^2}{e_l^T M_l^{-1}(\xi^*) e_l} \leq 1, \tag{7.3.11}$$

*with equality for the support points of* $\xi^*$.

*Proof* For a probability measure $\xi$ on the interval $[-1, 1]$ with $|M_m(\xi)| \neq 0$, define

$$\Phi(\xi) = \log \Psi_0^\beta(\xi) = -\sum_{l=1}^{m} \beta_l \log e_l^T M_l^{-1}(\xi) e_l.$$

Let

$$D_\Phi(\xi, \eta) = \frac{d}{d\alpha} \Phi((1 - \alpha)\xi + \alpha\eta)|_{\alpha=0+} \tag{7.3.12}$$

denote the *directional derivative* of the function $\Phi$ at $\xi$ in the direction of $\eta$. For a matrix $A = (a_{ij})$ we define its derivative by differentiating the elements, that is,

$$\frac{\partial}{\partial t} A = \left(\frac{\partial}{\partial t} a_{ij}\right)_{ij}.$$

It then follows for a nonsingular square matrix that

$$\frac{\partial}{\partial t} A^{-1} = -A^{-1} \frac{\partial}{\partial t} A \, A^{-1}.$$

This implies that for $l = 1, \ldots, m$,

$$\frac{d}{d\alpha} \log \left[ e_l^T \{(1 - \alpha)M_l(\xi) + \alpha M_l(\eta)\}^{-1} e_l \right] \Big|_{\alpha=0+}$$

$$= 1 - \frac{e_l^T M_l^{-1}(\xi) M_l(\eta) M_l^{-1}(\xi) e_l}{e_l^T M_l^{-1}(\xi) e_l},$$

and consequently the directional derivative of $\Phi$ at $\xi$ in the direction of $\eta$ is

$$D_\Phi(\xi, \eta) = -1 + \sum_{l=1}^{m} \beta_l \frac{e_l^T M_l^{-1}(\xi) M_l(\eta) M_l^{-1}(\xi) e_l}{e_l^T M_l^{-1}(\xi) e_l}. \tag{7.3.13}$$

If $\xi^*$ maximizes $\Psi_0^\beta$ or equivalently $\Phi$, then $D_\Phi(\xi^*, \eta) \leq 0$ for all $\eta$. If $\eta = \eta_x$

concentrates mass one at $x \in [-1, 1]$, then

$$0 \geq D_\Phi(\xi^*, \eta_x) = -1 + \sum_{l=1}^{m} \beta_l \frac{\left(e_l^T M_l^{-1}(\xi^*) f_l(x)\right)^2}{e_l^T M_l^-(\xi^*) e_l}, \tag{7.3.14}$$

which is equivalent to (7.3.11). Moreover integrating this inequality with respect to $d\xi^*(x)$ gives

$$\int_{-1}^{1} D_\Phi(\xi^*, \eta_x) d\xi^*(x) = 0$$

and shows that $D_\Phi(\xi^*, \eta_x)$ vanishes on the support of $\xi^*$. This proves the second assertion of the lemma. ∎

It is worthwhile to demonstrate at this point how concavity is used in the proof of the converse of Lemma 7.3.2. Integrating (7.3.11) and observing (7.3.13), and (7.3.14), it follows that

$$D_\Phi(\xi^*, \eta) \leq 0$$

for all probability measures $\eta$ on $[-1, 1]$. Now the concavity of $\Phi$ implies that

$$\Phi(\eta) - \Phi(\xi^*) \leq D_\Phi(\xi^*, \eta) \leq 0,$$

proving that $\xi^*$ maximizes $\Phi$ (or equivalently $\Psi_0^\beta$). A sufficient condition for the concavity of $\Phi$ is that all weights $\beta_l$ in the function $\psi_0^\beta$ are nonnegative.

**Theorem 7.3.3** *Let $\xi^*$ denote a symmetric probability measure on the interval $[-1, 1]$, with canonical moments of even order $p_2, \ldots, p_{2m} > 0$. The orthonormal polynomials $\{P_j(x, \xi^*)\}_{j=0}^{m}$ and $\{Q_j(x, \xi^*)\}_{j=0}^{m-1}$ with respect to the measures $d\xi^*(x)$ and $(1 - x^2) d\xi^*(x)$ satisfy the identity*

$$\sum_{l=1}^{m} \beta_l^* P_l^2(x, \xi^*) = 1 - (1 - x^2) \delta_{m-1}^* Q_{m-1}^2(x, \xi^*) \tag{7.3.15}$$

*where*

$$\beta_l^* = \prod_{j=1}^{l-1} \frac{q_{2j}}{p_{2j}} \left(1 - \frac{q_{2l}}{p_{2l}}\right), \qquad l = 1, \ldots, m-1,$$

$$\tag{7.3.16}$$

$$\beta_m^* = \prod_{j=1}^{m-1} \frac{q_{2j}}{p_{2j}} p_{2m}, \qquad \delta_{m-1}^* = \prod_{j=1}^{m-1} \frac{q_{2j}}{p_{2j}} q_{2m}.$$

*Proof*  Let $\bar{\xi}$ denote the upper principal representation of the moment point $(p_1,\ldots,p_{2m-1})$; that is, $\bar{\xi}$ is a symmetric probability measure with the same canonical moments $\bar{p}_j = p_j$ as $\xi^*$ up to the order $2m - 1$ and $\bar{p}_{2m} = 1$. By Corollary 2.3.6 and Remark 2.3.7 the orthonormal polynomials with respect to $\bar{\xi}$ and $\xi^*$ satisfy

$$P_l(x,\bar{\xi}) = P_l(x,\xi^*), \qquad l = 1,\ldots, m-1,$$

$$(7.3.17)$$

$$P_m(x,\bar{\xi}) = \sqrt{p_{2m}}\, P_m(x,\xi^*).$$

Now Theorem 6.2.6 shows that the probability measure $\bar{\xi}$ maximizes the function $\Psi_{0,\beta}$ for the weights $\beta = (\beta_1,\ldots,\beta_m)$ given by

$$\beta_l = \prod_{j=1}^{l-1} \frac{\bar{q}_{2j}}{\bar{p}_{2j}}\left(1 - \frac{\bar{q}_{2l}}{\bar{p}_{2l}}\right) = \begin{cases} \beta_l^* & \text{if } 1 \leq l \leq m-1, \\ \dfrac{1}{p_{2m}}\beta_m^* & \text{if } l = m. \end{cases} \qquad (7.3.18)$$

Here the equality is a consequence of the definition (7.3.16) and the fact that the canonical moments of the measure $\xi^*$ and $\bar{\xi}$ up to the order $2m - 1$ are identical. By Lemma 7.3.1 and 7.3.2 it follows that for the orthonormal polynomials $P_n(x,\bar{\xi})$ with respect to the measure $d\bar{\xi}(x)$,

$$1 \geq \sum_{l=1}^m \beta_l \frac{(e_l^T M_l^{-1}(\bar{\xi}) f_l(x))^2}{e_l^T M_l^{-1}(\bar{\xi}) e_l} = \sum_{l=1}^m \beta_l P_l^2(x,\bar{\xi}) = \sum_{l=1}^m \beta_l^* P_l^2(x,\xi^*)$$

whenever $x \in [-1, 1]$. Here we have used (7.3.17) and (7.3.18) in the last equality. Moreover the second part of Lemma 7.3.2 shows that there is equality on the support of the measure $\bar{\xi}$, which contains $m + 1$ points including $-1$ and $1$ (note that $\bar{p}_{2m} = 1$). By Theorem 2.2.3 (transferred to the interval $[-1, 1]$) these are given by the zeros of $(x^2 - 1)\bar{S}_{m-1}^2(x)$, where $\bar{S}_{m-1}(x)$ is the $(m-1)$th monic orthogonal polynomial with respect to the measure $(1 - x^2)d\bar{\xi}(x)$. By Corollary 2.3.6, $\bar{S}_{m-1}(x)$ is proportional to the $(m-1)$th orthonormal polynomial $Q_{m-1}(x,\xi^*)$ with respect to the measure $(1 - x^2)d\xi^*(x)$. Therefore the polynomials $\sum_{l=1}^m \beta_l^* P_l^2(x,\xi^*) - 1$ and $(x^2 - 1)Q_{m-1}^2(x,\xi^*)$ are of degree $2m$, nonpositive on the interval $[-1, 1]$, and equal to $0$ at the $m + 1$ support points of $\bar{\xi}$. Counting zeros with multiplicities shows that the polynomials must be proportional, and a comparison of the leading coefficients shows that

$$\sum_{l=1}^m \beta_l^* P_l^2(x,\xi^*) - 1 = \left(\prod_{j=1}^{m-1} p_{2j}^{-2}\right) x^{2m} + \ldots$$

$$= \delta_{m-1}^*(x^2 - 1)Q_{m-1}^2(x,\xi^*),$$

where Remark 2.3.7 is used in both equalities. This is equivalent to (7.3.15) and proves the assertion.  ∎

Note that Theorem 7.3.3 provides an identity for the sum of squares of orthogonal polynomials with respect to an arbitrary (symmetric) measure on the interval $[-1, 1]$. In the remaining part of this section we illustrate this identity in several examples.

**EXAMPLE 7.3.4**  *Arc-sine Distribution Revisited.* Let $\mu^T$ denote the arc-sine distribution on $[-1, 1]$, which has canonical moments $p_j = \frac{1}{2}$ ($j \in \mathbb{N}$). The orthonormal polynomials $Q_{m-1}(x, \mu^T)$ with respect to the measure $(1 - x^2)d\mu^T(x) = \sqrt{1 - x^2}\,dx/\pi$ are proportional to the Chebyshev polynomials of the second kind $U_{m-1}(x)$. From Remark 2.3.7 it follows that the leading coefficient is $\sqrt{2}2^{m-1}$, which shows that

$$Q_{m-1}(x, \mu^T) = \sqrt{2}U_{m-1}(x).$$

Similarly it can be shown that the orthonormal polynomials with respect to the arc-sine measure are given by

$$P_m(x, \mu^T) = \sqrt{2}T_m(x)$$

[see also the orthogonality relation (2.1.12) in Example 2.1.1]. Now $\beta_l^* = 0$ ($l = 1, \ldots, m - 1$), $\beta_m^* = \delta_{m-1}^* = \frac{1}{2}$ and (7.3.15) reduces to the "trigonometric identity" (7.3.1) for the Chebyshev polynomial of the first and second kind.

**EXAMPLE 7.3.5**  *Ultraspherical Polynomials.* Let $\mu^{(\alpha)}$ denote the probability measure on the interval $[-1, 1]$ with density

$$c_\alpha(1 - x^2)^{\alpha - 1/2}, \qquad \alpha > -\tfrac{1}{2}, \ \alpha \neq 0. \qquad (7.3.19)$$

The constant in (7.3.19) is given by

$$c_\alpha = \frac{\Gamma(2\alpha + 1)}{2^{2\alpha}[\Gamma(\alpha + \frac{1}{2})]^2} = \frac{\Gamma(\alpha + 1)}{\sqrt{\pi}\Gamma(\alpha + \frac{1}{2})},$$

which can be obtained from (2.1.14) and the duplication formula for the Gamma function in (2.3.18). The orthonormal polynomials with respect to the measure $d\mu^{(\alpha)}(x)$ are proportional to the ultraspherical polynomials $C_m^{(\alpha)}(x)$ defined by (2.1.18). The constant of proportionality can be obtained from (2.1.23) (which gives the coefficient of $x^m$ in $C_m^{(\alpha)}(x)$) and Remark 2.3.7. From Example 1.5.3 it follows that the canonical moments of $\mu^{(\alpha)}$ are given by

$$p_{2i}^{(\alpha)} = \frac{i}{2(i + \alpha)}, \qquad p_{2i-1} = \frac{1}{2} \qquad (i \in \mathbb{N}). \qquad (7.3.20)$$

If $\hat{C}_m^{(\alpha)}(x)$ denotes the monic version of $C_m^{(\alpha)}(x)$, this gives the representation (see

Remark 2.3.7)

$$P_l^2(x, \mu^{(\alpha)}) = \prod_{j=1}^{l} \left( q_{2j-2}^{(\alpha)} p_{2j}^{(\alpha)} \right)^{-1} \left[ \hat{C}_l^{(\alpha)}(x) \right]^2$$

$$= 2^{2l} \frac{\Gamma(2\alpha)\Gamma(l+1+\alpha)\Gamma(l+\alpha)}{\Gamma(l+1)\Gamma(l+2\alpha)\Gamma(\alpha)\Gamma(\alpha+1)} \left[ \hat{C}_l^{(\alpha)}(x) \right]^2 \qquad (7.3.21)$$

$$= \frac{\Gamma(2\alpha)\Gamma(l+1)(l+\alpha)}{\alpha\Gamma(l+2\alpha)} \left[ C_l^{(\alpha)}(x) \right]^2,$$

where the last equality follows from (2.3.19). Similarly the polynomial $Q_{m-1}(x, \mu^{(\alpha)})$ orthonormal with respect to the measure

$$(1 - x^2)d\mu^{(\alpha)}(x) = c_\alpha(1 - x^2)^{\alpha+1/2}dx$$

is proportional to $C_{m-1}^{(\alpha+1)}(x)$ and given by

$$Q_{m-1}^2(x, \mu^{(\alpha)}) = \frac{2\Gamma(m)\Gamma(2\alpha+1)(m+\alpha)}{\Gamma(m+2\alpha+1)} \left[ C_{m-1}^{(\alpha+1)}(x) \right]^2.$$

Finally the constants $\beta_l^*$ and $\delta_{m-1}^*$ in (7.3.16) can be calculated as

$$\beta_l^* = \begin{cases} -\dfrac{\Gamma(l+2\alpha)}{\Gamma(l+1)\Gamma(2\alpha)} & l = 1, \ldots, m-1, \\[2ex] \dfrac{m\Gamma(m+2\alpha)}{2(m+\alpha)\Gamma(m)\Gamma(2\alpha+1)}, & l = m, \end{cases}$$

and

$$\delta_{m-1}^* = \frac{\Gamma(m+2\alpha+1)}{2(m+\alpha)\Gamma(m)\Gamma(2\alpha+1)}.$$

Consequently we obtain from (7.3.15) the following identity for the sum of squares of ultraspherical polynomials

$$\left[ \frac{m}{2\alpha} C_m^{(\alpha)}(x) \right]^2 - \sum_{l=0}^{m-1} \frac{l+\alpha}{\alpha} \left[ C_l^{(\alpha)}(x) \right]^2 = (x^2 - 1) \left[ C_{m-1}^{(\alpha+1)}(x) \right]^2, \qquad (7.3.22)$$

which has a nice application in mathematical physics (see Dehesa, Van Assche, and Yáñez, 1997). We remark that (7.3.22) reduces to (7.3.1) when $\alpha \to 0$ which follows from (2.1.18) and (2.1.19).

**EXAMPLE 7.3.6** *The D-Optimal Design Revisited.* Let $\xi_m^D$ denote the probability measure with equal masses at the zeros of the polynomial $(x^2 - 1)P_m'(x)$, where $P_m(x)$ denotes the $m$th Legendre polynomial orthogonal with respect to the Lebesgue measure [see (2.1.20)]. It is shown in Theorem 5.5.3 and (5.5.10) that $\xi_m^D$ is the D-optimal design for a polynomial regression of degree $m$ and has canonical moments

$$p_{2j-1} = \frac{1}{2}, \quad p_{2j} = \frac{m-j+1}{2(m-j)+1}, \quad j = 1,\ldots, m. \tag{7.3.23}$$

By Theorem 5.5.2 and the preceding discussion $\xi_m^D$ is also G-optimal; that is, it minimizes

$$\max_{x\in[-1,1]} f_m^T(x)M_m^{-1}(\xi)f_m(x) \tag{7.3.24}$$

over the class of probability measures with $|M_m(\xi)| > 0$. Moreover the minimum value in (7.3.24) is $m + 1$ (see Theorem 5.5.2). Since the G- (or D-) optimal design $\xi_m^D$ for a regression of degree $m$ is often used in polynomial models of lower degree, it is reasonable to ask for the G-efficiency

$$\mathrm{eff}_k^G(\xi_m^D) = \frac{k+1}{\sup_{x\in[-1,1]} f_k^T(x)M_k^{-1}(\xi_m^D)f_k(x)} \tag{7.3.25}$$

of the optimal design for the polynomial of degree $m$ in the regression of degree $k \le m$.

Let $P_k(x, \xi_m^D)$ denote the $k$th orthonormal polynomial with respect to $d\xi_m^D(x)$, and put

$$g_k(x) = \left(P_0(x, \xi_m^D), P_1(x, \xi_m^D), \ldots, P_k(x, \xi_m^D)\right)^T = A_k f_k(x)$$

with an appropriate nonsingular $(k + 1) \times (k + 1)$ matrix. Then

$$f_k^T(x)M_k^{-1}(\xi_m^D)f_k(x) = g_k(x)^T g_k(x) = \sum_{l=0}^{k} P_l^2(x, \xi_m^D) \tag{7.3.26}$$

$(k = 0,\ldots,m)$. On the other hand, we obtain from Theorem 7.3.3 and (7.3.23) for $k < m$,

$$\sum_{l=1}^{k} P_l^2(x, \xi_m^D) + \frac{(m-k)^2}{2(m-k)-1} P_{k+1}^2(x, \xi_m^D) = m - m(1 - x^2)\delta_k^* Q_k^2(x, \xi_m^D). \tag{7.3.27}$$

Here $Q_k(x, \xi_m^D)$ is the $k$th orthonormal polynomial with respect to the measure

$(1 - x^2)d\xi_m^D(x)$. The identities (7.3.26) and (7.3.27) show that

$$f_{k+1}^T(x)M_{k+1}^{-1}(\xi_m^D)f_{k+1}(x) = \left(1 - \frac{2(m-k)-1}{(m-k)^2}\right)f_k^T(x)M_k^{-1}(\xi_m^D)f_k(x)$$

$$+ \frac{(m+1)(2(m-k)-1)}{(m-k)^2}$$

$$- c^*(1-x^2)Q_k^2(x,\xi_m^D)$$

$(0 \le k \le m-1)$ with a positive constant $c^*$. Now $f_0^T(x)M_0^{-1}(\xi_m^D)f_0(x) = 1$, and a simple induction with respect to $k$ yields that the left-hand side in (7.3.26) attains its maximum over the interval $[-1, 1]$ at the points $-1$ and $+1$. By (7.3.26) and (7.3.27) this maximum is given by

$$\max_{x\in[-1,1]} f_k^T(x)M_k^{-1}(\xi_m^D)f_k(x) = 1 + \left\{m - \frac{(m-k)^2}{2(m-k)-1}P_{k+1}^2(1,\xi_m^D)\right\}$$

(7.3.28)

$$= 1 + m\left\{1 - \left(\frac{m-k}{m}\right)^2\right\}.$$

The last equality follows from the representation

$$P_{k+1}^2(1,\xi_m^D) = \prod_{j=1}^{k}\frac{q_{2j}}{p_{2j}}\frac{1}{p_{2k+2}} = \frac{2(m-k)-1}{m},$$

where the first equality holds for the orthonormal polynomials with respect to any arbitrary symmetric probability measure (with canonical moments $p_j$). It can be proved by an induction argument combining Corollary 2.3.6 and Remark 2.3.7. As an immediate application we obtain the following result:

**Theorem 7.3.7** *The G-efficiency of the D-optimal design $\xi_m^D$ in a polynomial model of degree $k \le m$ is given by*

$$\text{eff}_k^G(\xi_m^D) = \frac{k+1}{m+1-\frac{(m-k)^2}{m}}, \qquad k = 0,\ldots,m.$$

Note that Theorem 7.3.7 provides an alternative proof of the equivalence between D- and G-optimality in the case of polynomial regression (just put $k = m$). As an example consider the cubic regression, namely $m = 3$, where the D-optimal design has masses $1/4$ at the points $-1, -1/\sqrt{5}, 1/\sqrt{5}$, and 1. If this design is used in a linear or quadratic regression, its G-efficiencies are given by 75% and 82%, respectively.

## 7.4  EXTREMAL PROBLEMS FOR POLYNOMIALS

In this section we discuss some extensions of an extremal property of the Chebyshev polynomials of the first kind. More precisely we consider the problem

$$\min_{a_0,\ldots,a_{m-1}} \sup_{x\in[-1,1]} \left| x^m - \sum_{j=0}^{m-1} a_j x^j \right| \tag{7.4.1}$$

of best approximation of the power $x^m$ by a (real) polynomial of degree $m - 1$. It is well known (see Natanson, 1955; Achieser, 1956; or Rivlin, 1990) that the minimum value in (7.4.1) is given by $1/2^{m-1}$ and that the "best" polynomial $x^m - \sum_{j=0}^{m-1} a_j x^j$ is given by $1/2^{m-1} T_m(x)$, where $T_m(x)$ is the Chebyshev polynomial of the first kind. We will present a new proof of this result based on a game-theoretic argument and the theory of canonical moments, which is motivated from the design theory developed in Chapter 5. This approach allows the treatment of more general extremal problems.

With the notation of Section 7.3, (7.4.1) can be rewritten as ($a \in \mathbb{R}^{m+1}$)

$$\inf_{|a^T e_m|^2=1} \sup_{x\in[-1,1]} (a^T \boldsymbol{f}_m(x))^2 = \inf_{|a^T e_m|^2=1} \sup_{\xi} \int_{-1}^{1} (a^T \boldsymbol{f}_m(x))^2 d\xi(x)$$

$$= \inf_{|a^T e_m|^2=1} \sup_{\xi} a^T M_m(\xi) a$$

$$= \sup_{\xi} \inf_{|a^T e_m|^2=1} a^T M_m(\xi) a \tag{7.4.2}$$

$$= \sup_{\xi} \inf_{a\in\mathbb{R}^{m+1}} \frac{a^T M_m(\xi) a}{(a^T e_m)^2} \tag{7.4.3}$$

$$= \sup_{\xi} (e_m^T M_m^{-1}(\xi) e_m)^{-1}$$

$$= \sup_{\xi} \frac{|M_m(\xi)|}{|M_{m-1}(\xi)|}. \tag{7.4.4}$$

with the convention that $(e_m^T M_m^{-1}(\xi) e_m)^{-1}$ is zero if $M_m(\xi)$ is singular. Here the equality in (7.4.2) follows from a game-theoretic argument and the fact that the kernel $w(a, \xi) = a^T M(\xi) a$ is convex in $a$ and concave (even linear) in $\xi$. Note that the calculation of the supremum in (7.4.3) can be restricted to the set of designs with nonsingular moment matrix of order $2m$ and that the equality between (7.4.3) and (7.4.4) is a consequence of Cauchy's inequality

$$(a^T e_m)^2 \le a^T M(\xi) a \cdot e_m^T M^{-1}(\xi) e_m. \tag{7.4.5}$$

Now (7.4.4) is the $D_1$-optimal design problem in a polynomial regression of degree $m$ which was solved in Theorem 5.5.4. If $\xi_m^{D_1}$ denotes the $D_1$-optimal design and

$$P_m(x) = \hat{a}^T f_m(x), \quad (\hat{a}^T e_m)^2 = 1, \tag{7.4.6}$$

is a optimal solution of (7.4.1), then there must be equality in (7.4.5), that is,

$$\hat{a} = cM^{-1}(\xi_m^{D_1})e_m, \tag{7.4.7}$$

where the constant $c$ is determined by $(\hat{a}^T e_m)^2 = 1$. Combining (7.4.6), (7.4.7), with Lemma 7.3.1, it follows that $P_m(x)$ is the $m$th monic orthogonal polynomial with respect to the measure $d\xi_m^{D_1}(x)$ (up to the sign). Now the canonical moments of $\xi_m^{D_1}$ up to the order $2m - 1$ are $\frac{1}{2}$ and coincide with the canonical moment of the arc-sine distribution. Therefore Corollary 2.3.6 shows that $P_m(x)$ is the $m$th monic orthogonal polynomial with respect to the arc-sine measure:

$$P_m(x) = \frac{1}{2^{m-1}} T_m(x).$$

This determines the solution of the extremal problem (7.4.1).

The solution of (7.4.1), based on a game-theoretic argument combined with the theory of canonical moments, allows the treatment of more general extremal problems that cannot be solved by the "classical" methods of approximation theory. To this end let $\mathbb{P}_j$ denote the set of polynomials of degree $j$, $I = \{i_1, \ldots, i_n\}$ denote a subset of $\{1, \ldots, m\}$ containing $m$, and define

$$P_I := \left\{ (P_j)_{j \in I} | P_j \in \mathbb{P}_j, \ j \in I, \ \sup_{x \in [-1,1]} \sum_{j \in I} P_j^2(x) \le 1 \right\}$$

as the set of all polynomials of degree $i_1, \ldots, i_n$ such that the sup norm of the sum of squares is bounded by 1 on the interval $[-1, 1]$. In the following, $m_l(P_l)$ denotes the leading coefficient of the polynomial $P_l \in \mathbb{P}_l$. We are interested in the (nonlinear) extremal problem

$$(\mathcal{P}_I) \qquad \max \left\{ \sum_{l \in I} \beta_l m_l^2(P_l) | (P_l)_{l \in I} \in P_I \right\},$$

where $\beta = (\beta_{i_1}, \ldots, \beta_{i_n})$ denotes a vector of positive weights with sum 1. Note that for $I = \{m\}$, the extremal problem $(\mathcal{P}_I)$ reduces to the problem of maximizing the highest coefficient among all polynomials of (precise) degree $m$ with sup norm bounded by 1. This is an alternative formulation of the "classical" Chebyshev approximation problem (7.4.1). Throughout this section the orthogonal polynomials with leading coefficient 1 corresponding to a probability measure will be denoted by $R_j(x, \xi)$ and their (squared) $L_2$-norm by

$$k_j(\xi) = \int_{-1}^{1} R_j^2(x, \xi)d\xi(x) = \frac{|M_j(\xi)|}{|M_{j-1}(\xi)|} = (e_j^T M_j^{-1}(\xi)e_j)^{-1}. \tag{7.4.8}$$

The main step for solving the extremal problem $(\mathcal{P}_I)$ is the following duality, which is the analogue in the game-theoretic argument in (7.4.2):

**Theorem 7.4.1** *If* $\Xi := \{\xi \in \Xi | |M_m(\xi)| > 0\}$ *denotes the set of all probability measures with nonsingular Hankel matrix* $M_m(\xi)$, *then the following duality holds:*

$$(\mathcal{P}_I) \qquad \max\left\{\sum_{l \in I} \beta_l m_l^2(P_l) | (P_j)_{j \in I} \in P_I\right\} = \min_{\xi \in \Xi} \max_{j \in I}\{\beta_j k_j^{-1}(\xi)\} \qquad (\mathcal{D}_I)$$

*and solutions of* $(\mathcal{P}_I)$ *and* $(\mathcal{D}_I)$ *exist.*
   *Moreover let* $\xi^*$ *be a solution of the problem* $(\mathcal{D}_I)$,

$$\mathcal{M}(\xi^*) = \left\{j \in I | \beta_j^{-1} k_j(\xi^*) = \min_{i \in I} \beta_i^{-1} k_i(\xi^*)\right\},$$

*and let* $\sqrt{1/k_j(\xi^*)}R_j(x, \xi^*)$ *denote the jth orthonormal polynomial with respect to the measure* $d\xi^*(x)$. *Then there exist constants* $\alpha_j \geq 0$ *with sum* 1 *satisfying*

$$\alpha_j = 0 \qquad if \ j \in I \setminus \mathcal{M}(\xi^*), \tag{7.4.9}$$

$$\sum_{j \in I} \alpha_j k_j^{-1}(\xi^*)R_j^2(x, \xi^*) \leq 1 \qquad for \ all \ x \in [-1, 1]. \tag{7.4.10}$$

*With this choice* $\{\sqrt{\alpha_j/k_j(\xi^*)}R_j(x, \xi^*)\}_{j \in I}$ *is a solution of the extremal problem* $(\mathcal{P}_I)$.

*Proof* Note that an optimal solution of $(\mathcal{D}_I)$ must have a nonsingular moment matrix $M_m(\xi)$. The proof of Theorem 7.4.1 is a consequence of the general duality 7.12 in Pukelsheim (1993, p. 172). To be precise, let $I = \{i_1, \ldots, i_n\}$, $i = n + \sum_{j=1}^{n} i_j$, $f_j(x) = (1, x, \ldots, x^j)^T$, and recall the definition of $M_j(\xi)$ in (7.3.3). In the following discussion we will collect all matrices $M_{i_1}(\xi), \ldots, M_{i_n}(\xi)$ in a matrix

$$M(\xi) = \begin{pmatrix} M_{i_1}(\xi) & & \\ & \ddots & \\ & & M_{i_n}(\xi) \end{pmatrix} \in \mathbb{R}^{i \times i}$$

and define two matrices by

$$K = \begin{pmatrix} \beta_{i_1}^{1/2} e_{i_1} & & \\ & \ddots & \\ & & \beta_{i_n}^{1/2} e_{i_n} \end{pmatrix} \in \mathbb{R}^{i \times n} \quad N = \begin{pmatrix} N_{i_1} & & \\ & \ddots & \\ & & N_{i_n} \end{pmatrix} \in \mathbb{R}^{i \times i},$$

where $e_j = (0, \ldots, 0, 1)^T \in \mathbb{R}^{j+1}$ is the $(j+1)$th unit vector $(j \in I)$, $N_{i_j}$ are nonnegative $(i_j + 1) \times (i_j + 1)$ matrices (i.e., $N_{i_j} \geq 0$), and all other entries in these matrices are 0. Defining $\lambda_{\min}(A)$ as the minimum eigenvalue of the nonnegative definite matrix $A \in \mathbb{R}^{n \times n}$, we obtain from Theorem 6.12 and 7.12 in Pukelsheim (1993, pp. 149, 172), that (note that $k_j(\xi) = [e_j^T M_j^{-1}(\xi) e_j]^{-1}$),

$$\max_{\xi \in \Xi} \min \{\beta_j^{-1} k_j(\xi) | j \in I\}$$

$$= \max_{\xi \in \Xi} \lambda_{\min}((K^T M^{-1}(\xi) K)^{-1})$$

$$= \min \Big\{ [\text{trace } (K^T N K)]^{-1} | N \in \mathbb{R}^{i \times i}, N \geq 0,$$

$$\quad \text{trace } (M(\xi) N) \leq 1 \; \forall \xi \in \Xi \Big\}$$

$$= \min \Big\{ \Big[ \sum_{j \in I} \beta_j e_j^T N_j e_j \Big]^{-1} | N_j \in \mathbb{R}^{(j+1) \times (j+1)}, N_j \geq 0 \; \forall j \in I, \tag{7.4.11}$$

$$\quad \sum_{j \in I} \text{trace } (M_j(\xi) N_j) \leq 1 \forall \xi \in \Xi \Big\}$$

$$= \min \Big\{ \Big[ \sum_{j \in I} \beta_j (e_j^T a_j)^2 \Big]^{-1} | a_j \in \mathbb{R}^{j+1}, \sum_{j \in I} (f_j^T(x) a_j)^2 \leq 1 \; \forall x \in [-1, 1] \Big\}$$

$$= \min \Big\{ \Big[ \sum_{j \in I} \beta_j m_j^2(P_j) \Big]^{-1} | (P_j)_{j \in I} \in P_I \Big\}.$$

In order to go from the third to the fourth line in (7.4.11), we have used the fact that

$$\sum_{j \in I} \text{trace } (M_j(\xi) N_j) = \sum_{j \in I} \int_{-1}^{1} f_j^T(x) N_j f_j(x) d\xi(x) \leq 1 \qquad \forall \xi \in \Xi$$

is equivalent to the inequality

$$\sum_{j \in I} f_j^T(x) N_j f_j(x) \leq 1 \qquad \forall x \in [-1, 1] \tag{7.4.12}$$

and the fact that the minimum value does not change if the matrices $N_j$ are replaced by matrices of the form $a_j a_j^T$ (see the following discussion). This proves the first part of the theorem. For the second part we discuss equality in (7.4.11), that is, equality in the general duality 7.12 of Pukelsheim (1993, p.172), and obtain

$$\sum_{j \in I} \text{trace}\, (M_j(\xi^*)N_j) = 1, \tag{7.4.13}$$

$$M_j(\xi^*)N_j = \frac{e_j e_j^T N_j}{e_j^T M_j^{-1}(\xi^*)e_j}, \qquad j \in I, \tag{7.4.14}$$

$$\sum_{j \in I} (\beta_j e_j^T N_j e_j) \min_{i \in I}\{(\beta_i e_i^T M_i^{-1}(\xi^*)e_i)^{-1}\}$$

$$= \sum_{j \in I} \frac{\beta_j e_j^T N_j e_j}{\max_{i \in I}\{(\beta_i e_i^T M_i^{-1}(\xi^*)e_i)\}} = \sum_{j \in I} \frac{e_j^T N_j e_j}{e_j^T M_j^{-1}(\xi^*)e_j} = 1. \tag{7.4.15}$$

Observing that $k_j^{-1}(\xi^*) = e_j^T M_j^{-1}(\xi^*)e_j$ ($j = 1, \ldots, m$), we obtain by straightforward calculation as a solution of (7.4.13) and (7.4.14), $N_j = \alpha_j a_j a_j^T$, where $a_j = \sqrt{k_j(\xi^*)}M_j^{-1}(\xi^*)\, e_j$ ($j \in I$), $\alpha_j \geq 0$ (because $N_j \geq 0$) and $\sum_{j \in I} \alpha_j = 1$. Finally it follows from (7.4.15) that $\alpha_j = 0$ whenever $j \notin M(\xi^*)$. By Lemma 7.3.1 the polynomials $P_l(x, \xi^*) = a_l^T f_l(x)$ are orthonormal with respect to the measure $d\xi^*(x)$, which yields for the monic orthogonal polynomials $R_l(x, \xi^*) = \sqrt{k_l(\xi^*)}a_l^T f_l(x)$ ($l = 1, \ldots, m$). Consequently a solution of the right-hand side of (7.4.11) is given by $\{\sqrt{\alpha_j/k_j(\xi^*)}R_j(x, \xi^*)\}_{j \in I}$, where $R_j(x, \xi^*)$ is the $j$th monic orthogonal polynomial with respect to the measure $d\xi^*(x)$ and the $\alpha_j$ have to satisfy

$$\sum_{j \in I} \alpha_j k_j^{-1}(\xi^*)R_j^2(x, \xi^*) = \sum_{j \in I} f_j(x)^T N_j f_j(x) \leq 1$$

for all $x \in [-1, 1]$. This completes the proof of Theorem 7.4.1. ■

Note that the dual problem $(\mathcal{D}_I)$ contains as a special case the optimization problem considered in Theorem 6.3.6 ($\beta_{i_j} = 1/n$, $j = 1, \ldots, n$). While from a statistical point of view the main interest are the support points and weights of the solution $\xi^*$ of $(\mathcal{D}_I)$, Theorem 7.4.1 shows that the orthogonal polynomials with respect to the optimal measure $d\xi^*(x)$ are needed for the solution of the extremal problem $(\mathcal{P}_I)$. The monic form of these polynomials is given by $R_0(x, \xi^*) = 1$, $R_1(x, \xi^*) = x + 1 - 2\zeta_1^*$,

$$R_m(x, \xi^*) = K\begin{pmatrix} & -4\zeta_1^*\zeta_2^* & \cdots & & -4\zeta_{2m-3}^*\zeta_{2m-2}^* & \\ x+1-2\zeta_1^* & & \cdots & & x+1-2\zeta_{2m-2}^* & -2\zeta_{2m-1}^* \end{pmatrix}$$

$$\tag{7.4.16}$$

where the continuants $K(\cdot)$ are defined in (3.2.19), $\zeta_j^* = q_{j-1}^* p_j^*, j \geq 1, (q_0^* = 1)$, and $p_1^*, p_2^*, \ldots$ denote the canonical moments of $\xi^*$. This follows from Corollary 2.3.5, part 2, because the right-hand side of (7.4.16) satisfies the same recurrence relation and initial conditions as the monic orthogonal polynomials with respect to the measure $d\xi^*(x)$. Remark 2.3.7 and (7.4.8) yield for the $L_2$-norm

$$k_m(\xi^*) = \int_{-1}^{1} R_m^2(x, \xi^*) d\xi^*(x) = 2^{2m} \prod_{j=1}^{m} \zeta_{2j-1}^* \zeta_{2j}^*. \tag{7.4.17}$$

**Theorem 7.4.2** *The solution $\xi^*$ of the dual problem $(\mathcal{D}_I)$ is uniquely determined by its canonical moments $(p_1^*, \ldots, p_{2m}^*)$ where $p_{2j-1}^* = \frac{1}{2}$ $(j = 1, \ldots, m)$, $p_{2m}^* = 1$, $p_{2(m-j)}^* = \frac{1}{2}$ if $m - j \notin I$ and*

$$p_{2(m-j)}^* = \max\left\{ 1 - \frac{\beta_m}{\beta_{m-j}} \prod_{i=m-j+1}^{m-1} (q_{2i}^* p_{2i}^*)^{-1}, \frac{1}{2} \right\} \tag{7.4.18}$$

*if $m - j \in I$.*

*Proof*   The result case can be proved by similar arguments as given in the proof of Theorem 6.3.6. To make this chapter self-contained, we provide an alternative proof that is directly based on the duality result of Theorem 7.4.1 and uses the identities for orthogonal polynomials derived in Section 7.3. For $m - j \in I$ let $\gamma_{m-j} = 1 - \beta_m/\beta_{m-j} \prod_{i=m-j+1}^{m-1}(q_{2i}^* p_{2i}^*)^{-1}$ $(\gamma_m = 1)$; then it is easy to see [observing (7.4.17) and (7.4.18)] that for $m - j \in I$,

$$\gamma_{m-j} \geq \tfrac{1}{2} \quad \text{if and only if} \quad \beta_m^{-1} k_m(\xi^*) = \beta_{m-j}^{-1} k_{m-j}(\xi^*),$$

$$\gamma_{m-j} < \tfrac{1}{2} \quad \text{if and only if} \quad \beta_m^{-1} k_m(\xi^*) < \beta_{m-j}^{-1} k_{m-j}(\xi^*).$$

Consequently we have for the set $\mathcal{M}(\xi^*)$ in Theorem 7.4.1,

$$m \in \mathcal{M}(\xi^*) = \{j \in I | \gamma_j \geq \tfrac{1}{2}\}, \tag{7.4.19}$$

$$p_{2j}^* = \tfrac{1}{2} \quad \text{if } j \notin \mathcal{M}(\xi^*). \tag{7.4.20}$$

In the following define weights $\alpha_1, \ldots, \alpha_m$ by

$$\alpha_j = \prod_{i=1}^{j-1} \frac{q_{2i}^*}{p_{2i}^*} \left( 1 - \frac{q_{2j}^*}{p_{2j}^*} \right). \tag{7.4.21}$$

These have sum 1 and are nonnegative, by the definition of $p_{2j}^*$ in (7.4.18). Additionally we have by (7.4.20) $\alpha_j = 0$ whenever $j \notin \mathcal{M}(\xi^*)$. From (7.4.17) it follows that the polynomials $k_l^{-1/2}(\xi^*) R_l(x, \xi^*)$ are orthonormal with respect to the measure $d\xi^*(x)$, and Theorem 7.3.3 shows (note that $p_{2m}^* = 1$) that

$$\sum_{j=1}^{m} \alpha_j k_j^{-1}(\xi^*) R_j^2(x, \xi^*) = \sum_{j \in \mathcal{M}(\xi^*)} \alpha_j k_j^{-1}(\xi^*) R_j^2(x, \xi^*) \leq 1 \qquad (7.4.22)$$

for all $x \in [-1, 1]$. In other words,

$$\{R_j^*(x)\}_{j \in I} := \left\{ \sqrt{\alpha_j / k_j(\xi^*)} R_j(x, \xi^*) \right\}_{j \in I} \in P_I, \qquad (7.4.23)$$

and by the definition of $\mathcal{M}(\xi^*)$ and $\sum_{j \in \mathcal{M}(\xi^*)} \alpha_j = 1$, it follows that

$$\sum_{j \in I} \beta_j m_j^2(R_j^*) = \sum_{j \in \mathcal{M}(\xi^*)} \beta_j m_j^2(R_j^*) = \frac{\beta_m}{k_m(\xi^*)} = \max\{\beta_j k_j^{-1}(\xi^*) | j \in I\}.$$

Therefore we have equality in Theorem 7.4.1 for $\{R_j^*\}_{j \in I} \in P_I, \xi^* \in \Xi$, and the assertion of the theorem follows.                                                                ∎

Observing the arguments at the end of the proof of the preceding theorem, the solution of $(\mathcal{P}_I)$ is obtained from (7.4.23) where the monic orthogonal polynomials with respect to the measure $d\xi^*(x)$ are given by (7.4.16) and the quantities $\alpha_j$ and $k_j(\xi^*)$ are obtained from (7.4.21) and (7.4.17), respectively. This provides a complete solution of the extremal problem $(\mathcal{P}_I)$. In the following we will discuss some interesting special cases and choose the $j$th weight $\beta_j$ proportional to $b^{-2j}$ for some positive constant $b$. This choice provides an extremal problem on the interval $[-b, b]$, that is,

$$(\mathcal{P}_I^{(b)}) \qquad\qquad \max\left\{ \sum_{l \in I} m_l^2(P_l) | (P_l)_{l \in I} \in P_I^{(b)} \right\},$$

where

$$P_I^{(b)} := \left\{ (P_j)_{j \in I} | P_j \in \mathbb{P}_j, j \in I, \sup_{x \in [-b,b]} \sum_{j \in I} P_j^2(x) \leq 1 \right\}.$$

A simple calculation shows that the solution of $(\mathcal{P}_I^{(b)})$ is obtained from a linear transformation of the solution of $(\mathcal{P}_I)$ for the weights $\beta_j$ proportional to $b^{-2j}$ ($j \in I$). In the following we consider two cases, namely $I = \{m - 1, m\}$ and $\{1, \ldots, m\}$. The last named case means that we are maximizing the sum of squares of the leading coefficients of $m$ polynomials of degree $1, \ldots, m$ subject to the condition that the sup norm over the interval $[-b, b]$ of the sum of squares is less or equal than 1. The other case considers the same problem for polynomials of degree $m - 1$ and $m$. It turns out that the solution of these problems can be expressed as a linear combination of the classical Chebyshev polynomials.

**Theorem 7.4.3** *Let $I = \{1, \ldots, m\}$, $T_k(x)$, $U_k(x)$ denote the kth Chebyshev polynomial of the first and second kind, respectively, and*

$$k = min \left\{ j \in \{1, \ldots, m\} | U_{2m-2i+1}\left(\frac{b}{2}\right) > 0 \quad \text{for } i = j, \ldots, m \right\}. \quad (7.4.24)$$

*The solution of the extremal problem $(\mathcal{P}_I^{(b)})$ is given by $\{R_l^*(x)\}_{l=1}^m$, where*

$$R_l^*(x) = \kappa_l \left[ T_k\left(\frac{x}{b}\right) U_{l-k}\left(\frac{x}{2}\right) - \frac{U_{m-k+1}\left(\frac{b}{2}\right)}{U_{m-k}\left(\frac{b}{2}\right)} T_{k-1}\left(\frac{x}{b}\right) U_{l-1-k}\left(\frac{x}{2}\right) \right] \quad (7.4.25)$$

*if $l = k, \ldots, m$, $R_l^*(x) = 0$ if $l = 1, \ldots, k-1$, and*

$$\kappa_l = \pm \frac{\sqrt{bU_{2m-2l+1}\left(\frac{b}{2}\right)}}{U_{m-k+1}\left(\frac{b}{2}\right)} \quad (l = k, \ldots, m).$$

*The maximum value of $(\mathcal{P}_I^{(b)})$ is $2^{2k-2}b^{-2k+1} U_{m-k}(b/2)/U_{m-k+1}(b/2)$.*

*Proof* By the preceding discussion the solution of the problem $(\mathcal{P}_I)$ is obtained from $(\mathcal{P}_I^{(b)})$ for weights $\beta_j$ proportional to $b^{-2j}$ ( $j = 1, \ldots, m$) and by a linear transformation of the optimal polynomials onto the interval $[-b, b]$. Consider the canonical moments $p_{2j}^*$ defined by (7.4.18) with $\beta_m/\beta_{m-j} = 2^{-2j}$. For $j = m, \ldots, 1$ define $\gamma_j(\xi^*) = 1 - b^{-2(m-j)} \prod_{i=j+1}^{m-1}(q_{2i}^* p_{2i}^*)^{-1}$ (here we put $\gamma_m(\xi^*) = 1$), then the recurrence relationship for the Chebyshev polynomials of the second kind (2.1.8), and a simple induction shows that

$$\gamma_j(\xi^*) = \frac{U_{m-j+1}\left(\frac{b}{2}\right)}{bU_{m-j}\left(\frac{b}{2}\right)} = \frac{U_{2m-2j+1}\left(\frac{b}{2}\right)}{2bU_{m-j}^2\left(\frac{b}{2}\right)} + \frac{1}{2}, \quad j = k, \ldots, m, \quad (7.4.26)$$

where the last identity follows from the trigonometric representation (2.1.6). The definition of $k$ in (7.4.24) and Theorem 7.4.2 provide for the canonical moments of the solution $\xi^*$ of the dual problem $(\mathcal{D}_I)$ $p_{2j}^* = \gamma_j(\xi^*)$ ( $j = k, \ldots, m$). The recursive relation of the Chebyshev polynomials of the second kind (2.1.8) implies that

$$q_{2j}^* p_{2j+2}^* = \frac{1}{b^2}, \quad j = k, \ldots, m-1. \quad (7.4.27)$$

If $k \geq 2$, then it follows that

$$\gamma_{k-1}(\xi^*) = 1 - b^{-2(m-k+1)} \prod_{i=k}^{m-1}(q_{2i}^* p_{2i}^*)^{-1} \leq \frac{1}{2}$$

and $b \leq 2$ [since, by Lemma 2.1.3 all zeros of $U_l(x)$ are located in the interval $(-1, 1)$]. This implies (by Theorem 7.4.2) that $p_{2k-2}^* = \frac{1}{2}$ and that $\gamma_{k-2} \leq 1 - 2b^{-2} \leq \frac{1}{2}$. Therefore the canonical moments of the solution $\xi^*$ of the dual problem $(\mathcal{D}_I)$ in Theorem 7.4.2 are given by

$$(\tfrac{1}{2},\tfrac{1}{2},\ldots,\tfrac{1}{2},p^*_{2k},\tfrac{1}{2},p^*_{2k+2},\tfrac{1}{2},\ldots,\tfrac{1}{2},p^*_{2m-2},\tfrac{1}{2},1) \tag{7.4.28}$$

where $p^*_{2j} = \gamma_j(\xi^*)$ $(j = k,\ldots,m)$ and $\gamma_j(\xi^*)$ is defined in (7.4.26). By Theorem 7.4.1 we have to find the orthonormal polynomials with respect to the measure $d\xi^*(x)$ and then consider a linear transformation. If $\xi^{(b)}$ denotes the linear transformation of $\xi^*$ on the interval $[-b, b]$, the monic orthogonal polynomials with respect to $d\xi^{(b)}(x)$ are

$$R_l(x, \xi^{(b)}) = b^l R_l\left(\frac{x}{b}, \xi^*\right) \tag{7.4.29}$$

where $R_l(x,\xi^*)$ is defined by (7.4.16) for the canonical moments in (7.4.28). Observing (7.4.27), we obtain for $R_l(x, \xi^{(b)})$,

$$K\left(\begin{array}{ccccccccc} \overbrace{-\dfrac{b^2}{2} \quad -\dfrac{b^2}{4}}^{k-1} \cdots -\dfrac{b^2}{4} & & -\dfrac{b^2}{2}p^*_{2k} - b^2 q^*_{2k}p^*_{2k+2} & \cdots & & -b^2 q^*_{2l-4}p^*_{2l-2} \\ x \quad x \quad \ldots & x & x & x \quad \ldots \quad x & & x \end{array}\right)$$

$$= K\left(\begin{array}{ccccccccc} \overbrace{-\dfrac{b^2}{2} \quad -\dfrac{b^2}{4}}^{k-1} \cdots -\dfrac{b^2}{4} & & -\dfrac{b^2}{2}p^*_{2k} & -1 & \cdots & -1 \\ x \quad x \quad \ldots & x & x & x \quad \ldots \quad x & & x \end{array}\right)$$

$$= 2^{-k+1}b^k\left[T_k\left(\frac{x}{b}\right)U_{l-k}\left(\frac{x}{2}\right) - \frac{U_{m-k+1}(\frac{b}{2})}{U_{m-k}(\frac{b}{2})}T_{k-1}\left(\frac{x}{b}\right)U_{l-1-k}\left(\frac{x}{2}\right)\right]. \tag{7.4.30}$$

Here we have used Lemma 4.4.1, (7.4.26) and the recursive relations (2.1.7) and (2.1.8) for the Chebyshev polynomials of the first and second kind, which imply that

$$b^k 2^{-k+1}T_k\left(\frac{x}{b}\right) = K\left(\begin{array}{cccc} \overbrace{-\dfrac{b^2}{2} \quad -\dfrac{b^2}{4}}^{k-1} \cdots -\dfrac{b^2}{4} \\ x \quad x \quad \ldots \quad x \quad x \end{array}\right)$$

$$U_k\left(\frac{x}{2}\right) = K\left(\begin{array}{cccc} \overbrace{-1 \quad -1 \ldots -1}^{k-1} \\ x \quad x \quad x \quad \ldots \quad x \quad x \end{array}\right).$$

Note that the case $k = 1$ has to be considered separately but gives the corresponding result in (7.4.30) for $k = 1$. The $L_2$-norm of this polynomial is obtained from (7.4.17), (7.4.27), and (7.4.29) (note that $p_{2j}^* = \frac{1}{2}$, $j = 1, \ldots, k - 1$ and $p_{2j}^* = \gamma_j(\xi^*), j = k, \ldots, m$),

$$k_l(\xi^{(b)}) = b^{2l}\left(\frac{1}{4}\right)^{k-1} p_{2k}^* \prod_{j=k+1}^{l} q_{2j-2}^* p_{2j}^* = \frac{b^{2k-1}}{2^{2k-2}} \frac{U_{m-k+1}\left(\frac{b}{2}\right)}{U_{m-k}\left(\frac{b}{2}\right)},$$

while the quantities $\alpha_l$ in (7.4.21) are given by $\alpha_l = 0, l = 1, \ldots, k - 1$ and

$$\alpha_l = \frac{U_{m-l}\left(\frac{b}{2}\right)\left[U_{m-l+1}\left(\frac{b}{2}\right) - U_{m-l-1}\left(\frac{b}{2}\right)\right]}{U_{m-k}\left(\frac{b}{2}\right)U_{m-k+1}\left(\frac{b}{2}\right)} = \frac{U_{2m-2l+1}\left(\frac{b}{2}\right)}{U_{m-k}\left(\frac{b}{2}\right)U_{m-k+1}\left(\frac{b}{2}\right)}$$

$(l = k, \ldots, m)$. This proves the assertion.                                         ∎

Theorem 7.4.3 shows that the structure of the solution of $(\mathcal{P}_I^{(b)})$ changes completely with the length of the interval $[-b, b]$. If $b \leq \sqrt{2}$, we obtain from (7.4.24) $k = m$, and consequently the sum of the squared leading coefficients of the polynomials $R_1^*, \ldots, R_m^*$ is maximized for the choice $R_l^*(x) = 0$ $(1 \leq l \leq m - 1)$ and $R_m^*(x) = T_m(x/b)$ with maximum value $(2^{m-1}b^{-m})^2$. If $b > \sqrt{2}$, the situation changes completely. In this case the index $1 \leq k \leq m$ defined by (7.4.24) depends on $m$ and $b$. The solution of the problem $(\mathcal{P}_I^{(b)})$ is given by (7.4.25). Finally, if $b \geq 2$, it follows that $k = 1$ and (7.4.25) simplifies to a linear combination of Chebyshev polynomials of the second kind:

$$R_l^*(x) = \frac{\sqrt{U_{2m-2l+1}\left(\frac{b}{2}\right)}}{\sqrt{b}U_m\left(\frac{b}{2}\right)}\left[U_l\left(\frac{x}{2}\right) - \frac{U_{m+1}\left(\frac{b}{2}\right)}{U_{m-1}\left(\frac{b}{2}\right)}U_{l-2}\left(\frac{x}{2}\right)\right], \qquad l = 1, \ldots, m.$$

**EXAMPLE 7.4.4**  Let $m = 3$, then we have to distinguish three different cases for the solution of the extremal problem $\mathcal{P}_I^{(b)}$:

1. If $b \leq \sqrt{2}$, it follows that $k = 3$; the optimal polynomials are given by

$$R_1^*(x) = R_2^*(x) = 0, R_3^*(x) = \pm T_3\left(\frac{x}{b}\right),$$

   and the maximum is $16b^{-6}$.

2. If $\sqrt{2} \leq b \leq \sqrt{3}$, then $k = 2$; the optimal polynomials are

$$R_1^*(x) = 0, R_2^*(x) = \pm\frac{b\sqrt{b^2 - 2}}{b^2 - 1}T_2\left(\frac{x}{b}\right),$$

$$R_3^*(x) = \pm\frac{1}{2(b^2 - 1)}\left[b^2 T_3\left(\frac{x}{b}\right) - (b^2 - 2)T_1\left(\frac{x}{b}\right)\right],$$

and the maximum value is $4b^{-2}(b^2 - 1)^{-1}$.

3. If $b \geq \sqrt{3}$, then $k = 1$; the optimal polynomials are

$$R_1^*(x) = \pm \frac{\sqrt{b^4 - 4b^2 + 3}}{b^3 - 2b} x,$$

$$R_2^*(x) = \pm \frac{1}{b\sqrt{b^2 - 2}} \left[ U_2\left(\frac{x}{2}\right) - \frac{b^4 - 3b^2 + 1}{b^2 - 1} \right],$$

$$R_3^*(x) = \pm \frac{1}{b^3 - 2b} \left[ U_3\left(\frac{x}{2}\right) - \frac{b^4 - 3b^2 + 1}{b^2 - 1} U_1\left(\frac{x}{2}\right) \right],$$

and the maximum value is $(b^2 - 1)/[b^2(b^2 - 2)]$.

In the remaining part of this section we will consider the index set $I = \{m - 1, m\}$. Thus the problem is to maximize the sum of the squared coefficients

$$m_{m-1}^2(P_{m-1}) + m_m^2(P_m)) \tag{7.4.31}$$

over the set of all polynomials (of degree $m - 1$ and $m$), satisfying

$$P_{m-1}^2(x) + P_m^2(x) \leq 1 \qquad \text{for all } x \in [-b, b]. \tag{7.4.32}$$

The solution of this problem can be obtained by the same reasoning as in Theorem 7.4.3 for $k = m$ and $k = m - 1$, and a detailed proof is omitted.

**Theorem 7.4.5** *The polynomials* $(R_{m-1}^*(x), R_m^*(x))$ *maximizing* (7.4.31) *subject to the restriction* (7.4.32) *are given by*

$$\left(0, \pm T_m\left(\frac{x}{b}\right)\right) \qquad \text{if } b \leq \sqrt{2},$$

$$\left(\pm \frac{b\sqrt{b^2 - 2}}{b^2 - 1} T_{m-1}\left(\frac{x}{b}\right), \pm \frac{1}{2(b^2 - 1)} \left[b^2 T_m\left(\frac{x}{b}\right) - (b^2 - 2) T_{m-2}\left(\frac{x}{b}\right)\right]\right) \text{if } b \geq \sqrt{2}.$$

*The maximum values in* (7.4.31) *are* $2^{2m-2} b^{-2m}$ *if* $b \leq \sqrt{2}$, *and* $2^{2m-4} b^{-(2m-4)}(b^2 - 1)^{-1}$ *if* $b \geq \sqrt{2}$, *respectively.*

# CHAPTER 8

# Canonical Moments and Random Walks

## 8.1 INTRODUCTION

A *random walk* is a discrete time homogeneous Markov chain whose state space $E$ is a subset of consecutive integers and whose one-step transition probabilities

$$P_{ij} = Pr(X_{n+1} = j | X_n = i), \qquad i, j \in E,$$

form a Jacobi matrix, that is,

$$P_{ij} = 0 \qquad \text{whenever } |i - j| > 1.$$

In this chapter we investigate the relation between canonical moments and random walks on the nonnegative integers. Our approach is based on an integral representation for the *transition matrix*

$$P = (P_{ij})_{ij \in E} \tag{8.1.1}$$

and its powers $(P^n)_{n=1,2,\ldots}$. To understand the basic idea, consider for a moment a random walk with a finite state space $E = \{0, \ldots, N\}$. Corresponding to the transition matrix $P$ in (8.1.1), we define a system of polynomials by $Q_{-1}(x) = 0$, $Q_0(x) = 1$,

$$xQ_n(x) = u_n Q_{n+1}(x) + h_n Q_n(x) + d_n Q_{n-1}(x) \tag{8.1.2}$$

$(0 \le n < N)$, where $d_n = P_{n,n-1}$, $h_n = P_{n,n}$, and $u_n = P_{n,n+1}$ are the *one-step transition probabilities*. If $\mathbf{Q}(x) = (Q_0(x), \ldots, Q_N(x))^T$ denotes the vector of these polynomials, then (8.1.2) can be conveniently rewritten in matrix form

$$x\mathbf{Q}(x) = P\mathbf{Q}(x). \tag{8.1.3}$$

**220**

It can be shown that the eigenvalues $\lambda_0, \ldots, \lambda_N$ of the $(N+1) \times (N+1)$ matrix $P$ are simple and real. For $j = 0, \ldots, N$ define vectors $Y_j = \alpha_j Q(\lambda_j) \in \mathbb{R}^{N+1}$, where

$$\alpha_j = \left( \sum_{k=0}^{N} \pi_k Q_k^2(\lambda_j) \right)^{-1}, \qquad j = 0, \ldots, N, \tag{8.1.4}$$

and

$$\pi_k = \frac{u_0 \cdots u_{k-1}}{d_1 \cdots d_k}, \qquad k = 1, \ldots, N$$

($\pi_0 = 1$). It then follows from (8.1.3) that $Y_0, \ldots, Y_N$ are $N+1$ linearly independent right eigenvectors of the transition matrix $P$. Similarly it follows that

$$Z_j = (\pi_0 Q_0(\lambda_j), \ldots, \pi_N Q_N(\lambda_j))^T \qquad (j = 0, \ldots, N)$$

define $N+1$ linearly independent left eigenvectors of $P$, i.e. $\lambda_j Z_j^T = Z_j^T P$. Moreover the definition of $\alpha_j$ shows that $Z_j^T Y_j = 1$ ($j = 0, \ldots, N$). On the other hand, we have

$$\lambda_j Z_i^T Y_j = Z_i^T P Y_j = \lambda_i Z_i^T Y_j,$$

which implies that $Z_i^T Y_j = 0$ whenever $i \neq j$. This means that the left and right eigenvectors are biorthogonal and the matrix of *n-step transition probabilities* may be written as

$$P^n = (Y_0, \ldots, Y_N) \begin{pmatrix} \lambda_0^n & & \\ & \ddots & \\ & & \lambda_N^n \end{pmatrix} \begin{pmatrix} Z_0^T \\ \vdots \\ Z_N^T \end{pmatrix}. \tag{8.1.5}$$

The equation (8.1.5) follows, since both matrices have the same eigenvalues and (linearly independent) eigenvectors. Comparing the elements in (8.1.5) yields

$$P_{ij}^n = \pi_j \sum_{k=0}^{N} \alpha_k \lambda_k^n Q_i(\lambda_k) Q_j(\lambda_k), \qquad i, j = 0, \ldots, N, \tag{8.1.6}$$

and putting $i = j = n = 0$ shows that $\sum_{k=0}^{N} \alpha_k = 1$. Finally *Gerschgorin's theorem* proves that the eigenvalues of $P$ are located in the interval $[-1, 1]$. Consequently the measure $\psi$ with masses $\alpha_i$ at the points $\lambda_i$ ($i = 0, \ldots, N$) defines a distribution on the interval $[-1, 1]$, and (8.1.6) can be rewritten as

$$P_{ij}^n = \pi_j \int_{-1}^{1} x^n Q_i(x) Q_j(x) d\psi(x).$$

Moreover, putting $n = 0$, $i = j$, it follows that

$$\pi_j^{-1} = \int_{-1}^{1} Q_j^2(x)d\psi(x),$$

which provides an integral representation for the $n$-step transition probabilities of a random walk with a finite state space:

$$P_{ij}^n = \frac{\int_{-1}^{1} x^n Q_i(x)Q_j(x)d\psi(x)}{\int_{-1}^{1} Q_j^2(x)d\psi(x)}. \tag{8.1.7}$$

Finally we remark that a consideration of the last row in (8.1.3) shows that the support points of the measure $\psi$ (i.e., the eigenvalues $\lambda_0, \ldots, \lambda_N$ of the transition matrix) are given by the $N + 1$ (distinct) zeros of the polynomial

$$(x - h_N)Q_N(x) - d_N Q_{N-1}(x).$$

In general, for a random walk with an infinite state space, the existence of a distribution $\psi$ such that (8.1.7) is valid can be derived from a spectral representation of a linear operator acting on an appropriate Hilbert space (see Karlin and McGregor, 1959a). Consequently the distribution $\psi$ will be called the *spectral measure* or *random walk measure* of the random walk $\{X_n\}_{n \in \mathbb{N}_0}$. In Section 8.2 we will present an elementary proof of the existence of a (unique) spectral measure $\psi$ that is based on the theory of canonical moments. As a consequence a characterization is obtained for the distributions on the interval $[-1, 1]$ which are spectral measures of some random walk. Conversely, it is possible to determine all random walks corresponding to a given spectral measure such that the integral representation (8.1.7) is valid.

While these results are interesting from a theoretical point of view, the integral representation can also be used for a much deeper analysis of the random walk. This is illustrated by deriving explicit expressions for the generating functions of the $n$-step transition probabilities in terms of the orthogonal polynomials (8.1.2) and the Stieltjes transform of the spectral measure $\psi$. There are intimate relationships between random walks, orthogonal polynomials, continued fractions, and canonical moments. Many results in moment theory are extremely useful in the analysis of random walks on the nonnegative integers.

## 8.2  RANDOM WALK MEASURES

Let $\{X_n\}_{n \in \mathbb{N}_0}$ be a random walk on a subset $E = \{0, \ldots, N\}$ of consecutive nonnegative integers with transition probabilities

$$P_{ij} = \text{Pr}(X_{n+1} = j | X_n = i), \qquad i, j \in E, \tag{8.2.1}$$

where $P_{ij} = 0$ if $|i - j| > 1$, and denote the one-step up-, down-, and holding transition probabilities by $u_j = P_{j,j+1}$, $d_j = P_{j,j-1}$, and $h_j = P_{j,j}$, respectively. If the state space is infinite, we formally define $N = \infty$ and assume that $u_j > 0$, $d_{j+1} > 0$, $d_j + u_j + h_j \leq 1$ ($j \geq 0$), and $d_0 = 0$. If $N < \infty$, the state space is finite, and we assume that $u_j > 0$ ($j = 0, \ldots, N - 1$), $u_N = 0$, $d_j > 0$ ($j = 1, \ldots, N$), and $d_0 = 0$. The case $u_j + d_j + h_j < 1$ (for some $j \in E$) is interpreted as a permanent absorbing state $j^*$ that can be reached from state $j$ with probability $1 - d_j - u_j - h_j$. Corresponding to the random walk, we define recursively polynomials $Q_j(x)$ of degree $j$ by $Q_0(x) = 1$, $Q_{-1}(x) = 0$,

$$xQ_n(x) = u_n Q_{n+1}(x) + h_n Q_n(x) + d_n Q_{n-1}(x) \tag{8.2.2}$$

whenever $0 \leq n < N$.

**Definition and Theorem 8.2.1** *A measure $\psi$ on the interval $[-1, 1]$ is called a spectral measure or a random walk measure if and only if there exists a random walk $\{X_n\}_{n \in \mathbb{N}_0}$ on the nonnegative integers with one-step down-, up-, and holding transition probabilities $d_j, u_j, h_j$ ($d_0 = 0$) such that one of the following two equivalent conditions is satisfied:*

1. *The polynomials $Q_n(x)$ defined by (8.2.2) and the n-step transition probabilities $P_{ij}^n$ of $\{X_n\}_{n \in \mathbb{N}}$ satisfy the integral representation*

$$P_{ij}^n = \Pr(X_n = j | X_0 = i) = \frac{\int_{-1}^{1} x^n Q_i(x) Q_j(x) d\psi(x)}{\int_{-1}^{1} Q_j^2(x) d\psi(x)} \tag{8.2.3}$$

*for all $i, j \in E$, $n \in \mathbb{N}_0$.*
2. *The polynomials $Q_n(x)$ defined by (8.2.2) are orthogonal with respect to the measure $d\psi(x)$.*

*Proof* Note that condition 2 is obtained from condition 1 by putting $i = 0$ and $n < j$ in (8.2.3). To show the converse Equations (8.2.2) are written in the form

$$x\mathbf{Q}(x) = P\mathbf{Q}(x),$$

where $\mathbf{Q}(x)$ denotes the column of polynomials $Q_0(x)$, $Q_1(x), \ldots$ and $P = (P_{ij})_{i,j=1}^{N}$ is the matrix of one-step transition probabilities. The vector $\mathbf{Q}(x)$ is of dimension $N + 1$ if the state space is finite. A simple iteration gives

$$x^n \mathbf{Q}(x) = P^n \mathbf{Q}(x). \tag{8.2.4}$$

If the polynomials in (8.2.2) are orthogonal with respect to the measure $d\psi(x)$, then the representation (8.2.3) is obtained by multiplying the $(i + 1)$th row of (8.2.4) by $Q_j(x)$ and integrating with respect to the measure $d\psi(x)$.    ∎

Multiplying (8.2.2) by $Q_{n-1}(x)$ and integrating with respect to $d\psi(x)$ shows inductively that the $L_2$-norm of the polynomials $Q_j(x)$ is given by

$$\int_{-1}^{1} Q_n^2(x) d\psi(x) = \frac{d_1 \ldots d_n}{u_0 \ldots u_{n-1}} =: \frac{1}{\pi_n}, \qquad 1 \leq n \leq N. \qquad (8.2.5)$$

For later purposes we also define $\pi_0 = 1$.

**EXAMPLE 8.2.2**   Consider the symmetric random walk $\{X_n\}_{n \in \mathbb{N}_0}$ on the nonnegative integers with reflecting barrier at the origin, that is, $u_0 = 1$, $u_j = d_j = \frac{1}{2}$ $(j \geq 1)$. In this case the recursion (8.2.2) coincides with (2.1.7), which shows that the polynomials $Q_n(x)$ are given by the Chebyshev polynomials $T_n(x)$ of the first kind. Consequently the arc-sine distribution with density $1/(\pi\sqrt{1 - x^2})$ is the spectral measure of the random walk $\{X_n\}_{n \in \mathbb{N}_0}$. The $n$-step transition probabilities are obtained from (8.2.3) as

$$P_{ij}^n = Pr(X_n = j | X_0 = i) = \frac{\int_{-1}^{1} x^n T_i(x) T_j(x)(1 - x^2)^{-1/2} dx}{\int_{-1}^{1} T_j^2(x)(1 - x^2)^{-1/2} dx}$$

$$= \frac{\int_0^{\pi} \cos^n u \cos(iu) \cos(ju) du}{\int_0^{\pi} \cos^2(ju) du}, \qquad i, j \in \mathbb{N}_0,$$

where the second equality follows from the trigonometric representation of the Chebyshev polynomials of the first kind in (2.1.5).

Example 8.2.2 demonstrates the calculation of the spectral measure corresponding to the symmetric random walk from the recurrence relation of the polynomials $Q_j(x)$ in (8.2.2). However, Definition 8.2.1 provides several open questions. The existence of a spectral measure for a given random walk is not clear in general. Conversely, a distribution $\psi$ on the interval $[-1, 1]$ is not necessarily a spectral measure of a random walk on the nonnegative integers. Moreover, even if this question can be answered positively, it is not clear if a process $\{X_n\}_{n \in \mathbb{N}_0}$ corresponding to a given spectral measure is unique. We will investigate some of these questions in the subsequent discussion.

The polynomials in (8.2.2) are more conveniently put in monic form as $P_0(x) = 1$, $P_1(x) = x - h_0$, and for $0 \leq n \leq N$,

$$P_{n+1}(x) = (x - h_n)P_n(x) - u_{n-1}d_n P_{n-1}(x). \qquad (8.2.6)$$

Note that if $N < \infty$, $P_{N+1}(x)$ is formally defined here and is proportional to $(x - h_N)Q_N(x) - d_N Q_{N-1}(x)$.

It is important to note that the measure $\psi$ on the interval $[-1, 1]$ uniquely determines the polynomials in monic form and the polynomials are uniquely determined by the coefficients $h_n$ and $u_{n-1}d_n$. Thus, if the polynomials in (8.2.2) or (8.2.6) are orthogonal with respect to some distribution $\psi$ on the interval $[-1, 1]$, then Corollary 2.3.5 shows that for some set of canonical moments, we must have

$$h_0 = -1 + 2\zeta_1,$$

$$h_n = -1 + 2\zeta_{2n} + 2\zeta_{2n+1}, \qquad (8.2.7)$$

$$u_{n-1}d_n = 4\zeta_{2n-1}\zeta_{2n}, \qquad 1 \le n \le N.$$

All solutions of (8.2.7) for $\{d_i, u_i, h_i\}_{i \in E}$ give the same monic orthogonal polynomials and hence will have the same spectral measure.

These equations are more readily analyzed if there is no absorption or the transition probabilities add to one, since in this case the mapping from the set $\{d_i, u_i, h_i\}_{i \in E}$ to the canonical moments turns out to be one to one. This normalization is accomplished by dividing (8.2.2) by $Q_n(1)$. Define transition probabilities

$$u'_n = \frac{Q_{n+1}(1)}{Q_n(1)} u_n, \quad h'_n = h_n, \quad d'_n = \frac{Q_{n-1}(1)}{Q_n(1)} d_n, \qquad 0 \le n < N, \qquad (8.2.8)$$

where in the case of a finite state space $(N < \infty)$, $u'_N = 0$, $h'_N = h_N$ and $d'_N = d_N Q_{N-1}(1)/Q_N(1)$. The "standardized" polynomials

$$R_n(x) = \frac{Q_n(x)}{Q_n(1)} \qquad (n \in E) \qquad (8.2.9)$$

then satisfy the recursion $R_{-1}(x) = 0$, $R_0(x) = 1$,

$$xR_n(x) = u'_n R_{n+1}(x) + h'_n R_n(x) + d'_n R_{n-1}(x), \qquad 1 \le n < N. \qquad (8.2.10)$$

Clearly $u'_{n-1}d'_n = u_{n-1}d_n$. Since $R_n(1) = 1$, Equation (8.2.10) implies that $d'_n + h'_n + u'_n = 1$, which proves the following result:

**Lemma 8.2.3** *The transition probabilities $\{d'_n, u'_n, h'_n\}_{n \in E}$ defined in (8.2.8) satisfy (8.2.7), and $h'_0 + u'_0 = 1$, $d'_n + h'_n + u'_n = 1$ $(1 \le n < N)$. Moreover, if $N < \infty$, $h'_N + d'_N = 1$ if and only if $x = 1$ is a zero of the polynomial $P_{N+1}(x)$ defined in (8.2.6).*

Note that by the preceding discussion the random walks with one-step transition probabilities $\{d_n, u_n, h_n\}_{n \in E}$ and $\{d'_n, u'_n, h'_n\}_{n \in E}$ have the same spectral measure.

The system

$$h'_0 = -1 + 2\zeta_1,$$

$$u'_{n-1}d'_n = 4\zeta_{2n-1}\zeta_{2n}, \tag{8.2.11}$$

$$h'_n = -1 + 2\zeta_{2n} + 2\zeta_{2n+1}, \qquad 1 \le n \le N,$$

with $d'_0 = 0$, $d'_n + h'_n + u'_n = 1$ can be simplified by noting that

$$-1 + 2\zeta_{2n} + 2\zeta_{2n+1} = 1 - 2p_{2n-1}p_{2n} - 2q_{2n}q_{2n+1},$$

$$4\zeta_{2n-1}\zeta_{2n} = (2q_{2n-2}q_{2n-1})(2p_{2n-1}p_{2n}).$$

In this case a simple induction shows that if $d'_0 = 0$ and $d'_n + h'_n + u'_n = 1$ for $0 \le n < N$, then

$$u'_0 = 2q_1,$$

$$u'_n = 2q_{2n}q_{2n+1}, \tag{8.2.12}$$

$$d'_n = 2p_{2n-1}p_{2n}, \qquad 1 \le n \le N.$$

Thus the solution of (8.2.11) for $d'_n$, $u'_n$, $h'_n$ is unique if $d'_n + u'_n + h'_n = 1$ ($0 \le n < N$, $d'_0 = 0$). Conversely, the solution of (8.2.12) for the canonical moments $p_k$ is clearly unique, and an induction argument shows that

$$0 < p_{2n} \le d'_n, \tag{8.2.13}$$

$$\tfrac{1}{2} \le p_{2n-1} < 1.$$

Therefore for given transition probabilities $\{d'_j, u'_j, h'_j\}_{j \in E}$ ($d'_j + u'_j + h'_j = 1$, $d'_0 = 0$) the solution $\{p_n\}_{n \in \mathbb{N}}$ of (8.2.11) is a sequence of canonical moments, and there exists a corresponding distribution $\psi$ on the interval $[-1, 1]$.

**Theorem 8.2.4 (Karlin and McGregor)**   *Let $\{X_n\}_{n \ge 0}$ denote a random walk on the nonnegative integers with one-step transition probabilities $d_n$, $u_n$, $h_n$ satisfying $d_0 = 0$, $u_n + h_n + u_n \le 1$ ($n \in E$). Then there exist a unique spectral measure $\psi$ on the interval $[-1, 1]$ such that the representation (8.2.3) is valid. The canonical moments of this measure are given by*

$$q_{2n-1} = \frac{u'_{n-1}/2}{|1|} - \frac{d'_{n-1}/2}{|1|} - \frac{u'_{n-2}/2}{|1|} - \cdots - \frac{u'_0/2}{|1|}, \tag{8.2.14}$$

$$p_{2n} = \frac{d'_n/2}{|1|} - \frac{u'_{n-1}/2}{|1|} - \frac{d'_{n-1}/2}{|1|} - \cdots - \frac{u'_0/2}{|1|},$$

*for $1 \leq n \leq N$. Here the one-step transition probabilities $\{d'_n, u'_n, h'_n\}_{n \in E}$ are obtained from $\{d_n, u_n, h_n\}_{n \in E}$ by the transformations (8.2.8). Moreover for every spectral measure $\psi$ there exists a unique random walk $\{X_n\}_{n \in \mathbb{N}_0}$ on the nonnegative integers with one-step up-, down-, and holding transition probabilities $d'_n, u'_n, h'_n$ satisfying $d'_0 = 0$, $d'_n + u'_n + h'_n = 1$ $(0 \leq n < N)$.*

*Proof*  The assertion is established by showing that the sequence of polynomials in (8.2.2) determines a sequence of canonical moments. By Theorem 8.2.1 the corresponding distribution turns out to be the spectral measure of the process. Consider first the case $N = \infty$. Starting with $d_i$, $u_i$, $h_i$, one calculates $d'_i$, $u'_i$, $h'_i$ from (8.2.8) and arrives at the canonical moments in (8.2.14) by a simple induction. Since (8.2.13) holds, there exists a unique measure corresponding to these canonical moments. By Lemma 8.2.3 this measure is the (unique) spectral measure of the random walk $\{X_n\}_{n \in \mathbb{N}_0}$.

Conversely, for a given spectral measure (in terms of its canonical moments) the preceding discussion shows that the solution of (8.2.11) is unique if $d'_0 = 0$, $d'_n + u'_n + h'_n = 1$ $(n \geq 0)$. Thus there exists a unique random walk with no absorbing states corresponding to the spectral measure $\psi$.

The case $N < \infty$ requires further elaboration. It follows that $\psi$ must now be supported on the $N + 1$ zeros of the polynomial $P_{N+1}(x)$, which is proportional to $(x - h_N)Q_N(x) - d_N Q_{N-1}(x)$. From the theory in Section 2.2 this will be the case when $\zeta_{2N+1}\zeta_{2N+2} = 0$, or equivalently when $p_{2N} = 1$, $p_{2N+1} = 1$, $p_{2N+1} = 0$ or $p_{2N+2} = 0$. From Equations (8.2.11) and (8.2.12) for $n = N$ we have

$$1 - h'_N = 2p_{2N-1}p_{2N} + 2q_{2N}q_{2N+1},$$

$$d'_N = 2p_{2N-1}p_{2N},$$

showing that $d'_N + h'_N = 1$ if and only if $q_{2N}q_{2N+1} = 0$. Thus, if $p_{2N} = 1$ or $p_{2N} < 1$ and $p_{2N+1} = 1$, the measure $\psi$ is determined by its canonical moment sequence, which is obtained from (8.2.14) [note that (8.2.13) implies that $p_{2N} > 0$]. If

$$0 < 1 - h_N - d_N = 2q_{2N}q_{2N+1},$$

then $p_{2N} < 1$ and $p_{2N+1} < 1$. By (8.2.13), $p_{2N} > 0$, and we either have $p_{2N+1} = 0$ ($q_{2N+1} = 1$) or $0 < p_{2N+1}$ ($q_{2N+1} < 1$) and $p_{2N+2} = 0$. Thus in all cases the measure $\psi$ is defined by its canonical moment sequence                                         ∎

The following result characterizes a spectral measure in terms of its canonical moments. The proof follows from the previous discussion.

**Theorem 8.2.5**  *Let $\psi$ denote a probability measure on the interval $[-1, 1]$ with canonical moments $p_1, p_2, \ldots$ ..*

1. *$\psi$ is the spectral measure of a random walk if and only if*

$$2p_{2n-1}p_{2n} + 2q_{2n}q_{2n+1} \leq 1 \qquad (8.2.15)$$

*whenever $1 \leq n \leq N$.*

2. *If $\psi$ is a spectral measure of a random walk, then the canonical moments of odd order satisfy $p_{2n-1} \geq \frac{1}{2}$ whenever $1 \leq n \leq N$.*

Some further remarks when $N < \infty$ are in order. By Lemma 8.2.3 and the proof of Theorem 8.2.4, it follows that $h'_N + d'_N = 1$ if and only if $x = 1$ is a support point of $\psi$ or equivalently $q_{2N}q_{2N+1} = 0$. In this case $p_{2N} = 1$ or $p_{2N+1} = 1$. This is in agreement with the general theory of canonical moments, since in these cases the corresponding measure is the upper principal representation and must have mass at the upper end point $x = 1$.

A random walk with finite state space $E = \{0, \ldots, N\}$ is called recurrent if there are no absorbing states; otherwise, it is called transient. From the equations (8.2.7) it can be shown inductively that

$$u_0 \leq 2q_1,$$

$$u_n \leq 2q_{2n}q_{2n+1}, \qquad 1 \leq n \leq N-1,$$

$$d_n \geq 2p_{2n-1}p_{2n}, \qquad 1 \leq n \leq N.$$

If $q_{2N}q_{2N+1} = 0$, an induction in (8.2.7) from the top end shows the reverse inequalities. Equality must then occur and $d_n + u_n + h_n = 1, 0 \leq n \leq N$. Thus in the nonabsorbing or recurrent case the process is actually unique. Moreover, if there is a strict inequality $d_n + u_n + h_n < 1$ for some $n \in E$, then $h_N + d_N < 1$. It follows that either all of the processes corresponding to a given $\psi$ (with finite support) are recurrent (in which case the process is unique) or they are all transient. The recurrent processes correspond to $p_{2N} = 1$ or $p_{2N+1} = 1$, while the transient processes correspond to $p_{2N+1} = 0$ or $p_{2N+2} = 0$.

Finally the case $p_{2N} = 1$ is further specialized. It was shown in (8.2.13) that $p_{2n-1} \geq \frac{1}{2}$ for all $1 \leq n \leq N$. If $p_{2N} = 1$, one can again start with the value $n = N$ and show similarly that $p_{2n-1} \leq \frac{1}{2}$ so that $p_{2n-1} = \frac{1}{2}$ for $1 \leq n \leq N$. In this situation $h_n = h'_n = 0, 0 \leq n \leq N, u_0 = u'_0 = 1$ and $d_N = d'_N = 1$. The spectral measure $\psi$ is symmetric with $N + 1$ support points including $-1$ and $+1$.

**EXAMPLE 8.2.6**  Gambler's Ruin Problem. Assume that an individual (player A) has initial fortune $x$ and plays against an adversary (player B) with limited fortune $y$ in a series of contests. If in the game the fortune of player A is $k$, the probability of winning or losing one unit is $u_k$ or $d_k$ $(u_k + d_k = 1)$, respectively. A player is ruined if his entire fortune is lost and he loses an additional game. The corresponding random walk has finite state space $E = \{0, \ldots, N\}$, where $N = x + y > 0$, and the one-step transition probabilities are

$$P_{ii+1} = u_i, \qquad i = 0, \ldots, N-1,$$
$$P_{ii-1} = d_i, \qquad i = 1, \ldots, N$$

$(u_i + d_i = 1; \; i = 1, \ldots, N-1)$. Absorption occurs from state 0 and $N$ with respective probabilities $1 - u_0$ and $1 - d_N$ corresponding to the ruin of player A or player B. Since $h_i = 0 \; (i = 0, \ldots, N)$, it follows from the spectral representation (8.2.3) that all odd moments of the spectral measure $\psi$ vanish, which implies its symmetry.

The process is transient, and we obtain from the previous discussion that the canonical moment sequence of the spectral measure $\psi$ terminates with either $p_{2N+1} = 0$ or $p_{2N+2} = 0$. The case $p_{2N+1} = 0$ is impossible because of the symmetry of $\psi$. Let $z_N$ denote the largest zero of the polynomial $zQ_N(z) - d_N Q_{N-1}(z)$ (i.e., the largest support point of the spectral measure $\psi$), then $z_N < 1$ because $\psi$ is a lower principal representation. Probabilistically it is clear that the probability of no ruin at time $n$ converges to zero. The rate of convergence is determined by the largest support point $z_N$ of the spectral measure, namely

$$P(X_n \in \{0, \ldots, N\} \mid X_0 = x) = O(z_N^n) \qquad \text{as } n \to \infty.$$

This equality follows directly from the Karlin-McGregor representation (8.2.3) by observing that the integration with respect to the discrete (symmetric) measure $\psi$ reduces to a finite sum.

Consider now the "classical" *Gambler's ruin problem* where $u_j = u$, $d_j = d$, $u + d = 1$. The polynomials $Q_j(z)$ are given recursively by $Q_0(z) = 1$, $Q_1(z) = z/u$ and

$$uQ_{n+1}(z) = zQ_n(z) - dQ_{n-1}(z), \qquad n = 0, 1, \ldots, N-1.$$

A simple induction shows that

$$Q_n(z) = \left(\frac{d}{u}\right)^{n/2} U_n\left(\frac{z}{2\sqrt{ud}}\right), \qquad n = 0, \ldots, N,$$

where $U_n(z)$ is the Chebyshev polynomial of the second kind defined by the recurrence relation (2.1.8). Applying this relation one more time, we obtain

$$H_{N+1}(z) = zQ_N(z) - dQ_{N-1}(z) = d\left(\frac{d}{u}\right)^{(N-1)/2} U_{N+1}\left(\frac{z}{2\sqrt{ud}}\right),$$

and the trigonometric representation (2.1.6) for the Chebyshev polynomials of the second kind gives for the largest zero of $H_{N+1}(z)$,

$$z_N = 2\sqrt{ud}\cos\left(\frac{\pi}{N+2}\right).$$

This means that the probability of no absorption at time $n$ converges to 0 with the rate $(2\sqrt{ud}\cos(\pi/(N+2)))^n$.

In what follows let

$$P_{ij}(z) = \sum_{n=0}^{\infty} P_{ij}^n z^n, \qquad |z| < 1, \tag{8.2.16}$$

denote the *generating function* of the $n$-step transition probabilities from state $i$ to state $j$ for a random walk $\{X_n\}_{n\in\mathbb{N}_0}$. If $\psi$ is the corresponding spectral measure, then the Stieltjes transform of $\psi$ and $P_{00}(z)$ are related by

$$S(z, \psi) = \int_{-1}^{1} \frac{d\psi(x)}{z - x} = \int_{-1}^{1} \frac{1}{z} \sum_{n=0}^{\infty} \left(\frac{x}{z}\right)^n d\psi(x)$$

$$= \frac{1}{z} \sum_{n=0}^{\infty} P_{00}^n \frac{1}{z^n} = \frac{1}{z} P_{00}\left(\frac{1}{z}\right), \tag{8.2.17}$$

where (8.2.17) holds whenever $|z| > 1$, and we have used the integral representation (8.2.3) for the second equality. An application of Theorem 3.3.3, (3.2.11) and (8.2.7) yields the following continued fraction expansion for the generating function of the return probabilities to state 0:

**Corollary 8.2.7** *Let $\{X_n\}_{n\in\mathbb{N}_0}$ denote a random walk on the nonnegative integers and $\psi$ the corresponding spectral measure with canonical moments $p_1, p_2, \ldots$; then for all $|z| < 1$,*

$$P_{00}(z) = \sum_{n=0}^{\infty} P_{00}^n z^n = \frac{1}{z} S\left(\frac{1}{z}, \psi\right)$$

$$= \frac{1}{\left|1 + (1 - 2\zeta_1)z\right|} - \frac{4\zeta_1\zeta_2 z^2}{\left|1 + (1 - 2\zeta_2 - 2\zeta_3)z\right|} - \frac{4\zeta_3\zeta_4 z^2}{\left|1 + (1 - 2\zeta_4 - 2\zeta_5)z\right|} - \cdots$$

$$= \frac{1}{\left|1 - h_0 z\right|} - \frac{u_0 d_1 z^2}{\left|1 - h_1 z\right|} - \frac{u_1 d_2 z^2}{\left|1 - h_2 z\right|} - \cdots \tag{8.2.18}$$

*In the case of a finite state space ($N < \infty$), these expressions terminate at $\zeta_{2N-1}\zeta_{2N}$ and $u_{N-1}d_N$, respectively.*

**EXAMPLE 8.2.8** Consider the $(d, u, h)$ random walk $\{X_n\}_{n \in \mathbb{N}_0}$ on the non-negative integers, that is,

$$
\begin{aligned}
d'_j &= d && (j \geq 1), d'_0 = 0, \\
h'_j &= h && (j \geq 1), h'_0 = 1 - u, \\
u'_j &= u && (j \geq 0),
\end{aligned}
\tag{8.2.19}
$$

where $d, u, h$ are positive with $d + u + h = 1$. By Corollary 8.2.7 and Lemma 3.2.3 the Stieltjes transform of the corresponding spectral measure $\psi$ is given by

$$
\begin{aligned}
S(z, \psi) &= \frac{1}{\left| z - 1 + u \right.} - \frac{ud}{\left| z - h \right.} - \frac{ud}{\left| z - h \right.} - \cdots \\
&= \left[ z - 1 + u - \frac{1}{2} \left( z - h - \sqrt{(z-h)^2 - 4ud} \right) \right]^{-1} \\
&= \frac{z - h - 2d - \sqrt{(z-h)^2 - 4ud}}{2d(1 - z)}, \qquad |z| > 1,
\end{aligned}
\tag{8.2.20}
$$

where the sign of the square root is determined by

$$
\left| \frac{z - h}{2\sqrt{ud}} - \sqrt{\frac{(z-h)^2}{4ud} - 1} \right| < 1.
\tag{8.2.21}
$$

The generating function of the return probabilities to state 0 is given by $P_{00}(z) = z^{-1} S(z^{-1}, \psi)$. Applying Theorem 3.6.5 and 3.6.7 it now follows that $\psi$ has the density

$$
\frac{1}{2\pi d} \sqrt{\frac{4ud - (x - h)^2}{1 - x}} \qquad \text{if } (x - h)^2 < 4ud,
$$

while there is an additional jump of size $1 - u/d$ at the point $x = 1$ if $d \geq u$.

In what follows let

$$
f^n_{ij} = \Pr(X_n = j, X_\nu \neq j, \ \nu = 1, \ldots, n - 1 \mid X_0 = i)
\tag{8.2.22}
$$

denote the probability that starting from state $i$, the first transition to state $j$ occurs at the $n$th transition, and define

$$
F_{ij}(z) = \sum_{n=1}^{\infty} f^n_{ij} z^n, \qquad |z| < 1,
\tag{8.2.23}
$$

as the generating functions of these probabilities. The generating functions in (8.2.23) and (8.2.16) are related by the well-known identities

$$P_{ij}(z) = F_{ij}(z)P_{jj}(z), \qquad i \neq j,$$

$$(8.2.24)$$

$$P_{ii}(z) = 1 + F_{ii}(z)P_{ii}(z),$$

and they play an important role in the analysis of Markov chains. The random walk is called *recurrent* if and only if

$$F_{ii}(1) = \lim_{z \to 1^-} F_{ii}(z) = \sum_{n=1}^{\infty} f_{ii}^n = 1, \quad (i \in E)$$

which means that if the process starts from any state $i$, then the probability of returning to this state after some finite length of time is one. These quantities are either all one or all less than one. A nonrecurrent random walk is called *transient*. It is clear probabilistically that a positive probability of absorption (i.e., $d_j + h_j + u_j < 1$ for some $j$) makes the process transient. In the case of recurrence certain limits of $P_{ij}^n$ (as $n$ tends to infinity) exist, i.e.

$$\lim_{n \to \infty} P_{ij}^{nd+r} = \begin{cases} \lambda_j & \text{if } d = 2 \text{ and } (i+j+r) = 0 \mod 2, \\ 0 & \text{if } d = 2 \text{ and } (i+j+r) = 1 \mod 2, \\ \lambda_j & \text{if } d = 1, \end{cases}$$

where

$$\lambda_j = \lambda_j(d) = d \left[ \sum_{m=1}^{\infty} m f_{jj}^m \right]^{-1},$$

the right-hand side is interpreted as 0 if the sum diverges, $r \in \{0, 1\}$ is a fixed number, and $d \in \{1, 2\}$ denotes the *period* of the random walk. This means that $d = 2$ if and only if all holding probabilities vanish (*periodic case*) and $d = 1$ if there is at least one $i \in E$ with $h_i > 0$ (*aperiodic* case). A recurrent random walk is called *positive recurrent* if the expected first passage times are all finite, that is,

$$\sum_{m=1}^{\infty} m f_{jj}^m < \infty, \qquad j \in E;$$

otherwise, it is called *null recurrent*. In the aperiodic case $d = 1$ $\{\lambda_j\}_{j \in \mathbb{N}}$ is called the *stationary distribution* of the random walk $\{X_n\}_{n \in \mathbb{N}_0}$. From (8.2.24) it follows that the random walk is recurrent if and only if $\lim_{z \to 1^-} P_{00}(z) = \infty$, or equivalently the integral $\int_{-1}^{1} (1-x)^{-1} d\psi(x)$ diverges. The process $\{X_n\}_{n \in \mathbb{N}_0}$ is positive recurrent if and only if $\lambda_0 = \lim_{n \to \infty} P_{00}^{2n}$ is positive. The dominated convergence theorem and the integral representation $P_{00}^{2n} = \int_{-1}^{1} x^{2n} d\psi(x)$ imply that this is the case if and only if $\psi$ has a jump at $x = 1$ or $x = -1$. If $\psi$ has no jump at $x = 1$, then

$$\psi(-1) = \lim_{n \to \infty} \left( -\int_{-1}^{1} x^{2n+1} d\psi(x) + \int_{-1^-}^{1} x^{2n+1} d\psi(x) \right)$$

$$= -\lim_{n \to \infty} P_{00}^{2n+1} \leq 0,$$

so there is also no jump at $x = -1$. If $\{X_n\}_{n \in \mathbb{N}_0}$ is positive recurrent, then

$$\lambda_j = \lim_{n \to \infty} P_{jj}^{nd} = \frac{d \cdot \psi(1)}{\int_{-1}^{1} Q_j^2(x) d\psi(x)} = d\psi(1)\pi_j', \qquad (8.2.25)$$

where

$$\pi_0' = 1, \qquad \pi_k' = \frac{u_0' u_1' \ldots u_{k-1}'}{d_1' d_2' \ldots d_k'}, \qquad 1 \leq k \leq N. \qquad (8.2.26)$$

The equality in (8.2.25) follows from $Q_j(1) = 1$ because there is no absorption in the recurrent case. Consider first the aperiodic random walk ($d = 1$), which occurs if and only if $h_i > 0$ for some $i \in E$. In this case $\lim_{n \to \infty} P_{ij}^n$ exists, which implies that the spectral measure $\psi$ has no jump at $x = -1$. On the other hand, if $d = 2$ (i.e., $h_i = 0$ for all $i \in E$), then $P_{00}^{2n+1} = 0$ ($n \in \mathbb{N}$) and the Karlin-McGregor representation shows that $\psi$ is symmetric. In both cases the second equality in (8.2.25) follows from Lebesgue's dominated convergence theorem. Now (8.2.25) implies ($\sum \lambda_j = d$) that the size of the jump at the point $x = 1$ is

$$\psi(1) = \left( \sum_{j \in E} \pi_j' \right)^{-1}. \qquad (8.2.27)$$

We summarize some of these observations in the following lemma:

**Lemma 8.2.9** *If $\psi$ denotes the spectral measure of a random walk $\{X_n\}_{n \in \mathbb{N}_0}$ on the nonnegative integers, then*

$\{X_n\}_{n \in \mathbb{N}_0}$ *is recurrent if and only if* $\int_{-1}^{1} \dfrac{d\psi(x)}{1 - x}$ *diverges,*

$\{X_n\}_{n \in \mathbb{N}_0}$ *is positive recurrent if and only if the spectral measure $\psi$ has a jump at the point $x = 1$.*

In the case of a finite state space, it is probabilistically clear that (positive) recurrence is equivalent to no absorption. The following result provides a well-known characterization of this property in terms of the transition probabilities of the process:

**Theorem 8.2.10**  *If* $\{X_n\}_{n \in \mathbb{N}_0}$ *is a random walk with infinite state space and one-step transition probabilities* $\{d'_j, h'_j, u'_j\}_{j \geq 0}$ *adding to one,* $d'_0 = 0$, *then*

$$\{X_n\}_{n \in \mathbb{N}_0} \text{ is recurrent if and only if } \sum_{j=0}^{\infty} \frac{1}{u'_j \pi'_j} \text{ diverges}$$

$$\{X_n\}_{n \in \mathbb{N}_0} \text{ is positive recurrent if and only if } \begin{cases} \displaystyle\sum_{j=0}^{\infty} \frac{1}{u'_j \pi'_j} \text{ diverges} \\ \displaystyle\sum_{j=0}^{\infty} \pi'_j \text{ converges} \end{cases}$$

*Proof*  For the first part of the proof assume that the random walk is transient, which means that $P_{00}(1) < \infty$. By Corollary 8.2.7, Equations (3.2.11), (3.2.15), (3.2.16), and Lemma 3.2.5, we obtain

$$2P_{00}(1) = \left|\frac{1}{1 - \zeta_1}\right| - \left|\frac{\zeta_1 \zeta_2}{1 - \zeta_2 - \zeta_3}\right| - \left|\frac{\zeta_3 \zeta_4}{1 - \zeta_4 - \zeta_5}\right| - \cdots$$

$$= \left|\frac{1}{1}\right| - \left|\frac{\zeta_1}{1}\right| - \left|\frac{\zeta_2}{1}\right| - \cdots \tag{8.2.28}$$

$$= 1 + \sum_{k=1}^{\infty} \frac{p_1 \cdots p_k}{q_1 \cdots q_k} = 1 + \frac{p_1}{q_1} + \sum_{k=1}^{\infty} \left(\frac{p_1 \cdots p_{2k}}{q_1 \cdots q_{2k}} + \frac{p_1 \cdots p_{2k+1}}{q_1 \cdots q_{2k+1}}\right)$$

$$= \frac{2}{u'_0} + 2\sum_{k=1}^{\infty} \frac{d'_1 \cdots d'_k}{u'_0 u'_1 \cdots u'_k}(q_{2k+1} + p_{2k+1}) = 2\sum_{k=0}^{\infty} \frac{1}{u'_k \pi'_k},$$

where we have used (8.2.12) for the equality from the third to the fourth line. This proves the first part. For the second part we note that by Lemma 8.2.9 positive recurrence is equivalent to the fact that the spectral measure $\psi$ has a jump at $x = 1$, and then we apply (8.2.27).  ∎

**REMARK 8.2.11**  The derivation of Theorem 8.2.10 proves a special case of a well-known relation between orthogonal polynomials and the corresponding measure of orthogonality. To explain the general statement, we let $P_n(x) = \sqrt{\pi_n}Q_n(x)$ denote the orthonormal polynomial with respect to the spectral measure $\psi$. Then (8.2.27) can be represented as

$$\psi(1) = \left[\sum_{n=0}^{\infty} P_n^2(1)\right]^{-1} = \left[\sum_{n=0}^{\infty} \pi'_n\right]^{-1}.$$

The general version of this result is as follows: Let $\alpha$ be a measure on the real line with existing moments $\{c_n\}_{n\in\mathbb{N}}$ such that the corresponding moment problem is determinate (i.e., $\alpha$ is the unique measure with given moments $\{c_n\}_{n\in\mathbb{N}}$); then the mass of $\alpha$ at any point $x \in \mathbb{R}$ can be represented as

$$\alpha(x) = \left[\sum_{j=0}^{\infty} P_j^2(x)\right]^{-1},$$

where the polynomial $P_n(x)$ is the $n$th orthonormal polynomial with respect to the measure $d\alpha(x)$ (see Shohat and Tamarkin (1943)).

**EXAMPLE 8.2.12** Consider the $(d, u, h)$ random walk on the nonnegative integers of Example 8.2.8. If $d \geq u$, the process is recurrent (which is well known) and the Stieltjes transform diverges at $z = 1$ [see (8.2.20) and (8.2.21)]. If $d > u$, the random walk $\{X_n\}_{n\in\mathbb{N}_0}$ is positive recurrent and the spectral measure $\psi$ has a jump at $x = 1$. The size of this jump is given by

$$\psi(1) = \left[\sum_{j=0}^{\infty} \pi_j'\right]^{-1} = \left[\sum_{j=0}^{\infty}\left(\frac{u}{d}\right)^j\right]^{-1} = 1 - \frac{u}{d},$$

which was already proved in Example 8.2.8 using the Stieltjes inversion formula.

## 8.3  PROCESSES WITH THE SAME SPECTRAL MEASURE

By the discussion in Section 8.2, all solutions $\{d_j, u_j, h_j\}_{j\in E}$ of

$$h_0 = -1 + 2\zeta_1,$$
$$h_k = -1 + 2\zeta_{2k} + 2\zeta_{2k+1}, \tag{8.3.1}$$
$$u_{k-1}d_k = 4\zeta_{2k-1}\zeta_{2k}, \qquad 1 \leq k \leq N,$$

correspond to random walks with the same spectral measure. To describe the set of all transition probabilities $\{d_n, u_n, h_n\}_{n\in E}$ satisfying the system of equations in (8.3.1), define (for a given set of canonical moments)

$$\kappa_j := \left|\frac{2\zeta_{2j+1}\zeta_{2j+2}}{1 - \zeta_{2j+2} - \zeta_{2j+3}}\right| - \left|\frac{\zeta_{2j+3}\zeta_{2j+4}}{1 - \zeta_{2j+4} - \zeta_{2j+5}}\right| - \cdots \tag{8.3.2}$$

$(j = 0, 1, \ldots)$ and

$$S_k = 1 + \sum_{\ell=k}^{\infty}\prod_{i=k}^{\ell}\frac{p_i}{q_i}, \qquad k = 1, 2, \ldots.$$

Note that in the case of a finite state space ($N < \infty$), some care is necessary in these definitions. If the sequence of canonical moments terminates with $p_l = 0$ for $l \in \{2N+1, 2N+2\}$, then the continued fraction and the series terminate. If $p_l = 1$ for $l \in \{2N, 2N+1\}$, a simple induction shows that $\kappa_j = 2q_{2j}q_{2j+1}$. In this case we formally define $S_j = \infty$ ($j = 1, 2, \ldots$) and $1/\infty = 0$. The reason for this convention becomes clear from the following result:

**Theorem 8.3.1** *Let $\psi$ denote a spectral measure on the interval $[-1, 1]$ with canonical moments $p_1, p_2, p_3, \ldots$. A random walk with one-step down, up, and holding transitions probabilities $d_n, u_n, h_n$ has $\psi$ as its spectral measure if and only if $\{u_n, d_n, h_n\}_{n \in E}$ satisfies (8.3.1) and*

$$2\left(q_1 - \frac{1}{S_1}\right) = \kappa_0 \le u_0 \le 2q_1, \tag{8.3.3}$$

$$2\left(q_{2j}q_{2j+1} - \frac{q_{2j}}{S_{2j+1}}\right) = \kappa_j \le u_j \le 2\left\{1 - \zeta_{2j} - \zeta_{2j+1} - \frac{2\zeta_{2j-1}\zeta_{2j}}{u_{j-1}}\right\}$$

$$\le 2q_{2j}q_{2j+1}, \tag{8.3.4}$$

*holds for all $0 \le j < N$. Moreover, if $u_{j_0} = \kappa_{j_0}$ for some $j_0 \in E$, then $u_j = \kappa_j$ for all $j_0 \le j < N$.*

*Proof* By the discussion of Section 8.2, $\psi$ is the spectral measure of the random walk if and only if (8.3.1) is satisfied, and it remains to show the bounds for the upward transition probabilities. We first verify the right-hand inequality for $u_j$. Note that the first two equations in (8.3.1) imply that $u_0 \le 1 - h_0 = 2q_1$ and

$$d_j + u_j \le 1 - h_j = 2(1 - \zeta_{2j} - \zeta_{2j+1}),$$

so by the third equation in (8.3.1),

$$u_j \le 2\left(1 - \zeta_{2j} - \zeta_{2j+1} - \frac{2\zeta_{2j-1}\zeta_{2j}}{u_{j-1}}\right), \qquad 1 \le j < N. \tag{8.3.5}$$

The choice $u_0 = 2q_1$ and successively using equality in this bound produces the maximal solution for $u_j$ which is $2q_{2j}q_{2j+1}$ for all $1 \le j < N$. Note that by Theorem 8.2.5 the odd canonical moments of the spectral measure satisfy $p_{2i+1} \ge \frac{1}{2}$, which shows $u_j \le 1$.

To prove the left-hand inequality on $u_j$, we write

$$u_j = \frac{4\zeta_{2j+1}\zeta_{2j+2}}{d_{j+1}} \ge \frac{4\zeta_{2j+1}\zeta_{2j+2}}{1 - h_{j+1}} = \frac{2\zeta_{2j+1}\zeta_{2j+2}}{1 - \zeta_{2j+2} - \zeta_{2j+3}} =: \kappa_{j,1}. \tag{8.3.6}$$

Combining (8.3.6) with (8.3.5) and replacing $j$ by $j + 1$ yields

$$\frac{2\zeta_{2j+3}\zeta_{2j+4}}{1 - \zeta_{2j+4} - \zeta_{2j+5}} \leq u_{j+1} \leq 2\left\{1 - \zeta_{2j+2} - \zeta_{2j+3} - \frac{2\zeta_{2j+1}\zeta_{2j+2}}{u_j}\right\}. \qquad (8.3.7)$$

Looking at the extremes in (8.3.7) produces

$$u_j \geq 2\left\{\left|\frac{\zeta_{2j+1}\zeta_{2j+2}}{1 - \zeta_{2j+2} - \zeta_{2j+3}}\right| - \left|\frac{\zeta_{2j+3}\zeta_{2j+4}}{1 - \zeta_{2j+4} - \zeta_{2j+5}}\right|\right\} =: \kappa_{j,2}.$$

Repeating this argument shows that for all $0 \leq j < N$ with $2j + 2k < 2N$,

$$u_j \geq 2\left\{\left|\frac{\zeta_{2j+1}\zeta_{2j+2}}{1 - \zeta_{2j+2} - \zeta_{2j+3}}\right| - \cdots - \left|\frac{\zeta_{2j+2k-1}\zeta_{2j+2k}}{1 - \zeta_{2j+2k} - \zeta_{2j+2k+1}}\right|\right\} =: \kappa_{j,k}. \qquad (8.3.8)$$

It is easy to see that $\kappa_{j,k}$ is increasing with $k$ which proves that in the case $N = \infty$, this continued fraction converges and its limit is also a lower bound for $u_j$. To verify the value for $\kappa_j$ in (8.3.3), note that by an even contraction and Lemma 3.2.5,

$$\frac{1}{1 - \zeta_1 - \kappa_0/2} = S_1,$$

which gives the equality in (8.3.3). To complete the proof, we verify the value for $\kappa_1$; the other cases are treated similarly by induction. From (8.3.2) we have

$$\kappa_1 = 2\left\{1 - \zeta_2 - \zeta_3 - \frac{\zeta_1\zeta_2}{\kappa_0/2}\right\} = 2\left\{q_2q_3 + p_1p_2 - \frac{p_1q_1p_2}{q_1 - 1/S_1}\right\},$$

and $S_j = 1 + (p_j/q_j)S_{j+1}$ implies that

$$p_1p_2 - \frac{p_1q_1p_2}{q_1 - 1/S_1} = p_1p_2 - \frac{p_1p_2}{1 - 1/(q_1 + p_1S_2)} = p_1p_2\left(1 - \frac{q_1 + p_1S_2}{p_1(S_2 - 1)}\right)$$

$$= -\frac{p_2}{(S_2 - 1)} = -\frac{p_2}{(\frac{p_2}{q_2}S_3)} = -\frac{q_2}{S_3}.$$

Combining these identities yields

$$\kappa_1 = 2\left\{q_2q_3 - \frac{q_2}{S_3}\right\},$$

which is the required representation in the case $j = 1$.

Finally, if $\kappa_j = u_j$ for $j = j_0$, then (8.3.2) and (8.3.4) for $j = j_0 + 1$ yield

$$\kappa_{j_0+1} \leq u_{j_0+1} \leq 2\left[1 - \zeta_{2j_0+2} - \zeta_{2j_0+3} - \frac{2\zeta_{2j_0+1}\zeta_{2j_0+2}}{\kappa_{j_0}}\right] = \kappa_{j_0+1},$$

which shows that there is equality also for $j = j_0 + 1 < N$.                    ∎

**Theorem 8.3.2**    *Let* $\{X_n\}_{n \in \mathbb{N}_0}$ *denote a random walk on the nonnegative integers, and let* $\psi$ *be the corresponding spectral measure.* $\{X_n\}_{n \in \mathbb{N}_0}$ *is the unique random walk with spectral measure* $\psi$ *if and only if it is recurrent.*

*Proof*    If there is to be a unique random walk then the two bounds for $u_0$ in (8.3.3) must be equal. Otherwise, one can construct an infinite class of processes with the same spectral measure by choosing $u_0 \in [\kappa_0, 2q_1]$ arbitrarily, solving successively (8.3.1) for $d_j$ and putting

$$u_j = 2\left[1 - \zeta_{2j} - \zeta_{2j+1} - \frac{2\zeta_{2j-1}\zeta_{2j}}{u_{j-1}}\right] \qquad (1 \leq j < N).$$

Now $\kappa_0 = 2q_1$ if and only if $S_1 = \infty$, and the proof of Theorem 8.3.1 shows $u_0 = 2q_1$, and $u_j = 2q_{2j}q_{2j+1}$ ($d_j = 2p_{2j-1}p_{2j}$, $j = 1, \ldots, N$, $d_0 = 0$). On the other hand, $\{d_j, u_j, h_j\}_{j \in E}$ satisfies (8.3.1), which implies that $d_j + u_j + h_j = 1$ ($1 \leq j < N$). Recurrence is equivalent to $P_{00}(1) = \infty$. From (8.2.18) and Lemma 3.2.5, we observe that $2P_{00}(1) = S_1 = \infty$, which proves the assertion.                    ∎

**Corollary 8.3.3**    *A random walk* $\{X_n\}_{n \in \mathbb{N}_0}$ *on the nonnegative integers is determined by its* $n$*-step transition probabilities* $\{P_{00}^n\}_{n=0,1,\ldots}$ *if and only if it is recurrent.*

**EXAMPLE 8.3.4**    Consider the $(d, u, h)$ random walk $\{X_n\}_{n \in \mathbb{N}_0}$ on the nonnegative integers of Example 8.2.8. If $u > d$, the process is transient, and by Theorem 8.3.1, (8.3.1), (3.2.11), and Lemma 3.2.3 the upward transition probabilities of a random walk with the same spectral measure satisfy

$$u_j \geq \frac{ud}{1-h} - \frac{ud}{1-h} - \frac{ud}{1-h} - \cdots = \frac{u+d-|d-u|}{2} = d,$$

for all $j \geq 0$, while the holding probabilities are given by

$$h_j = -1 + 2\zeta_{2j} + \zeta_{2j+1} = h_j' = \begin{cases} 1 - u & \text{if } j = 0, \\ h & \text{if } j \geq 1. \end{cases}$$

Thus we obtain from (8.3.1)

$$d \leq u_0 \leq 2[1 - \zeta_1] = 1 - h_0' = u,$$

$$d \leq u_j \leq 2\left[1 - \zeta_{2j} - \zeta_{2j+1} - \frac{2\zeta_{2j-1}\zeta_{2j}}{u_{j-1}}\right]$$

$$= u + d - ud/u_{j-1}, \qquad (8.3.9)$$

$$d_j = \frac{ud}{u_{j-1}} \qquad (j \geq 1),$$

$$h_0 = 1 - u, \quad h_j = h \qquad (j \geq 1),$$

and every random walk $\{Y_n\}_{n \in \mathbb{N}_0}$ on the nonnegative integers with one-step transition probabilities $\{d_n, u_n, h_n\}_{n \geq 0}$ satisfying (8.3.9) has the same $n$-step transition probabilities $P_{00}^n$ as $\{X_n\}_{n \in \mathbb{N}_0}$.

For example, a two parameter class of such processes is obtained by putting $u_2 = d$, which by Theorem 8.3.1 implies that $u_j = d$ $(j \geq 2)$ and $d_j = u$ $(j \geq 3)$. The holding probabilities are $h_0 = 1 - u, h_j = h (j \geq 1)$, while $(u_0, u_1)$ vary in the two-dimensional set

$$\left\{(s, t) | d \leq s \leq u, \ d \leq t \leq d + u - \frac{du}{s}\right\}.$$

Similarly, by putting $u_{j_0} = d$, we obtain a $j_0$-dimensional class of random walks with the same $n$-step transition probabilities as $\{X_n\}_{n \in \mathbb{N}_0}$.

The "extremal" random walks should also be singled out. The maximal value for $u_j$ is $u$, and the resulting random walk is given by (8.2.19). The minimal value is given by $u_j = d$, and the resulting process has

$$d_j = u \qquad (j \geq 1),$$

$$h_j = h \quad h_0 = 1 - u \qquad (j \geq 1), \qquad (8.3.10)$$

$$u_j = d \qquad (j \geq 0).$$

Note that the process in (8.2.19) is transient if $u > d$ and has no absorbing state, while the process in (8.3.10) would be recurrent except for absorption from zero, since $u_0 + h_0 = 1 - (u - d) < 1$.

## 8.4   GENERATING FUNCTIONS

In this section we will derive representations for the generating functions of the $n$-step transition probabilities of a random walk on the nonnegative integers in terms of the Stieltjes transform of the spectral measure $\psi$ and the corresponding orthogonal polynomials. The first result expresses the Stieltjes transform in terms of a series of the polynomials $Q_j(x)$ defined by (8.2.2).

**Lemma 8.4.1**    *The Stieltjes transform of the spectral measure $\psi$ satisfies*

$$S(z, \psi) = \sum_{n \in E} \left[ u_n \pi_n Q_n(z) Q_{n+1}(z) \right]^{-1}, \qquad z \in \mathbb{C} \setminus [-1, 1],$$

*where the polynomials $Q_j(x)$ ($j \in E$) are defined recursively by (8.2.2).*

*Proof*    By Corollary 8.2.7,

$$S(z, \psi) = \frac{1}{|z - h_0|} - \frac{u_0 d_1}{|z - h_1|} - \frac{u_1 d_2}{|z - h_2|} - \cdots, \qquad (8.4.1)$$

where the continued fraction on the right-hand side terminates at $u_{N-1} d_N$ in the case of a finite state space. Define

$$\frac{P_{n-1}^{(1)}(z)}{P_n(z)} := \frac{1}{|z - h_0|} - \frac{u_0 d_1}{|z - h_1|} - \cdots - \frac{u_{n-2} d_{n-1}}{|z - h_{n-1}|} \qquad (8.4.2)$$

as the $n$th convergent of the continued fraction in (8.4.1). Then it follows from (3.2.2) and (3.2.3) that the polynomials in the denominator can be obtained recursively by (8.2.6), while the polynomials in the numerator are given by $P_{-1}^{(1)}(z) = 0$, $P_0^{(1)}(z) = 1$,

$$P_{j+1}^{(1)}(z) = (z - h_{j+1}) P_j^{(1)}(z) - u_j d_{j+1} P_{j-1}^{(1)}(z) \quad 0 \le j < N. \qquad (8.4.3)$$

From (8.2.6) we obtain that $P_j(z)$ is the monic version of $Q_j(z)$ defined in (8.2.2), i.e.

$$P_j(z) = \left( \prod_{i=0}^{j-1} u_i \right) Q_j(z), \qquad (8.4.4)$$

and (3.2.14) represents the continued fraction in (8.4.1) as an infinite series

$$\int_{-1}^1 \frac{d\psi(x)}{z - x} = \frac{1}{P_1(z)} + \frac{u_0 d_1}{P_1(z) P_2(z)} + \frac{u_0 d_1 u_1 d_2}{P_2(z) P_3(z)} + \cdots$$

$$= \sum_{n \in E} [P_n(z) P_{n+1}(z)]^{-1} \left( \prod_{j=1}^n u_{j-1} d_j \right)$$

$$= \sum_{n \in E} \left[ u_n \pi_n Q_n(z) Q_{n+1}(z) \right]^{-1},$$

which is the assertion of the lemma.                                                    ∎

The identity (8.4.4) relates the monic orthogonal polynomials with respect to the spectral measure to the polynomials $Q_j(x)$ defined by the recurrence relation (8.2.2). In the following we need a similar relation for the numerator polynomials $P_j^{(1)}(z)$ given in (8.4.3). To this end we consider the *first associated polynomials*

$$Q_j^{(1)}(z) = u_0 \int_{-1}^{1} \frac{Q_{j+1}(z) - Q_{j+1}(x)}{z - x} \, d\psi(x).$$

A simple induction shows that these polynomials satisfy the recurrence relation $Q_{-1}^{(1)}(z) = 0$, $Q_0^{(1)}(z) = 1$,

$$u_{j+1} Q_{j+1}^{(1)}(z) = (z - h_{j+1}) Q_j^{(1)}(z) - d_{j+1} Q_{j-1}^{(1)}(z), \tag{8.4.5}$$

which is obtained from (8.2.2) by replacing the transition probabilities $u_j, d_j, h_j$ in the $j$th step by $u_{j+1}, d_{j+1}, h_{j+1}$, respectively. It is now easy to see that

$$P_j^{(1)}(z) = \left( \prod_{i=1}^{j} u_i \right) Q_j^{(1)}(z) = \int_{-1}^{1} \frac{P_{j+1}(z) - P_{j+1}(x)}{z - x} \, d\psi(x). \tag{8.4.6}$$

**Theorem 8.4.2** *For all $i, j \in E$ the generating functions of the n-step transition probabilities of a random walk on the nonnegative integers are given by*

$$P_{ij}(z) = \frac{\pi_j}{z} Q_i\left(\frac{1}{z}\right) Q_j\left(\frac{1}{z}\right) \sum_{\substack{n \geq max\{i,j\} \\ n \in E}} \left[ u_n \pi_n Q_n\left(\frac{1}{z}\right) Q_{n+1}\left(\frac{1}{z}\right) \right]^{-1} \tag{8.4.7}$$

*If $i \leq j$, then*

$$P_{ij}(z) = \pi_j Q_i\left(\frac{1}{z}\right) \left[ Q_j\left(\frac{1}{z}\right) \int_{-1}^{1} \frac{d\psi(x)}{1 - xz} - \frac{1}{u_0 z} Q_{j-1}^{(1)}\left(\frac{1}{z}\right) \right]. \tag{8.4.8}$$

*Here $\pi_j$ is defined by (8.2.5), $Q_k(x)$ is the kth orthogonal polynomial with respect to spectral measure $d\psi(x)$ recursively defined by (8.2.2), and $Q_k^{(1)}(x)$ is the kth first associated polynomial recursively defined in (8.4.5).*

*Proof* Without loss of generality we assume that $i \leq j$. From the integral representation (8.2.3) and (8.2.5) we obtain for the generating function

$$P_{ij}(z) = \sum_{n=0}^{\infty} P_{ij}^n z^n = \pi_j \int_{-1}^{1} \frac{Q_i(x) Q_j(x)}{1 - zx} \, d\psi(x), \qquad i, j \in E \tag{8.4.9}$$

($|z| < 1$), and the proof is now performed in three steps.

1. We show that it is sufficient to prove (8.4.7) in the case $i = 0$.
2. We prove (8.4.7) for $i = 0$.
3. We prove (8.4.8).

First, since for fixed $|z| < 1$,

$$H_{i-1}\left(x, \frac{1}{z}\right) = \frac{Q_i(x) - Q_i(\frac{1}{z})}{\frac{1}{z} - x}$$

is a polynomial of degree $i - 1 < j$ in the variable $x$, we obtain from the orthogonality of the polynomials $Q_k(x)$ with respect to the spectral measure $d\psi(x)$ and (8.4.9) that

$$P_{ij}(z) = \frac{\pi_j}{z} \int_{-1}^{1} \frac{Q_i(x) Q_j(x)}{\frac{1}{z} - x} d\psi(x)$$

$$= \frac{\pi_j}{z} \left[ \int_{-1}^{1} Q_j(x) H_{i-1}\left(x, \frac{1}{z}\right) d\psi(x) + Q_i\left(\frac{1}{z}\right) \int_{-1}^{1} \frac{Q_j(x)}{\frac{1}{z} - x} d\psi(x) \right]$$

$$= \pi_j Q_i\left(\frac{1}{z}\right) \int_{-1}^{1} \frac{Q_j(x)}{1 - zx} d\psi(x) = Q_i\left(\frac{1}{z}\right) P_{0j}(z). \tag{8.4.10}$$

Thus, if (8.4.7) is correct for $i = 0, j \in E$, then it follows from (8.4.10) that it also holds in the general case $i, j \in E$, $i \leq j$.

Next we show the assertion for $i = 0$, $j \in E$. The case $j = 0$ follows from Lemma 8.4.1, since

$$P_{00}(z) = \frac{1}{z} S\left(\frac{1}{z}, \psi\right) = \frac{\pi_0}{z} \sum_{n \in E} \left[ u_n \pi_n Q_n\left(\frac{1}{z}\right) Q_{n+1}\left(\frac{1}{z}\right) \right]^{-1}.$$

The general case is obtained by induction. To this end we define the so-called "functions of the second kind"

$$Q_k^*(z) = \int_{-1}^{1} \frac{Q_k(x)}{z - x} d\psi(x), \qquad k \in E, z \in \mathbb{C} \setminus [-1, 1], \tag{8.4.11}$$

where the case $k = 0$ corresponds to the Stieltjes transform of the spectral measure, i.e. $Q_0^*(z) = S(z, \psi)$. Observing the recurrence relation (8.2.2), we obtain

$$u_j Q_{j+1}^*(z) + h_j Q_j^*(z) + d_j Q_{j-1}^*(z)$$

$$= \int_{-1}^1 \frac{u_j Q_{j+1}(z) + h_j Q_j(z) + d_j Q_{j-1}(z)}{z - x} \, d\psi(x)$$

$$= \int_{-1}^1 \frac{x Q_j(x)}{z - x} \, d\psi(x) = z \int_{-1}^1 \frac{Q_j(x)}{z - x} \, d\psi(x) - \int_{-1}^1 Q_j(x) d\psi(x)$$

$$= \begin{cases} z Q_j^*(z) & \text{if } j \geq 1, \\ z Q_0^*(z) - 1 & \text{if } j = 0, \end{cases}$$

where the last line follows from the orthogonality of the polynomials $Q_j(x)$ with respect to the spectral measure $d\psi(x)$. Thus the functions of the second kind $Q_j^*(x)$ satisfy the same recursive relations as the polynomials $Q_j(x)$ but with different initial conditions, that is,

$$Q_0^*(z) = \int_{-1}^1 \frac{d\psi(x)}{z - x}, \quad z Q_0^*(z) = u_0 Q_1^*(z) + h_0 Q_0^*(z) + 1,$$

$$(8.4.12)$$

$$u_j Q_{j+1}^*(z) = (z - h_j) Q_j^*(z) - d_j Q_{j-1}^*(z), \qquad j \geq 1.$$

Now assume that the relation (8.4.7) holds for $i = 0$ and $j \in E$. Then it follows for $j + 1 \in E$,

$$P_{0,j+1}(z) = \frac{\pi_{j+1}}{z} Q_{j+1}^*\left(\frac{1}{z}\right) = \frac{\pi_{j+1}}{z u_j}\left\{\left(\frac{1}{z} - h_j\right) Q_j^*\left(\frac{1}{z}\right) - d_j Q_{j-1}^*\left(\frac{1}{z}\right)\right\}$$

$$= \frac{\pi_{j+1}}{u_j}\left\{\frac{1}{\pi_j}\left(\frac{1}{z} - h_j\right) P_{0,j}(z) - \frac{d_j}{\pi_{j-1}} P_{0,j-1}\left(\frac{1}{z}\right)\right\},$$

where we used (8.4.9), (8.4.11), and the recursive relation (8.4.12) for the functions of the second kind. The induction hypotheses and the recursive relation (8.2.2) then give

$$P_{0,j+1}(z) = \frac{\pi_{j+1}}{z u_j}\left[\left(\frac{1}{z} - h_j\right) Q_j\left(\frac{1}{z}\right) \sum_{n \geq j; n \in E}\left\{u_n \pi_n Q_n\left(\frac{1}{z}\right) Q_{n+1}\left(\frac{1}{z}\right)\right\}^{-1}\right.$$

$$\left. - d_j Q_{j-1}\left(\frac{1}{z}\right) \sum_{n \geq j-1; n \in E}\left\{u_n \pi_n Q_n\left(\frac{1}{z}\right) Q_{n+1}\left(\frac{1}{z}\right)\right\}^{-1}\right]$$

$$= \frac{\pi_{j+1}}{z u_j}\left[u_j Q_{j+1}\left(\frac{1}{z}\right) \sum_{n \geq j; n \in E}\left\{u_n \pi_n Q_n\left(\frac{1}{z}\right) Q_{n+1}\left(\frac{1}{z}\right)\right\}^{-1}\right.$$

$$-d_j \left\{ u_{j-1} \pi_{j-1} Q_j \left( \frac{1}{z} \right) \right\}^{-1} \Bigg]$$

$$= \frac{\pi_{j+1}}{z} Q_{j+1} \left( \frac{1}{z} \right) \sum_{n \ge j+1; n \in E} \left\{ u_n \pi_n Q_n \left( \frac{1}{z} \right) Q_{n+1} \left( \frac{1}{z} \right) \right\}^{-1},$$

which is the assertion (8.4.7) for $i = 0$ and $j + 1$. Note that the case $i = 0, j = 1$ has to be treated separately but follows by exactly the same arguments.

We finally show (8.4.8). From Lemma 8.4.1 we rewrite (8.4.7) as (note that $i \le j$)

$$P_{ij}(z) = \frac{\pi_j}{z} Q_i \left( \frac{1}{z} \right) Q_j \left( \frac{1}{z} \right) \left[ \int_{-1}^1 \frac{d\psi(x)}{\frac{1}{z} - x} - \sum_{n=0}^{j-1} \left\{ u_n \pi_n Q_n \left( \frac{1}{z} \right) Q_{n+1} \left( \frac{1}{z} \right) \right\}^{-1} \right].$$

$$(8.4.13)$$

The same argument as in the proof of Lemma 8.4.1 yields ($w = 1/z$)

$$\sum_{n=0}^{j-1} \left\{ u_n \pi_n Q_n(w) Q_{n+1}(w) \right\}^{-1}$$

$$= \frac{1}{\left| w - h_0 \right.} - \frac{u_0 d_1}{\left| w - h_1 \right.} - \cdots - \frac{u_{j-2} d_{j-1}}{\left| w - h_{j-1} \right.} = \frac{P_{j-1}^{(1)}(w)}{P_j(w)},$$

where $P_j(w)$ and $P_{j-1}^{(1)}(w)$ are defined in (8.2.6) and (8.4.3). The assertion now follows from (8.4.4), (8.4.6), and (8.4.13).    ∎

The following theorem gives a corresponding statement for the generating function of the first return probabilities. The proof is a direct consequence of Theorem 8.4.2 and the identities (8.2.24) and is therefore omitted.

**Theorem 8.4.3** *The generating function* (8.2.23) *of the first return probabilities* $f_{ij}^{(n)}$ *satisfies*

$$F_{ij}(z) = \frac{Q_i \left( \frac{1}{z} \right)}{Q_j \left( \frac{1}{z} \right)}, \qquad i < j,$$

$$F_{jj}(z) = 1 - \left[ \pi_j Q_j \left( \frac{1}{z} \right) \left\{ Q_j \left( \frac{1}{z} \right) \int_{-1}^1 \frac{d\psi(x)}{1 - zx} - \frac{1}{u_0 z} Q_{j-1}^{(1)} \left( \frac{1}{z} \right) \right\} \right]^{-1},$$

$$F_{ij}(z) = \frac{Q_i \left( \frac{1}{z} \right) \int_{-1}^1 \frac{d\psi(x)}{1 - zx} - \frac{1}{u_0 z} Q_{i-1}^{(1)} \left( \frac{1}{z} \right)}{Q_j \left( \frac{1}{z} \right) \int_{-1}^1 \frac{d\psi(x)}{1 - zx} - \frac{1}{u_0 z} Q_{j-1}^{(1)} \left( \frac{1}{z} \right)}, \qquad i > j.$$

**EXAMPLE 8.4.4**   Consider a random walk with state space $\mathbb{N}_0$ and transition probabilities

$$u_i = \frac{i+1}{2i+1}, \quad d_i = \frac{i}{2i+1}, \quad h_i = 0 \quad (i \in \mathbb{N}_0). \tag{8.4.14}$$

(Note that this chain is null recurrent, by Theorem 8.2.10.) The orthogonal polynomials $Q_i(x)$ satisfying the recursive relation (8.2.2) are the Legendre polynomials $P_i(x)$ [see (2.3.20), $\lambda = \frac{1}{2}$, and (2.1.20)], which are orthogonal with respect to the uniform distribution $\lambda_U$ on the interval $[-1, 1]$. By (2.1.23) these polynomials are given by

$$P_n(x) = \frac{1}{2^n} \sum_{m=0}^{\lfloor n/2 \rfloor} (-1)^m \binom{n}{m} \binom{2n-2m}{n} x^{n-2m} \tag{8.4.15}$$

where we used the duplication formula for the Gamma function in (2.3.18). Consequently the Stieltjes transform of the spectral measure $\lambda_U$ is obtained as

$$S(z, \lambda_U) = \frac{1}{2} \int_{-1}^{1} \frac{dx}{z-x} = \frac{1}{2} \log \frac{z+1}{z-1}.$$

Finally the polynomials $Q_j^{(1)}(z)$ defined in (8.4.5) are the associated Legendre polynomials of order 1 (see Chihara, 1978, pp. 201–202), which can be represented as

$$P_n^{(1)}(z) = \sum_{k=0}^{n} \frac{P_k(z) P_{n-k}(z)}{k+1}.$$

Thus the generating function of the $n$-step transition probabilities are given by

$$P_{ij}(z) = \frac{2j+1}{z} P_i\left(\frac{1}{z}\right) \left[ \frac{1}{2} P_j\left(\frac{1}{z}\right) \log \frac{1+z}{1-z} - \sum_{k=0}^{j-1} \frac{P_k(z) P_{j-1-k}(z)}{k+1} \right]$$

if $i \leq j$ and by $P_{ij}(z) = \frac{2j+1}{2i+1} P_{ji}(z)$ if $i \geq j$ (note that $\pi_n = 2n + 1$ for all $n \in \mathbb{N}_0$ and that $u_0 = 1$). The generating functions for the first return probabilities are given by Theorem 8.4.3. For example, we have for $i < j$,

$$F_{ij}(z) = \frac{P_i\left(\frac{1}{z}\right)}{P_j\left(\frac{1}{z}\right)} \tag{8.4.16}$$

which allows a simple calculation for the moments of the first passage time $T_{ij}$, where $T_{ij}$ is the minimum number of steps the random walk takes to reach state $j$ when it starts in state $i < j$. For example, the first two (factorial) moments are calculated from (8.4.16),

$$E[T_{ij}] = \frac{P_j'(1)P_i(1) - P_j(1)P_i'(1)}{P_j^2(1)},$$

$$E[T_{ij}(T_{ij} - 1)] = -2E[T_{ij}]\left[1 - \frac{P_j'(1)}{P_j(1)}\right] + \frac{P_i''(1)P_j(1) - P_j''(1)P_i(1)}{P_j^2(1)},$$

whenever $i < j$ (note that these formulas do not depend on the special spectral measure). Now (2.1.19) and (2.1.20) yield $P_n(1) = 1$, $P_n'(1) = n(n+1)/2$, $P_n''(1) = (n-1)n(n+1)(n+2)/8$, and a straightforward calculation gives the expectation and the variance of the first passage time $T_{ij}$,

$$E[T_{ij}] = \tfrac{1}{2}\{(j)_2 - (i)_2\}, \quad \mathrm{Var}(T_{ij}) = \tfrac{1}{8}\{(j-1)_4 - (i-1)_4\} \qquad (i < j),$$

where $(j)_k = j(j+1)\ldots(j+k-1)$ denotes the Pochhammer symbol.

We conclude this section with a more detailed discussion of the finite state space $E = \{0, \ldots, N\}$ ($u_N = 0$), which yields a further simplification of the representation formulas for the generating functions. To this end we observe from (8.4.1) that in the case $N < \infty$ the Stieltjes transform of the spectral measure has a finite continued fraction expansion

$$S(z, \psi) = \int_{-1}^{1} \frac{d\psi(x)}{z - x} = \frac{1}{|z - h_0|} - \frac{u_0 d_1}{|z - h_1|} - \cdots - \frac{u_{N-1} d_N}{|z - h_N|} = \frac{P_N^{(1)}(z)}{P_{N+1}(z)},$$

where $P_{N+1}(z)$ and $P_N^{(1)}(z)$ are defined recursively by (8.2.6) and (8.4.3). Thus it follows from (8.4.4) and (8.4.6) that

$$P_{N+1}(z) = \left(\prod_{i=0}^{N-1} u_i\right)[(z - h_N)Q_N(z) - d_N Q_{N-1}(z)],$$

$$P_N^{(1)}(z) = \left(\prod_{i=1}^{N-1} u_i\right)[(z - h_N)Q_{N-1}^{(1)}(z) - d_N Q_{N-2}^{(1)}(z)],$$

and we obtain by straightforward algebra the following auxiliary result.

**Lemma 8.4.5** *In the random walk with finite state space $E = \{0, \ldots, N\}$, the generating functions of the n-step transition probabilities $P_{ij}^n$ are*

$$P_{ij}(z) = \frac{\pi_j Q_i(\tfrac{1}{z})}{u_0 z}\left\{ \frac{Q_j(\tfrac{1}{z})[(\tfrac{1}{z} - h_N)Q_{N-1}^{(1)}(\tfrac{1}{z}) - d_N Q_{N-2}^{(1)}(\tfrac{1}{z})]}{[(\tfrac{1}{z} - h_N)Q_N(\tfrac{1}{z}) - d_N Q_{N-1}(\tfrac{1}{z})]} - Q_{j-1}^{(1)}\left(\tfrac{1}{z}\right) \right\}$$

*if $i \leq j$, and by $P_{ij}(z) = (\pi_j/\pi_i)P_{ji}(z)$ if $i \geq j$, where the polynomials $Q_j(z)$ and $Q_j^{(1)}(z)$ are defined recursively by (8.2.2) and (8.4.5).*

**Theorem 8.4.6** *In the random walk with finite state space $E = \{0, \ldots, N\}$ the generating functions of the n-step transition probabilities $P_{ij}^n$ are given by*

$$P_{ij}(z) = \left(\prod_{l=0}^{j-1} u_l\right) \frac{1}{z} Q_i\left(\frac{1}{z}\right) \prod_{l=0}^{N} \left(\frac{1}{z} - x_l\right)^{-1} H\left(\frac{1}{z}, j+1, N\right)$$

*whenever $i \leq j$. Here $x_0, \ldots, x_N$ denote the support points of the spectral measure $\psi$, the polynomials $Q_j(z)$ are defined recursively by (8.2.2), $H(w, N+1, N) = 1$, $H(w, i, i) = w - h_i$, and for $i < k$,*

$$H(w, i, k) = K\left(\begin{matrix} -u_i d_{i+1} & & & -u_{k-1} d_k \\ z - h_i & z - h_{i+1} & \cdots & z - h_k \end{matrix}\right).$$

*Proof* Let $x_0, \ldots, x_N$ denote the support points of the spectral measure $\psi$. These points are the $N+1$ zeros of the equation $(z - h_N)Q_N(z) - d_N Q_{N-1}(z) = 0$, and from (8.4.4) we have

$$\left(\frac{1}{z} - h_N\right) Q_N\left(\frac{1}{z}\right) - d_N Q_{N-1}\left(\frac{1}{z}\right) = \left(\prod_{l=0}^{N-1} u_l\right)^{-1} \prod_{l=0}^{N} \left(\frac{1}{z} - x_l\right). \quad (8.4.17)$$

Now $Q_j(w) = \prod_{i=0}^{j-1} u_i^{-1} H(w, 0, j-1)$, $Q_{j-1}^{(1)}(w) = \prod_{i=1}^{j-1} u_i^{-1} H(w, 1, j-1)$, and it follows that

$$\begin{aligned} M(w) &= Q_j(w)[(w - h_N)Q_{N-1}^{(1)}(w) - d_N Q_{N-2}^{(1)}(w)] \\ &\quad - Q_{j-1}^{(1)}(w)[(w - h_N)Q_N(w) - d_N Q_{N-1}(w)] \\ &= \rho_j[H(w, 0, j-1)H(w, 1, N) - H(w, 1, j-1)H(w, 0, N)] \\ &= \rho_j u_0 d_1[H(w, 1, j-1)H(w, 2, N) - H(w, 2, j-1)H(w, 1, N)], \end{aligned}$$

where

$$\rho_j = \left(\prod_{l=0}^{j-1} u_l\right)^{-1} \left(\prod_{l=1}^{N-1} u_l\right)^{-1}$$

and the last equality follows by an expansion of the two determinants $H(w, 0, N)$ and $H(w, 0, j-1)$ with respect to the first column. Repeating this argument yields

$$M(w) = \prod_{l=1}^{j} \frac{d_l}{u_l} \left( \prod_{l=j+1}^{N-1} u_l \right)^{-1} H(w, j+1, N),$$

and the assertion now follows directly from Lemma 8.4.5 and (8.4.17).  ∎

Note that Theorem 8.4.6 yields a recursion for the generating functions

$$P_{ij}(z) = \prod_{l=j}^{j+k-1} u_l^{-1} \frac{H(\frac{1}{z}, j+1, N)}{H(\frac{1}{z}, j+k+1, N)} P_{i,j+k}(z), \qquad 0 \le i \le j \le j+k \le N,$$

which gives in the special case $k = N - j$,

$$P_{ij}(z) = \left( \prod_{l=j}^{N-1} u_l \right)^{-1} H\left( \frac{1}{z}, j+1, N \right) P_{iN}(z), \qquad 0 \le i \le j \le N.$$

For example, the generating function of the probabilities $P_{i,N-2}^n$, $P_{i,N-1}^n$, $P_{i,N}^n$ in a random walk with finite state space $E = \{0, \ldots, N\}$ are given by ($0 \le i \le N$)

$$P_{iN}(z) = \frac{\frac{1}{z} Q_i(\frac{1}{z})}{(\frac{1}{z} - h_N) Q_N(\frac{1}{z}) - d_N Q_{N-1}(\frac{1}{z})}$$

$$P_{i,N-1}(z) = \frac{(\frac{1}{z} - h_N)}{u_{N-1}} P_{iN}(z) = \frac{\frac{1}{z}(\frac{1}{z} - h_N) Q_i(\frac{1}{z})}{u_{N-1} \left[ (\frac{1}{z} - h_N) Q_N(\frac{1}{z}) - d_N Q_{N-1}(\frac{1}{z}) \right]},$$

$$P_{i,N-2}(z) = \frac{(\frac{1}{z} - h_N)(\frac{1}{z} - h_{N-1}) - u_{N-1} d_N}{u_{N-2} u_{N-1}} P_{iN}(z)$$

$$= \frac{\frac{1}{z} \left[ (\frac{1}{z} - h_N)(\frac{1}{z} - h_{N-1}) - u_{N-1} d_N \right] Q_i(\frac{1}{z})}{u_{N-2} u_{N-1} \left[ (\frac{1}{z} - h_N) Q_N(\frac{1}{z}) - d_N Q_{N-1}(\frac{1}{z}) \right]}.$$

**EXAMPLE 8.4.7**   *The Generalized Ehrenfest Urn* (*Krafft and Schaefer, 1993*). Consider two urns, say, A and B containing N balls. The chain is in state $i$ if there are $i$ balls in urn A. At each time $n$ one ball is chosen with equal probabilities. If it is in urn A, it is placed in urn B, with probability $t \in (0, 1]$ and returned to urn A with probability $1 - t$. If it is in urn B it is put into urn A with probability $s \in (0, 1]$ and into urn B with probability $1 - s$. For $s = t = 1$ we obtain the classical Ehrenfest urn model (e.g., see Kemperman, 1961) which was used as a model of heat exchange between two isolated bodies (Ehrenfest, 1907). For $s = t \in (0, 1]$ this model was also considered by Johnson and Kotz (1977). The state space of the corresponding random walk is $E = \{0, \ldots, N\}$, and the transition probabilities are

$$u_i = \left(1 - \frac{i}{N}\right)s,$$

$$d_i = \frac{i}{N}\,t, \qquad (8.4.18)$$

$$h_i = 1 - u_i - d_i = \frac{(1-s)N + i(s-t)}{N}$$

$(i = 0, \ldots, N)$. The recurrence relation for the monic form of the polynomial $Q_j(x)$ is given by $P_0(x) = 1$, $P_{-1}(x) = 0$,

$$P_{n+1}(x) = (x - h_n)P_n(x) - u_{n-1}d_nP_{n-1}(x), \qquad n = 0, \ldots, N,$$
$$\qquad (8.4.19)$$
$$= \left(x - \frac{(1-s)N + n(s-t)}{N}\right)P_n(x) - \left(1 - \frac{n-1}{N}\right)st\frac{n}{N}\,P_{n-1}(x).$$

The monic orthogonal polynomials with respect to a measure on an interval $[a, b]$ are given by (3.6.5) and (3.6.6), and a comparison with (8.4.19) yields that the spectral measure corresponding to the random walk in (8.4.18) is supported in the interval

$$[a, b] = [1 - (s+t), 1] \subseteq [-1, 1]$$

and has canonical moments (calculated with respect to the interval $[1 - s - t, 1]$)

$$p_{2j} = \frac{j}{N}, \quad p_{2j-1} = \frac{t}{s+t}, \qquad j = 1, \ldots, N.$$

Example 2.3.8 showed that these are the canonical moments of the Binomial distribution transferred to the interval $[1 - s - t, 1]$. Consequently the spectral measure puts masses

$$\psi(\{x_j\}) = \binom{N}{j}\left(\frac{s}{t+s}\right)^j\left(\frac{t}{t+s}\right)^{N-j}, \qquad j = 0, \ldots, N \qquad (8.4.20)$$

at the points

$$x_j = 1 - \frac{s+t}{N}\,j, \qquad j = 0, \ldots, N. \qquad (8.4.21)$$

The orthogonal polynomials with respect to the binomial distribution on the set $\{0, \ldots, N\}$ are the Krawtchouk polynomials $k_n(x, p, N)$ defined in Example 2.3.8. Now a transformation onto the interval $[1 - s - t, 1]$ and a comparison of the leading coefficients identifies the polynomials $Q_j(x)$ as

$$Q_j(x) = k_j\left(\frac{N(1-x)}{s+t}, \frac{s}{s+t}, N\right), \qquad j = 0, \ldots, N.$$

From the integral representation, we thus obtain for the $n$-step transition probabilities

$$P_{k,\ell}^n = \binom{N}{\ell}\left(\frac{s}{t}\right)^\ell \sum_{j=0}^N \binom{N}{j} q_1^j p_1^{N-j} k_k(j, q_1, N) k_\ell(j, q_1, N)\left(1 - \frac{s+t}{N}j\right)^n,$$

where $p_1 = 1 - q_1 = t/(s+t)$, and we have used the fact that

$$\int_{-1}^1 Q_\ell^2(x) d\psi(x) = \left\{\binom{N}{\ell}\left(\frac{s}{t}\right)^\ell\right\}^{-1}, \qquad \ell = 0, \ldots, N,$$

which follows from (8.2.5). The expected first hitting times can be calculated from Theorem 8.4.3. For example, if $0 \le i < j \le N$, the generating functions of the first return probabilities are

$$F_{ij}(z) = \frac{k_i\left(\frac{N(1-1/z)}{s+t}, \frac{s}{s+t}, N\right)}{k_j\left(\frac{N(1-1/z)}{s+t}, \frac{s}{s+t}, N\right)} \qquad (i < j).$$

From Example 2.3.8 it follows that

$$k_n'(0, p, N) := \frac{d}{dx} k_n(x, p, N)\big|_{x=0} = -\sum_{k=1}^n \frac{(-n)_k}{(-N)_k \, k}\left(\frac{1}{p}\right)^k$$

$(n = 1, \ldots, N)$, and we obtain for the expected first passage times in the generalized Ehrenfest urn

$$E[T_{ij}] = F_{ij}'(1)$$

$$= \frac{N}{s+t} \frac{k_j(0, \frac{s}{s+t}, N)k_i'(0, \frac{s}{s+t}, N) - k_i(0, \frac{s}{s+t}, N)k_j'(0, \frac{s}{s+t}, N)}{k_j^2(0, \frac{s}{s+t}, N)}$$

$$= \frac{N}{s+t} \sum_{k=1}^j \left(\frac{s+t}{s}\right)^k \frac{1}{k(-N)_k} [(-j)_k - (-i)_k], \qquad i < j.$$

## 8.5   SYMMETRIC RANDOM WALKS

Random walks with holding probabilities $h_n \equiv 0$ for all $n \in E$ are of some special interest. If $h_0 = -1 + 2\zeta_1 = 0$, then $p_1 = \frac{1}{2}$, and $h_k = -1 + 2\zeta_{2k} + 2\zeta_{2k+1} = 0$, implies inductively that $p_{2k+1} \equiv \frac{1}{2}$ whenever the canonical moments are defined. Conversely, if all canonical moments of odd order are $\frac{1}{2}$, then $h_k \equiv 0$ for all $k \in E$. By Corollary 1.3.4 this in turn is equivalent to the spectral measure $\psi$ being symmetric. Consequently a random walk with vanishing holding probabilities $h_j = 0$ ($j \in E$) is called *symmetric*.

Moreover from Theorem 8.2.5 we see that every symmetric probability measure $\psi$ is the spectral measure of at least one random walk on the nonnegative integers. Equation (8.2.7) then reduces to

$$u_{i-1}d_i = q_{2i-2}p_{2i}, \qquad i \in E \ (q_0 := 1). \tag{8.5.1}$$

**EXAMPLE 8.5.1**   *Chain Sequences.* Assume for a moment that the state space is infinite, i.e. $E = \mathbb{N}_0$. Equations of the form (8.5.1) are closely related to the theory of chain sequences. See, for example, Wall (1948) or Chihara (1978). A sequence $a_1, a_2, a_3, \ldots$ is called a *chain sequence* if there exists another sequence $g_0, g_1 \ldots$ such that

$$(1 - g_{i-1})g_i = a_i, \qquad i \geq 1,$$

where $0 \leq g_i \leq 1$. We will discuss only the case where $0 < g_i < 1$. The sequence $\{g_i\}_{i\geq 0}$ is called a *parameter sequence* for the chain sequence. Any chain sequence has a *maximal* and *minimal parameter sequence* $\{M_i\}_{i\geq 0}$ and $\{m_i\}_{i\geq 0}$, respectively. This means that whenever $\{g_i\}_{i\geq 0}$ is a parameter sequence of $\{a_i\}_{i\geq 1}$, then $m_i \leq g_i \leq M_i$ holds for all $i \geq 0$. The minimal sequence $\{m_i\}_{i\geq 0}$ is clearly given by choosing $m_0 = 0$ and recursively calculating the other $m_i$. In general, the quantity $g_0$ cannot be chosen too large; otherwise; the remaining $g_i$ will not be in the interval $(0, 1)$. For the calculation of the maximal parameter sequence $\{M_i\}_{i\geq 0}$; define

$$M_i = 1 - \frac{a_{i+1}|}{|1} - \frac{a_{i+2}|}{|1} - \cdots. \tag{8.5.2}$$

From Lemma 3.2.5 we see that this expression is equal to

$$M_i = 1 - (1 - g_i)\left(1 - \frac{1}{T_{i+1}}\right), \tag{8.5.3}$$

where $T_{i+1}$ is defined as

$$T_{i+1} = 1 + \sum_{\ell=i+1}^{\infty} \prod_{k=i+1}^{\ell} \frac{g_k}{1 - g_k}$$

with the convention that this is $+\infty$ if the series diverges. Using (8.5.3) and $T_i = 1 + g_i/(1 - g_i)T_{i+1}$, it is easily seen that $(1 - M_{i-1})M_i = a_i$ and that

$g_i \le M_i$, in which case the sequence $\{M_i\}_{i \ge 0}$ is the maximal parameter sequence.

The minimal parameter sequence of the right-hand side in (8.5.1) is of course the sequence of canonical moments of even order, namely $d_0' = 0$, $d_i' = p_{2i}$ ($i \in \mathbb{N}$). From (8.5.2) and (8.5.3) we see that the maximal sequence $\{M_i^*\}_{i \ge 0}$ is given by

$$M_i^* = 1 - \frac{q_{2i}p_{2i+2}}{\lfloor 1 \rfloor} - \frac{q_{2i+2}p_{2i+4}}{\lfloor 1 \rfloor} - \cdots$$

$$= p_{2i} + q_{2i}\left(1 + \sum_{\ell=i+1}^{\infty} \prod_{k=i+1}^{\ell} \frac{p_{2k}}{q_{2k}}\right)^{-1}.$$

If $\{X_n\}_{n \in \mathbb{N}}$ is a random walk on $\mathbb{N}_0$ with symmetric spectral measure $\psi$ (or equivalently $h_j \equiv 0$ for all $j$), then its one-step transition probabilities satisfy $p_{2i} = d_i' \le d_i \le M_i^*$ ($i \ge 0$) and $u_{i-1}d_i = q_{2i-2}p_{2i}$ ($i \ge 1$).

Throughout this section we will consider symmetric random walks with transition probabilities adding to one, namely $u_0 = 1$, $u_j + d_j = 1$, $1 \le j < N$ ($d_N = 1$ if $N < \infty$). If $\psi$ denotes the spectral measure of such a process, then the return probabilities to the state 0 can be represented as

$$P_{00}^n = \begin{cases} 0 & \text{if } n \text{ is odd}, \\ \int_{-1}^{1} x^{2m}d\psi(x) = \int_0^1 x^m d\psi^*(x) & \text{if } n = 2m \text{ is even}, \end{cases} \tag{8.5.4}$$

where the distribution $\psi^*$ is defined on the interval $[0,1]$ and related to the (symmetric) spectral measure $\psi$ by the transformation

$$\psi^*([0, x^2]) = \psi([-x, x]) \qquad (0 \le x \le 1). \tag{8.5.5}$$

The discussion in Example 8.5.1 shows that the correspondence between the distribution functions on the interval $[0, 1]$ and the symmetric random walks with no absorption is one to one.

**Theorem 8.5.2** *If $\{X_n\}_{n \ge 0}$ is a symmetric random walk with no absorption (i.e., $u_0 = 1$, $u_j + d_j = 1$  $1 \le j \le N$), then*

$$P_{00}^{2m} = \int_0^1 x^m d\psi^*(x), \tag{8.5.6}$$

*where $\psi^*$ is the measure on the interval $[0, 1]$ corresponding to the canonical moment sequence $(d_1, d_2, d_3, \ldots)$.*

*Proof*   If the canonical moments of $\psi$ are denoted by $p_j$, we obtain by symmetry

$p_{2j-1} = \frac{1}{2}$. Since there is no absorption, we have to solve (8.5.1) with $u_0 = 1$, $(q_0 = 1)$, $u_i + d_i = 1$ $(1 \leq i < N)$. Example 8.5.1 shows that this gives the minimal parameter sequence, $p_{2i} = d_i$, whenever $1 \leq i \leq N$. By Theorem 1.3.5 the canonical moments of $\psi^*$ are the even canonical moments of $\psi$, which proves the theorem. ∎

**EXAMPLE 8.5.3 (Kemperman)** As a first application of Theorem 8.5.2 we will provide a probabilistic proof of Theorem 2.4.4 which gives the inverse map from the set of canonical moment sequences onto the set of ordinary moment sequences. To this end let $\psi^*$ denote a distribution on the interval $[0, 1]$ with moments $c_n = \int_0^1 x^n d\psi^*(x)$ and canonical moments $p_n$ $(n = 1, 2, \ldots)$. In Section 2.4 we showed that the moments can be calculated as $c_n = S_{n,n}$ if $S_{0,j} \equiv 1$, $S_{i,j} = 0$ for $i > j$ and

$$S_{i,j} = S_{i,j-1} + \zeta_{j-i+1}S_{i-1,j} \qquad (i \leq j). \tag{8.5.7}$$

Moreover

$$S_{n,n} = \sum_{k=0}^{\lfloor n/2 \rfloor} \left( \prod_{j=1}^{n-2k} \zeta_j \right) S_{k,n-k}^2. \tag{8.5.8}$$

We will now prove these identities probabilistically. For the sake of simplicity assume that $\psi^*$ has infinite support. By Theorem 8.5.2 the moments of $\psi^*$ are the $2n$-step transition probabilities $P_{00}^{2n}$ of the random walk $\{X_n\}_{n \geq 0}$ with infinite state space and one-step transition probabilities $d_j = p_j$, $u_j = q_j$ $(j \geq 1)$, $u_0 = 1$, $d_0 = 0$. Define $S_{i,i} = P_{00}^{2i}$ $(i \geq 1)$, $S_{i,j} = 0$ $(i > j)$, and

$$S_{i,j} = \frac{1}{u_0 u_1 \ldots u_{j-i-1}} P_{0,j-i}^{i+j}, \qquad 0 \leq i < j, \tag{8.5.9}$$

where $P_{ij}^n$ denotes the probability of a transition from state $i$ to state $j$ in $n$ steps. From Theorem 8.5.2 we have $c_n = P_{00}^{2n} = S_{n,n}$, and (8.5.9) implies that

$$S_{0,k} = \frac{P_{0k}^k}{u_0 u_1 \ldots u_{k-1}} = 1. \tag{8.5.10}$$

Conditioning on the last transition, we obtain from (8.5.9)

$$S_{i,j} = \frac{1}{u_0 u_1 \ldots u_{j-i-1}} \left\{ u_{j-i-1} P_{0,j-i-1}^{i+j-1} + d_{j-i+1} P_{0,j-i+1}^{i+j-1} \right\}$$

$$= S_{i,j-1} + \zeta_{j-i+1}S_{i-1,j} \qquad (j \geq i),$$

where $\zeta_{j-i+1} = u_{j-i}d_{j-i+1} = q_{j-i}p_{j-i+1}$ $(j \geq i)$. This is the recurrence relation

(2.4.7) for the moments in terms of the canonical moments derived in Section 2.4. For the second identity note that

$$
c_n = P_{00}^{2n} = \sum_{k=0}^{n} P_{0k}^n P_{k0}^n = \sum_{k=0}^{n} \pi_k^{-1} (P_{0k}^n)^2, \tag{8.5.11}
$$

where $\pi_k$ is defined in (8.2.5). For the last equality in (8.5.11) we used the well-known relation $\pi_i P_{ij}^n = \pi_j P_{ji}^n$ which is obvious from the integral representation (8.2.3) and (8.2.5). Now $P_{0k}^n$ vanishes unless $k = n - 2i$ and $u_j = q_j = 1 - p_j = 1 - d_j \ (j \geq 1)$ yields

$$
c_n = \sum_{k=0}^{\lfloor n/2 \rfloor} \left( \prod_{j=1}^{n-2k} u_{j-1} d_j \right) \left( \frac{P_{0,n-2k}^n}{u_0 u_1 \ldots u_{n-2k-1}} \right)^2 = \sum_{k=0}^{\lfloor n/2 \rfloor} \left( \prod_{j=1}^{n-2k} \zeta_j \right) S_{k,n-k}^2,
$$

where we have used (8.5.9) for the last equality. This provides a probabilistic proof of Theorem 2.4.4. Other formulas similar to (8.5.8) can also be derived probabilistically by using

$$
c_n = P_{00}^{2n} = \sum_{k=0}^{n} P_{0k}^r P_{k0}^{2n-r}
$$

for a fixed $r$ in place of (8.5.11).

**Theorem 8.5.4** (*Kemperman*) *The moments* $c_1, c_2, \ldots$ *and canonical moments of a probability measure* $\psi^*$ *on the interval* $[0, 1]$ *are related by the formula*

$$
c_n = \sum_{q=1}^{n} \sum_{\substack{k_1 + \cdots + k_q = n \\ k_1 > 0, \ldots, k_q > 0}} \beta_q(k_1, \ldots, k_q) \zeta_1^{k_1} \zeta_2^{k_2} \cdots \zeta_q^{k_q}, \tag{8.5.12}
$$

*where*

$$
\beta_q(k_1, \ldots, k_q) = \begin{cases} \prod_{j=1}^{q-1} \binom{k_j + k_{j+1} - 1}{k_{j+1}} & \text{if} \quad q \geq 2, \\ 1 & \text{if} \quad q = 1, \end{cases} \tag{8.5.13}
$$

*and the summation in* (8.5.12) *is only performed over the terms for which the canonical moments are well defined.*

*Proof* If $\{X_n\}_{n \geq 0}$ is the random walk with one-step transition probabilities $d_j = p_j \ (j \geq 1), d_0 = 0, u_j = q_j \ (j \geq 1), u_0 = 1$, then Theorem 8.5.2 shows that the $n$th moment of $\psi^*$ is precisely the probability that the process returns to state 0 after $2n$ steps. Let $\beta_q(k_1, \ldots, k_q)$ denote the number of paths of length $2n$ that start and end at 0 and go up exactly to state $q$, with $k_j$ "up-crossings" from state $j-1$ to $j, j, = 1, \ldots, q$. Since the process returns to zero there are exactly the same number $k_j$ of "down-crossings" from $j$ to $j-1$ and $\sum_{j=1}^{q} k_j = n$. The form of Equation (8.5.12) is then clear since the probability associated with each path must contain products of terms of the form $(q_{j-1}p_j)^{k_j}, j = 1, \ldots, q$. It remains to show the representation (8.5.13). Since $\beta_1(k_1) = 1$, this assertion can be established by showing the recursion

$$\beta_{q+1}(k_1, \ldots, k_{q+1}) = \binom{k_q + k_{q+1} - 1}{k_{q+1}} \beta_q(k_1, \ldots, k_q) \qquad (8.5.14)$$

Now there are $k_q$ transitions from state $q-1$ to state $q$ (and back) and $k_{q+1}$ crossings from $q$ to $q+1$ (and back). Thus one has to "insert $k_{q+1}$ crossings (to $q+1$ and back) into $k_q$ up-crossings to $q$ and the adjacent down-crossing to $q-1$. This can be done in

$$\binom{k_q + k_{q+1} - 1}{k_{q+1}}$$

ways. This proves (8.5.14) and the assertion follows.                               ■

In the remaining part of this section we will discuss a possibility for calculating return probabilities in random walks with "switched" one-step transition probabilities. To this end let $\{\tilde{X}_n\}_{n\geq 0}$ denote the random walk on $E$ which is obtained from $\{X_n\}_{n\geq 0}$ by switching $u_i$ and $d_i$ for all $1 \leq i \leq N$, i.e.

$$\tilde{u}_i = P(\tilde{X}_{n+1} = i+1 \mid \tilde{X}_n = i) = P(X_{n+1} = i-1 \mid X_n = i) = d_i,$$

$$\tilde{d}_i = P(\tilde{X}_{n+1} = i-1 \mid \tilde{X}_n = i) = P(X_{n+1} = i+1 \mid X_n = i) = u_i$$

$(P(\tilde{X}_{n+1} = 1 \mid \tilde{X}_n = 0) = 1)$. In the following discussion we call $\{\tilde{X}_n\}_{n\geq 0}$ a *dual* random walk of $\{X_n\}_{n\geq 0}$, and denote the corresponding one- and $n$-step transition probabilities by $\tilde{u}_j, \tilde{d}_j, \tilde{P}_{ij}^n, \tilde{f}_{ij}^n$.

**Theorem 8.5.5**    *Let $\{X_n\}_{n\geq 0}$ denote a symmetric random walk with no absorption, and $\{\tilde{X}_n\}_{n\geq 0}$ denote the dual walk obtained by switching the transition probabilities for $1 \leq i \leq N$. The probabilities for a return to state 0 in both processes are related by*

$$\tilde{f}_{00}^{2n} = P_{00}^{2n-2} - P_{00}^{2n}, \qquad n = 1, 2, \ldots,$$

*or*

$$P_{00}^{2n} = 1 - \sum_{j=1}^{n} \tilde{f}_{00}^{2j}, \qquad n = 0, 1, \ldots.$$

*Proof* The equivalence between the two expressions is obvious. To prove the first expression, we consider for $\{X_n\}_{n\geq 0}$ and $\{\tilde{X}_n\}_{n\geq 0}$ the generating function of the return probabilities to state 0 and obtain from Corollary 8.2.7,

$$P_{00}(z) = \sum_{n=0}^{\infty} P_{00}^{2n} z^{2n} = \frac{1}{\big|} \frac{1}{1} - \frac{d_1 z^2}{\big|} \frac{1} - \frac{u_1 d_2 z^2}{\big|} \frac{1} - \frac{u_2 d_3 z^2}{\big|} \frac{1} - \cdots,$$

$$\tilde{P}_{00}(z) = \sum_{n=0}^{\infty} \tilde{P}_{00}^{2n} z^{2n} = \frac{1}{\big|} \frac{1}{1} - \frac{u_1 z^2}{\big|} \frac{1} - \frac{d_1 u_2 z^2}{\big|} \frac{1} - \frac{d_2 u_3 z^2}{\big|} \frac{1} - \cdots$$

$$(8.5.15)$$

$(|z| < 1)$. Now (8.5.15) and Theorem 3.4.2 transferred to the interval $[-1, 1]$ yield

$$P_{00}(z)\tilde{P}_{00}(z) = \frac{1}{1 - z^2}. \qquad (8.5.16)$$

Observing that $\tilde{F}_{00}(z) = 1 - 1/\tilde{P}_{00}(z)$, this gives

$$(1 - z^2) \sum_{n=0}^{\infty} P_{00}^{2n} z^{2n} = (1 - z^2) P_{00}(z) = 1 - \tilde{F}_{00}(z) = 1 - \sum_{n=1}^{\infty} \tilde{f}_{00}^{2n} z^{2n},$$

and the assertion follows by comparing coefficients. ∎

**EXAMPLE 8.5.6** Consider the random walk $\{X_n\}_{n\geq 0}$ with one-step transition probabilities $u_0 = 1$,

$$u_n = \frac{u^{n-1} + d^{n-1}}{u^n + d^n} ud \quad d_n = \frac{u^{n+1} + d^{n+1}}{u^n + d^n}, \qquad n = 1, 2, \ldots, \qquad (8.5.17)$$

where $0 < u < 1$ and $d = 1 - u$. The dual process $\{\tilde{X}_n\}_{n\geq 0}$ has transition probabilities ($\tilde{u}_0 = 1$)

$$\tilde{u}_n = \frac{u^{n+1} + d^{n+1}}{u^n + d^n}, \quad \tilde{d}_n = \frac{u^{n-1} + d^{n-1}}{u^n + d^n} ud, \qquad n = 1, 2, \ldots. \qquad (8.5.18)$$

The generating function of the return probabilities of $\{\tilde{X}_n\}_{n\geq 0}$ are obtained from

(8.5.15) and Lemma 3.2.3 as

$$\tilde{P}_{00}(z) = \sum_{n=0}^{\infty} \tilde{P}_{00}^{2n} z^{2n} = \frac{1}{\lfloor 1} - \frac{2udz^2}{\lfloor 1} - \frac{udz^2}{\lfloor 1} - \frac{udz^2}{\lfloor 1} - \cdots$$

$$= \frac{1}{\sqrt{1 - 4\,udz^2}} = \sum_{n=0}^{\infty} \binom{2n}{n} (ud)^n z^{2n},$$

$$\tilde{F}_{00}(z) = \sum_{n=1}^{\infty} \tilde{f}_{00}^n z^{2n} = 1 - \frac{1}{\tilde{P}_{00}(z)} = \sum_{n=1}^{\infty} \binom{2n}{n} \frac{(ud)^n}{2n-1} z^{2n}.$$

Comparing coefficients yields for the return probabilities of $\{\tilde{X}_n\}_{n\geq 0}$,

$$\tilde{P}_{00}^{2n} = \binom{2n}{n} (ud)^n, \qquad\qquad (8.5.19)$$

$$\tilde{f}_{00}^{2n} = \binom{2n}{n} \frac{(ud)^n}{2n-1}. \qquad\qquad (8.5.20)$$

Now Theorem 8.5.5 provides a method for calculating the corresponding probabilities of $\{X_n\}_{n\geq 0}$, that is,

$$P_{00}^{2n} = 1 - \sum_{m=1}^{n} \tilde{f}_{00}^{2m} = 1 - \sum_{m=1}^{n} \binom{2m}{m} \frac{(ud)^m}{2m-1},$$

$$f_{00}^{2n} = \tilde{P}_{00}^{2n-2} - \tilde{P}_{00}^{2n} = \binom{2n}{n} (ud)^n \left\{ \frac{n}{2(2n-1)} \frac{1}{ud} - 1 \right\}.$$

There does not seem to be any simple way for deriving these formulas. So without exploiting the duality relation, it is not evident why there should be such simple formulas for the return probabilities to state 0.

# CHAPTER 9

# The Circle and Trigonometric Functions

## 9.1 INTRODUCTION

This chapter is devoted to the trigonometric functions on the circle or $[-\pi, \pi)$. Thus instead of using the powers $1, t, \ldots, t^n$ defined on some interval $[a, b]$, our concern here is with the system of functions

$$1, \quad \cos\theta, \quad \sin\theta, \quad \cos 2\theta, \quad \sin 2\theta, \ldots, \quad \cos m\theta, \quad \sin m\theta \quad (9.1.1)$$

for $\theta$ in $[-\pi, \pi)$ or some interval of length $2\pi$. It should be mentioned that the functions in (9.1.1) are periodic and defined on the circle, and although we will write $[-\pi, \pi)$, on numerous occasions, there are no special "end points." For a given probability measure on $[-\pi, \pi)$ the corresponding *trigonometric moments* are given by $\alpha_0 = 1$,

$$\alpha_k = \int_{-\pi}^{\pi} \cos k\theta d\sigma(\theta), \quad \beta_k = \int_{-\pi}^{\pi} \sin k\theta d\sigma(\theta), \quad k = 1, \ldots, m. \quad (9.1.2)$$

These can be alternatively expressed by

$$\gamma_k = \int_{-\pi}^{\pi} e^{-ik\theta} d\sigma(\theta) \quad k = 0, \pm 1, \ldots, \pm m. \quad (9.1.3)$$

It is clear that $\overline{\gamma}_k = \gamma_{-k}$ and $\gamma_k = \alpha_k - i\beta_k, k = 1, \ldots, m$. The quantities $\alpha_k$ and $\beta_k$ can be expressed as $\alpha_k = (\gamma_k + \overline{\gamma}_k)/2$ and $\beta_k = (\overline{\gamma}_k - \gamma_k)/(2i)$.

The theory corresponding to much of the material in early chapters has a corresponding analogue for the functions in (9.1.1), and in some respects the theory for the trigonometric functions is even richer than for the powers $1, t, \ldots, t^n$.

The moment space for the system of functions (9.1.1), together with the canonical representations of the moment points, is described in Section 9.2, while the corresponding canonical moments are discussed in Section 9.3. The *Szegö polynomials* are investigated in Section 9.4, and some basic transforms and continued fractions are introduced in Section 9.5. The canonical moments of measures on the circle appear quite naturally in discrete-time stationary random processes, and some simple aspects of this theory are described in Sections 9.6 and 9.7. Finally Section 9.8 has some applications to the design of experiments for linear regression models with trigonometric functions (9.1.1).

## 9.2 MOMENT SPACES AND CANONICAL REPRESENTATIONS

The moment space $M_{2m}^c$ using the functions in (9.1.1) is defined as

$$M_{2m}^c = \{(\alpha_1, \beta_1, \ldots, \alpha_m, \beta_m) | \sigma \in \mathcal{P}_{[-\pi,\pi)}\}, \tag{9.2.1}$$

where $\mathcal{P}_{[-\pi,\pi)}$ is the set of all probability measures on the circle and $\alpha_k$, $\beta_k$ are defined in (9.1.2). As before this set can be shown to be convex and compact with interior points, and it is the convex hull of the set of $2m$ dimensional points

$$(\cos\theta, \; \sin\theta, \ldots, \; \cos m\theta, \; \sin m\theta)$$

as $\theta$ varies over the interval $[-\pi, \pi)$. The following theorem is the same as Theorem 1.4.1:

**Theorem 9.2.1** *The point $(\alpha_1, \beta_1, \ldots, \alpha_m, \beta_m)$ is contained in $M_{2m}^c$ if and only if for every $(d_0, \ldots, d_m, e_1, \ldots, e_m)$ such that*

$$d_0 + \sum_{k=1}^m (d_k \cos k\theta + e_k \sin k\theta) \geq 0 \qquad \text{for all } \theta \tag{9.2.2}$$

*it follows that* $d_0 + \sum_{k=1}^m (d_k \alpha_k + e_k \beta_k) \geq 0.$

*Moreover $(\alpha_1, \beta_1, \ldots, \alpha_m, \beta_m) \in \text{Int } M_{2m}^c$ if and only if the same condition holds with the last inequality being strict if $(d_0, \ldots, d_m, e_1, \ldots, e_m) \neq 0$.*

**Definition 9.2.2** *Any probability measure $\sigma$ satisfying (9.1.2) is called a representation of the moment point $(\alpha_1, \beta_1, \ldots, \alpha_m, \beta_m)$. The index $I(\sigma)$ of a representation is defined as the number of points in the support of $\sigma$.*

Our first task is to investigate representations of index $I(\sigma) = m + 1$. To do this, we first consider the boundary $\partial M_{2m}^c$ of the moment space $M_{2m}^c$. A point $(\alpha_1, \beta_1, \ldots, \alpha_m, \beta_m) \in \partial M_{2m}^c$ if and only if there exists a corresponding supporting hyperplane, i.e. there exists a vector $(d_0, d_1, \ldots, d_m, e_1, \ldots, e_m) \neq 0$

such that

$$g(\theta) = d_0 + \sum_{k=1}^{m}(d_k \cos k\theta + e_k \sin k\theta) \geq 0 \qquad \text{for all } \theta$$

and

$$d_0 + \sum_{k=1}^{m} d_k\alpha_k + e_k\beta_k = 0.$$

In this case the support of any $\sigma$ representing $(\alpha_1, \beta_1, \ldots, \alpha_m, \beta_m)$ must be contained in the set of zeros of $g(\theta)$. An expression like $g(\theta)$ is called a *trigonometric polynomial*. It is said to be of exact degree $m$ if at least one of $d_m$ or $e_m$ is nonzero. The following three lemmas will be stated without proof:

**Lemma 9.2.3**
1. *Any nontrivial nonnegative trigonometric polynomial $g(\theta)$ of degree $m$ has at most $m$ distinct zeros.*
2. *For any set of $r \leq m$ distinct points $\theta_1, \theta_2, \ldots, \theta_r$, there exists a nonnegative trigonometric polynomial $g(\theta)$ of degree at most $m$ vanishing precisely at $\theta_1, \ldots, \theta_r$.*

**Lemma 9.2.4**   *The moment point $(\alpha_1, \beta_1, \ldots, \alpha_m, \beta_m)$ is contained in $\partial M_{2m}^c$ if and only if it has a unique representation $\sigma$ of index $I(\sigma) \leq m$.*

**Lemma 9.2.5**   *Let $(\alpha_1, \beta_1, \ldots, \alpha_m, \beta_m) \in \text{Int } M_{2m}^c$. For each $\theta \in [-\pi, \pi)$, there exists a unique representation $\sigma_\theta$ of $(\alpha_1, \beta_1, \ldots, \alpha_m, \beta_m)$ that includes $\theta$ and $I(\sigma_\theta) = m + 1$.*

To characterize which $2m$-dimensional points $(\alpha_1, \beta_1, \ldots, \alpha_m, \beta_m)$ actually lie in $M_{2m}^c$ an analogue of the basic Theorem 1.4.2 is needed.

**Theorem 9.2.6**   *Let $g(\theta)$ be a nonnegative trigonometric polynomial (with real coefficients). Then there exists an ordinary polynomial $h(z)$ (with possibly complex coefficients) of the same degree such that*

$$g(\theta) = |h(e^{i\theta})|^2. \tag{9.2.3}$$

*If $g(\theta) \not\equiv 0$, then $h(z)$ can be chosen so that $h(z) \neq 0$ for $|z| < 1$ and $h(0) > 0$, and in this case it is uniquely defined. If $g(\theta)$ is a cosine polynomial, then $h(\theta)$ is a polynomial with real coefficients.*

*Proof*   See Szegö (1939 p. 3–4) or Grenander and Szegö (1958, pp. 20–21). For any polynomial

$$\rho(z) = \sum_{i=0}^{n} \rho_i z^i$$

with complex coefficients, let

$$\bar{p}(z) = \sum_{i=0}^{n} \bar{p}_i z^i$$

and then define the *reciprocal polynomial* $\rho^*(z)$ by

$$\rho^*(z) = z^n \bar{\rho}(z^{-1}) = \sum_{i=0}^{n} \bar{\rho}_{n-i} z^i. \tag{9.2.4}$$

Note that for $z \neq 0$,

$$\rho^*(\bar{z}^{-1}) = \overline{z^{-n} \rho(z)},$$

so $\bar{z}^{-1}$ is a root of $\rho^*$ if $z$ is a root of $\rho$.

If $g(\theta)$ is a trigonometric polynomial of degree $m$, then by expressing $\sin k\theta$ and $\cos k\theta$ in exponential form, $g(\theta)$ can be written as

$$g(\theta) = d_0 + \sum_{k=1}^{m} (d_k \cos k\theta + e_k \sin k\theta)$$

$$= \frac{1}{2} \sum_{k=0}^{m} (\xi_k e^{ik\theta} + \bar{\xi}_k e^{-ik\theta}),$$

where $\xi_k = d_k - ie_k$, $k = 0, 1, \ldots, m$, $e_0 = 0$. This is then rewritten as (note that $g$ is a nonnegative polynomial)

$$g(\theta) = |z^{-m} G(z)|, \qquad z = e^{i\theta},$$

where

$$G(z) = \frac{1}{2} \sum_{k=0}^{m} (\xi_k z^{m+k} + \bar{\xi}_k z^{m-k})$$

is a polynomial of degree $2m$.

Since $G^*(z) = G(z)$, the zeros of $G(z)$ must be symmetrical with respect to the unit circle; that is, if $G(z_0) = 0$, then $z_0^* = 1/\bar{z}_0$ is also a zero of $G$. Moreover both zeros must be of the same multiplicity. Further, since $g(\theta) \geq 0$, any zero $z_j = e^{i\theta_j}$ of $G$ must be of even multiplicity. Finally, if $g(\theta)$ is of exact degree $m$, then $\xi_m \neq 0$ and $z = 0$ is not a zero of $G$. This shows that

$$G(z) = a^2 \prod_{i=1}^{q} (z - z_i)(z - z_i^*) \prod_{j=1}^{r} (z - e^{i\theta_j})^2,$$

where $2q + 2r = 2m$.

For any $z_0$,

$$\frac{1 - \bar{z}_0 z}{z - z_0}$$

has absolute value one for $z = e^{i\theta}$. This implies that $|z - z_i| = |z_i||z - z_i^*|$ for $z = e^{i\theta}$, and the form (9.2.3) follows where the degree of $h$ is the same as the degree of $g$, namely $m$. Further representations of the form (9.2.3) can be obtained by replacing $h(z)$ by

$$\frac{1 - \bar{z}_0 z}{z - z_0} h(z).$$

In this way all the zeros of $h(z)$ with $|z| < 1$ can be removed. The uniqueness statement for $h$ and the implication when $g(\theta)$ is a cosine polynomial are omitted.                                                                                       ∎

The following theorem is an immediate consequence of Theorem 9.2.1 and 9.2.6:

**Theorem 9.2.7**
1. $(\alpha_1, \beta_1, \ldots, \alpha_m, \beta_m) \in M_{2m}^c$ if and only if

$$\sum_{i=0}^{m} \sum_{j=0}^{m} d_i \bar{d}_j \gamma_{i-j} \geq 0 \qquad \text{for all complex } d_0, \ldots, d_m. \tag{9.2.5}$$

2. $(\alpha_1, \beta_1, \ldots, \alpha_m, \beta_m) \in Int\, M_{2m}^c$ if and only if the inequality in (9.2.5) is strict unless $d_i = 0$, $i = 0, \ldots, m$.

The matrix

$$T_m = \begin{pmatrix} \gamma_0 & \gamma_1 & \cdots & \gamma_m \\ \gamma_{-1} & \gamma_0 & \cdots & \gamma_{m-1} \\ \vdots & \vdots & & \vdots \\ \gamma_{-m} & \gamma_{-m+1} & \cdots & \gamma_0 \end{pmatrix} \tag{9.2.6}$$

is called a *Toeplitz matrix*. The determinant of $T_m$ is denoted by $\Delta_m$.

**Corollary 9.2.8**
1. $(\alpha_1, \beta_1, \ldots, \alpha_m, \beta_m) \in M_{2m}^c$ if and only if $\Delta_i \geq 0$, $i = 0, 1, \ldots, m$.
2. $(\alpha_1, \beta_1, \ldots, \alpha_m, \beta_m) \in Int\, M_{2m}^c$ if and only if $\Delta_i > 0$, $i = 0, 1, \ldots, m$.

**REMARK 9.2.9**   Theorem 9.2.6 can be used to provide an alternative proof of Theorem 1.4.2; see Szegö (1959, thm. 1.21.1). The statement of Theorem 1.4.2 for a nonnegative polynomial $P$ of degree $n$ on the interval $[-1, 1]$ is that $P(x)$ can be written in the form

$$P(x) = \begin{cases} P_m^2(x) + (1 - x^2)Q_{m-1}^2(x), & n = 2m, \\ (1 + x)P_m^2(x) + (1 - x)Q_m^2(x), & n = 2m + 1, \end{cases}$$

where $P_k$, $Q_k$ are polynomials of degree $\leq k$. By Theorem 9.2.6 we can write

$$P(\cos\theta) = |h(e^{i\theta})|^2 = |e^{-in\theta/2}h(e^{i\theta})|^2, \tag{9.2.7}$$

where $h$ is of degree $n$ and has real coefficients. It is shown in Example 2.1.1 that

$$\cos k\theta, \quad \frac{\sin(k+1)\theta}{\sin\theta}, \quad \frac{\cos(k+(1/2))\theta}{\cos(\theta/2)}, \quad \frac{\sin(k+(1/2))\theta}{\sin(\theta/2)},$$

are polynomials of degree $k$ in $\cos\theta$. Therefore it follows by straightforward algebra that

$$e^{-in\theta/2}h(e^{i\theta})$$

$$= \begin{cases} P_m(\cos\theta) + i\sin\theta Q_{m-1}(\cos\theta), & \text{if } n = 2m, \\ \sqrt{2}\cos\left(\frac{\theta}{2}\right)P_m(\cos\theta) + i\sqrt{2}\sin\left(\frac{\theta}{2}\right)Q_m(\cos\theta), & \text{if } n = 2m + 1, \end{cases}$$

where $P_k$ and $Q_k$ are polynomials of degree $k$. Using (9.2.7), the desired result follows.

## 9.3  CANONICAL MOMENTS ON THE CIRCLE

The canonical moments of a given fixed measure $\sigma$ on the circle are defined somewhat analogous to the ordinary power case. The measure $\sigma$ determines the infinite sequence of trigonometric moments

$$\gamma_k = \int_{-\pi}^{\pi} e^{-ik\theta}d\sigma(\theta), \quad k = 0, \pm 1, \pm 2, \ldots$$

For fixed arbitrary $m$ the problem now involves the range $C_{m+1}$ of the $(m+1)$th moment

$$w = \int_{-\pi}^{\pi} e^{-i(m+1)\theta}d\eta(\theta) \tag{9.3.1}$$

as $\eta$ varies over those measures in $\mathcal{P}[-\pi, \pi)$ such that

$$\gamma_k = \int_{-\pi}^{\pi} e^{-ik\theta} d\eta(\theta), \qquad k = 1, 2, \ldots, m.$$

For this to be nontrivial we assume that $\Delta_m > 0$ or $(\alpha_1, \beta_1, \ldots, \alpha_m, \beta_m) \in \text{Int } M_{2m}^c$. By Corollary 9.2.8 the range of $w$ in (9.3.1) is determined by the inequality

$$\Delta_{m+1}(w) = \begin{vmatrix} \gamma_0 & \gamma_1 & \cdots & \gamma_m & w \\ \gamma_{-1} & \gamma_0 & \cdots & \gamma_{m-1} & \gamma_m \\ \vdots & \vdots & & \vdots & \vdots \\ \bar{w} & \gamma_{-m} & \cdots & \gamma_{-1} & \gamma_0 \end{vmatrix} \geq 0. \tag{9.3.2}$$

Using Sylvester's identity [Lemma (1.5.2)], the inequality in (9.3.2) is equivalent to

$$\begin{vmatrix} \Delta_m & R_m(w) \\ \overline{R_m(w)} & \Delta_m \end{vmatrix} \geq 0, \tag{9.3.3}$$

where $R_0(w) = w$ and for $m \geq 1$,

$$R_m(w) = \begin{vmatrix} \gamma_1 & \gamma_2 & \cdots & \gamma_m & w \\ \gamma_0 & \gamma_1 & \cdots & \gamma_{m-1} & \gamma_m \\ \vdots & \vdots & & \vdots & \vdots \\ \gamma_{-m+1} & \gamma_{-m+2} & \cdots & \gamma_0 & \gamma_1 \end{vmatrix}. \tag{9.3.4}$$

Writing $R_m(w) = (-1)^m w \Delta_{m-1} + R_m(0)$ $(\Delta_{-1} = 1)$, (9.3.3) can be simplified to

$$\left| w + (-1)^m \frac{R_m(0)}{\Delta_{m-1}} \right|^2 \leq \left( \frac{\Delta_m}{\Delta_{m-1}} \right)^2. \tag{9.3.5}$$

Thus the range $C_{m+1}$ of the $(m+1)$th moment is a circle with center

$$s_{m+1} = (-1)^{m+1} \frac{R_m(0)}{\Delta_{m-1}}$$

and radius $r_{m+1} = \Delta_m / \Delta_{m-1}$.

**Definition 9.3.1** *Given $\sigma \in \mathcal{P}[-\pi, \pi)$ and corresponding moments $\gamma_k = \alpha_k - i\beta_k$, $k \geq 1$; let*

$$N(\sigma) = \min\{k \in \mathbb{N} | (\alpha_1, \beta_1, \ldots, \alpha_k, \beta_k) \in \partial M_{2k}^c\}.$$

*The canonical moments $a_1, \ldots, a_N$ of the measure $\sigma$ on the circle are defined by*

$$a_{m+1} = \frac{\gamma_{m+1} - s_{m+1}}{r_{m+1}}, \qquad 0 \le m < N(\sigma). \qquad (9.3.6)$$

Note that $a_1 = \gamma_1$, $|a_{N(\sigma)}| = 1$ and that the canonical moments $a_m$ are left undefined for $m > N(\sigma)$. If $\sigma$ provides a moment point whose sections are in the interior of $M_{2m}^c$ for all $m$, then $|a_m| < 1$ for all $m$; otherwise, the canonical moment sequence is finite in length. As in the power case the canonical moment sequence uniquely determines the measure $\sigma$. Checking Equations (9.3.3) and (9.3.5) the canonical moments can be alternatively expressed as

$$a_{m+1} = \frac{(-1)^m R_m(\gamma_{m+1})}{\Delta_m} \qquad (9.3.7)$$

$$= \frac{(-1)^m}{\Delta_m} \begin{vmatrix} \gamma_1 & \gamma_2 & \cdots & \gamma_{m+1} \\ \gamma_0 & \gamma_1 & \cdots & \gamma_m \\ \vdots & \vdots & & \vdots \\ \gamma_{-m+1} & \gamma_{-m+2} & \cdots & \gamma_1 \end{vmatrix}, \qquad 0 \le m < N(\sigma).$$

### EXAMPLE 9.3.2

1. For the uniform measure on $[-\pi, \pi)$ with constant density $\sigma'(\theta) = 1/(2\pi)$, we have

$$\gamma_k = \int_{-\pi}^{\pi} e^{-ik\theta} \frac{d\theta}{2\pi} = 0, \qquad k \ge 1.$$

From the definition in (9.3.7) it follows readily that the canonical moments of $\sigma$ on the circle are given by $a_k = 0$ for all $k \ge 1$.

2. If the measure $\sigma$ has the density $\sigma'(\theta) = \frac{1}{2\pi}(1 + \cos\theta)$ on $[-\pi, \pi)$, then $\gamma_0 = 1$, $\gamma_1 = \frac{1}{2}$ and $\gamma_k = 0$, $k \ge 2$. Evaluating the determinants $\Delta_m$ and $R_m(\gamma_{m+1})$, it follows that

$$\Delta_m = \frac{(m+2)}{2^{m+1}} \quad \text{and} \quad a_m = \frac{(-1)^{m+1}}{m+1}.$$

**REMARK 9.3.3** Suppose that the measure $\sigma$ on the circle or $[-\pi, \pi)$ is symmetric about zero, or more precisely that $\sigma([-\theta, 0]) = \sigma([0, \theta])$ for all $0 \le \theta < \pi$. In this case there is a close connection between the canonical moments $a_k$ on the circle and the quantities $p_k$ defined for the polynomial case on the interval $[-1, 1]$. Clearly, if $\sigma$ is symmetric, then the trigonometric moments $\gamma_k = \int_{-\pi}^{\pi} \cos(k\theta) d\sigma(\theta)$ are all real, and the canonical moments $a_k, k \ge 0$, are also real.

There is a one-to-one mapping between the symmetric measures $\sigma$ on the circle and the measures $\xi$ on $[-1, 1]$ defined by projecting $\sigma$ onto $[-1, 1]$ by the mapping $x = \cos\theta$. We will show that through this correspondence the canonical moments of $\sigma$ and $\xi$ are related by $[m < N(\sigma) = N(\xi)]$

$$a_{m+1}(\sigma) = 2p_{m+1}(\xi) - 1. \tag{9.3.8}$$

Let $c_0, c_1, \ldots, c_m$ denote the ordinary moments of the corresponding $\xi$. It is shown in Example 2.1.1 that $T_m(x) = \cos(m\theta)$, $x = \cos\theta$ is a polynomial of degree $m$ with leading coefficient $2^{m-1}$, which is called the $m$th Chebyshev polynomial of the first kind. The symmetry of $\sigma$ implies that

$$\gamma_m(\sigma) = \int_{-\pi}^{\pi} \cos m\theta \, d\sigma(\theta) = \int_{-1}^{1} T_m(x) \, d\xi(x)$$

$$= \sum_{k=0}^{\lfloor m/2 \rfloor} (-1)^k \frac{m\Gamma(m-k)}{\Gamma(k+1)\Gamma(m-2k+1)} 2^{m-2k-1} c_{m-2k}(\xi) \tag{9.3.9}$$

$$= 2^{m-1} c_m(\xi) - m 2^{m-3} c_{m-2}(\xi) + \cdots,$$

where the second equality is obtained from (2.1.11). From (9.3.9) it follows that specifying the $m$ values $\gamma_1, \ldots, \gamma_m$ for a symmetric measure $\sigma$ on the circle is equivalent to specifying the moments $c_1, c_2, \ldots, c_m$ for the corresponding "projection" $\xi$ on the interval $[-1, 1]$.

The range of $w$ in (9.3.1) is a circle $C_{m+1}$ with center in $[-1, 1]$. Let $\gamma_{m+1}^+$ and $\gamma_{m+1}^-$ be the upper and lower limits of the intersection of $C_{m+1}$ and $[-1, 1]$. Since the center is in $[-1, 1]$, it follows that

$$\gamma_{m+1}^+ = \sup_{\eta} \int_{-\pi}^{\pi} \cos((m+1)\theta) \, d\eta(\theta),$$

where the sup is taken over all measures with the same values $\gamma_1, \ldots, \gamma_m$ as $\sigma$. This supremum is attained by a symmetric $\eta$, as can be seen by replacing $\eta$ by its "symmetrization" $\eta_s(\theta) = \frac{1}{2}[\eta(\theta) + \eta(-\theta)]$. If $c_1, \ldots, c_m$ denote the moments of the measure on $[-1, 1]$ corresponding to $\gamma_1, \ldots, \gamma_m$ via the "projection" $x = \cos\theta$, then we obtain from (9.3.9),

$$\gamma_{m+1}^+ = \sup_{\mu} \int_{-1}^{1} T_{m+1}(x) \, d\mu(x)$$

$$= 2^m c_{m+1}^+ - (m+1)2^{m-2} c_{m-1} + \cdots.$$

Here $\mu$ is any measure on $[-1, 1]$ with moments $c_1, \ldots, c_m$ and $c_{m+1}^+ = c_{m+1}(\mu_m^+)$, where $\mu_m^+$ is the corresponding upper principal representation. A similar result holds for $\gamma_{m+1}^-$. In this case the center and the radius of the circle

$C_{m+1}$ are given by $s_{m+1} = (\gamma_{m+1}^+ + \gamma_{m+1}^-)/2$, and $r_{m+1} = (\gamma_{m+1}^+ - \gamma_{m+1}^-)/2$ and it follows from (9.3.6) that

$$a_{m+1} = \frac{(\gamma_{m+1} - (\gamma_{m+1}^+ + \gamma_{m+1}^-)/2)}{(\gamma_{m+1}^+ - \gamma_{m+1}^-)/2}$$

$$= 2\left(\frac{c_{m+1} - c_{m+1}^-}{c_{m+1}^+ - c_{m+1}^-}\right) - 1 = 2p_{m+1} - 1.$$

Using the relation (9.3.8), some simple properties of $p_k$ can be derived from properties of $a_k$. For example, if we rotate (counterclockwise) the measure $\sigma$ through the angle $\theta_0$ to give $d\mu(\theta) = d\sigma(\theta - \theta_0)$, then the resulting moments $\gamma_k'$ satisfy $\gamma_k' = e^{-ik\theta_0}\gamma_k$. Writing down the definition of $a_k'$, one can extract factors of $e^{il\theta_0}$ from various rows and columns in the determinants involved to show that the canonical moments of $\mu$ and $\sigma$ are related by the transformation

$$a_k' = e^{-ik\theta_0} a_k, \qquad k = 1, 2, \ldots. \tag{9.3.10}$$

Using (9.3.10), with $\theta_0 = \pi$, we can see the result on the $p_k$ of reflecting a measure $\xi$ on $[-1, 1]$ which is given in Theorem 1.3.3. That is, if $\xi$ is on $[-1, 1]$ and $d\xi'(x) = d\xi(-x)$, the resulting transformation on the circle rotates $\theta$ through an angle $\theta_0 = \pi$, in which case $a_k' = (-1)^k a_k$. By (9.3.8) the corresponding $p_k$ then satisfy $p_{2i}' = p_{2i}$ and $p_{2i+1}' = 1 - p_{2i+1}$.

One of the very useful results involves expressing the determinant of the Toeplitz matrix $\Delta_m$ in terms of the canonical moments $a_k$.

**Lemma 9.3.4**

$$\frac{\Delta_m}{\Delta_{m-1}} = \prod_{k=1}^{m}(1 - |a_k|^2), \qquad 1 \le m \le N(\sigma). \tag{9.3.11}$$

*Proof* Recall the definition in (9.3.4), and let $\wedge_{m+1} = R_m(\gamma_{m+1})$. Then (9.3.7) gives

$$a_{m+1} = \frac{(-1)^m \wedge_{m+1}}{\Delta_m} \tag{9.3.12}$$

To prove (9.3.11), it suffices to show that

$$\frac{\Delta_{m+1}/\Delta_m}{\Delta_m/\Delta_{m-1}} = 1 - |a_{m+1}|^2$$

or

$$\Delta_m^2 |a_{m+1}|^2 = \Delta_m^2 - \Delta_{m+1}\Delta_{m-1}.$$

From (9.3.12)

$$\Delta_m^2 |a_{m+1}|^2 = \wedge_{m+1}\overline{\wedge}_{m+1},$$

so we require

$$\Delta_{m+1}\Delta_{m-1} = \Delta_m^2 - \wedge_{m+1}\overline{\wedge}_{m+1}.$$

However, this follows from Sylvester's identity in Lemma 1.5.2.                    ∎

**Corollary 9.3.5**

$$\Delta_m = \prod_{k=1}^{m}(1 - |a_k|^2)^{m-k+1}, \qquad 1 \le m \le N(\sigma). \tag{9.3.13}$$

## 9.4   ORTHOGONAL POLYNOMIALS ON THE CIRCLE

Given a measure $\sigma$ with corresponding trigonometric moments $\gamma_0, \gamma_1\, \gamma_2, \ldots$, the basic set of monic polynomials is defined by $\varphi_0(z) \equiv 1$ and

$$\varphi_k(z) = \frac{1}{\Delta_{k-1}} \begin{vmatrix} \gamma_0 & \gamma_1 & \cdots & \gamma_{k-1} & 1 \\ \gamma_{-1} & \gamma_0 & \cdots & \gamma_{k-2} & z \\ \vdots & \vdots & & \vdots & \vdots \\ \gamma_{-k} & \gamma_{-k+1} & \cdots & \gamma_{-1} & z^k \end{vmatrix} \tag{9.4.1}$$

for $1 \le k \le N(\sigma)$, that is, as long as $\Delta_{k-1} > 0$. They are orthogonal with respect to the measure $\sigma$ in the sense that

$$\int_{-\pi}^{\pi} \varphi_k(e^{i\theta})\overline{\varphi_\ell(e^{i\theta})}d\sigma(\theta) = \begin{cases} 0, & \ell \ne k, \\ \dfrac{\Delta_k}{\Delta_{k-1}}, & \ell = k. \end{cases} \tag{9.4.2}$$

The polynomials $\varphi_k$ are called *Szegö polynomials with respect to the measure* $\sigma$. The reciprocal polynomials $\varphi_k^*(z)$ [see (9.2.4)] can be expressed as

$$\varphi_k^*(z) = \frac{1}{\Delta_{k-1}} \begin{vmatrix} \gamma_0 & \gamma_{-1} & \cdots & \gamma_{-k+1} & z^k \\ \gamma_1 & \gamma_0 & \cdots & \gamma_{-k+2} & z^{k-1} \\ \vdots & \vdots & & \vdots & \vdots \\ \gamma_k & \gamma_{k-1} & \cdots & \gamma_1 & 1 \end{vmatrix}, \tag{9.4.3}$$

and they satisfy the relations

$$\int_{-\pi}^{\pi} \varphi_k^*(e^{i\theta})\overline{e^{i\ell\theta}}\,d\sigma(\theta) = \begin{cases} \Delta_k/\Delta_{k-1} & \ell = 0, \\ 0, & \ell = 1, \ldots, k. \end{cases} \tag{9.4.4}$$

In terms of the polynomials $\varphi_m$ and $\varphi_{m-1}^*$ the $m$th canonical moment $a_m$ can be written as

$$a_m = -\overline{\varphi_m(0)} \tag{9.4.5}$$

or

$$a_m = \frac{\Delta_{m-2}}{\Delta_{m-1}} \int_{-\pi}^{\pi} \varphi_{m-1}^*(e^{i\theta})e^{-im\theta}\,d\sigma(\theta). \tag{9.4.6}$$

The two sequences $\varphi_k$ and $\varphi_k^*$ can be defined recursively by $\varphi_0(z) \equiv 1 \equiv \varphi_0^*(z)$, and for $n \geq 1$

$$\varphi_n(z) = z\varphi_{n-1}(z) - \bar{a}_n\varphi_{n-1}^*(z),$$
$$\varphi_n^*(z) = \varphi_{n-1}^*(z) - a_n z\varphi_{n-1}(z). \tag{9.4.7}$$

These equations follow, as in Theorem 2.3.1, by Sylvester's identity. In each case one starts with the determinant expressions in (9.4.1) or (9.4.3) and uses the pivot block obtained by deleting the first and last column and row. Note that the leading coefficient of $\varphi_n$ and $\varphi_n^*$ are 1 and $-a_n$, respectively. This shows that the degree of $\varphi_n$ is $n$, while $\varphi_n^*$ is of degree $n$ if and only if $a_n \neq 0$.

The polynomials $\varphi_n$ and $\varphi_n^*$ satisfy certain minimization properties that will be used in Section 9.6. To explain these properties in more detail, we let

$$P_n(z) = \sum_{k=0}^{n} d_k z^k$$

denote an arbitrary polynomial of degree at most $n$. Since $\varphi_k(z)$ is of exact degree $k$, then $P_n(z)$ can be written as

$$P_n(z) = \sum_{k=0}^{n} e_k \varphi_k(z),$$

and by (9.4.2),

$$\int_{-\pi}^{\pi} |P_n(e^{i\theta})|^2\,d\sigma(\theta) = \int_{-\pi}^{\pi} \left| \sum_{k=0}^{n} e_k \varphi_k(e^{i\theta}) \right|^2 d\sigma(\theta)$$
$$= \sum_{k=0}^{n} |e_k|^2 \int_{-\pi}^{\pi} |\varphi_k(e^{i\theta})|^2\,d\sigma(\theta).$$

Since $\varphi_n$ has leading coefficient 1, we have $d_n = e_n$, and it follows that

$$\min_{d_n=1} \int_{-\pi}^{\pi} |P_n(e^{i\theta})|^2 d\sigma(\theta) = \min_{d_n=1} \int_{-\pi}^{\pi} |d_0 + d_1 e^{i\theta} + \cdots + d_n e^{in\theta}|^2 d\sigma(\theta)$$

(9.4.8)

$$= \int_{-\pi}^{\pi} |\varphi_n(e^{i\theta})|^2 d\sigma(\theta) = \frac{\Delta_n}{\Delta_{n-1}}.$$

Moreover

$$\min_{d_0=1} \int_{-\pi}^{\pi} |P_n(e^{i\theta})|^2 d\sigma(\theta)$$

$$= \min_{d_0=1} \int_{-\pi}^{\pi} |d_0 + d_1 e^{i\theta} + \cdots + d_n e^{in\theta}|^2 d\sigma(\theta)$$

$$= \min_{d_0=1} \int_{-\pi}^{\pi} |\overline{d}_0 + \overline{d}_1 e^{-i\theta} + \cdots + \overline{d}_n e^{-in\theta}|^2 |e^{in\theta}|^2 d\sigma(\theta)$$

$$= \min_{d_0=1} \int_{-\pi}^{\pi} |\overline{d}_0 e^{in\theta} + \overline{d}_1 e^{i(n-1)\theta} + \cdots + \overline{d}_n|^2 d\sigma(\theta)$$

$$= \min_{d_n=1} \int_{-\pi}^{\pi} |P_n(e^{i\theta})|^2 d\sigma(\theta) = \int_{-\pi}^{\pi} |\varphi_n(e^{i\theta})|^2 d\sigma(\theta). = \frac{\Delta_n}{\Delta_{n-1}}.$$

Since $\varphi_n^*(0) = 1$ and

$$\int_{-\pi}^{\pi} |\varphi_n^*(e^{i\theta})|^2 d\sigma(\theta) = \int_{-\pi}^{\pi} |\varphi_n(e^{i\theta})|^2 d\sigma(\theta) = \frac{\Delta_n}{\Delta_{n-1}}.$$

[see (9.4.4)], it follows that

$$\min_{d_0=1} \int_{-\pi}^{\pi} |P_n(e^{i\theta})|^2 d\sigma(\theta) = \int_{-\pi}^{\pi} |\varphi_n^*(e^{i\theta})|^2 d\sigma(\theta).$$

(9.4.9)

It is of some interest to determine where the zeros of $\varphi_n(z)$ are located.

**Theorem 9.4.1**   *If $\Delta_n > 0$, then the polynomial $\varphi_n(z)$ has no zeros in the region $|z| \geq 1$, and therefore the polynomial $\varphi_n^*(z)$ has no zeros in the region $|z| \leq 1$.*

*Proof*   The proof will make use of the *Christoffel-Darboux formula*

$$\sum_{i=0}^{n} \frac{\varphi_i(z)\overline{\varphi_i(w)}}{k_i} = \frac{\varphi_n^*(z)\overline{\varphi_n^*(w)} - z\overline{w}\varphi_n(z)\overline{\varphi_n(w)}}{k_n(1 - z\overline{w})},$$

where $k_0 = 1$ and $k_n = \int_{-\pi}^{\pi} |\varphi_n(e^{i\theta})|^2 d\sigma(\theta) = \Delta_n/\Delta_{n-1}$. The proof of this identity uses induction and the recurrence formula (9.4.7). If $\varphi_n(z_0) = 0$ and $|z_0| = 1$, then $\varphi_n^*(z_0) = 0$ and

$$\sum_{i=0}^{n} \frac{|\varphi_i(z_0)|^2}{k_i} = 0. \tag{9.4.10}$$

Note, however, that $\varphi_0(z) = 1 = k_0$, so (9.4.10) is impossible. If $\varphi_n(z_0) = 0$ and $|z_0| > 1$, then

$$\sum_{i=0}^{n} \frac{|\varphi_i(z_0)|^2}{k_i} = \frac{|\varphi_n^*(z_0)|^2}{k_n(1 - |z_0|^2)} \leq 0,$$

which is also impossible. ∎

Now consider a sequence $\gamma_0, \gamma_1, \ldots, \gamma_m, \gamma_{m+1}$ of trigonometric moments from some $\sigma$ such that the determinants of the Toeplitz matrices satisfy $\Delta_m > 0$ and $\Delta_{m+1} = 0$. In this case $\gamma_{m+1} = w$ is on the boundary of the circle $C_{m+1}$ described in Section 9.3. Let $\varphi_{m+1}(z) = \varphi_{m+1}(z; w)$ in order to emphasize the dependence on $w$, and denote the corresponding measure by $\sigma_w$. Then by (9.4.2),

$$\int_{-\pi}^{\pi} |\varphi_{m+1}(e^{i\theta}; w)|^2 d\sigma_w(\theta) = \frac{\Delta_{m+1}}{\Delta_m} = 0, \tag{9.4.11}$$

in which case the support of $\sigma_w$ for each $w$ must be contained in the set of zeros of $\varphi_{m+1}(z; w)$. Since $\Delta_m > 0$, it follows from Corollary 9.2.8 and Lemma 9.2.4 that $\varphi_{m+1}(z; w)$ must have $m + 1$ distinct zeros satisfying $|z| = 1$.

**Theorem 9.4.2** *If $\Delta_m > 0$ and $\Delta_{m+1} = 0$, then the support of the corresponding measure on the unit circle $\sigma$ is given by the zeros of $\varphi_{m+1}(z)$ or $\varphi_{m+1}^*(z)$.*

## 9.5 TRANSFORMS AND CONTINUED FRACTIONS

For the circle case the analogue of the Stieltjes transform is given by

$$F(z) = F(z; \sigma) = \int_{-\pi}^{\pi} \frac{e^{i\theta} + z}{e^{i\theta} - z} d\sigma(\theta), \qquad |z| \neq 1. \tag{9.5.1}$$

This represents two analytic functions, one for $|z| < 1$ and one for $|z| > 1$. Expanding the integrand in (9.5.1) for $|z| < 1$ gives

$$\frac{e^{i\theta} + z}{e^{i\theta} - z} = 1 + 2 \sum_{k=1}^{\infty} z^k e^{-ik\theta}$$

so that

$$F(z) = 1 + 2\sum_{k=1}^{\infty} z^k \gamma_k, \qquad |z| < 1. \tag{9.5.2}$$

For $|z| > 1$ we obtain

$$\frac{e^{i\theta} + z}{e^{i\theta} - z} = -1 - 2\sum_{k=1}^{\infty} e^{ik\theta} z^{-k},$$

and hence

$$F(z) = -1 - 2\sum_{k=1}^{\infty} \overline{\gamma}_k z^{-k}, \qquad |z| > 1.$$

For example, if $\sigma(\theta)$ is the uniform measure, then by Example 9.3.2, $\gamma_k = 0$ for $k \geq 1$ and

$$F(z) = \begin{cases} 1 & \text{if } |z| < 1, \\ -1 & \text{if } |z| > 1. \end{cases}$$

The function $F(z)$ is associated with the continued fraction

$$1 - \frac{2|}{|1} + \frac{1|}{|-a_1 z} + \frac{(1 - |a_1|^2)z|}{|-\overline{a}_1} + \frac{1|}{|-a_2 z} + \frac{(1 - |a_2|^2)z|}{|-\overline{a}_2} + \cdots. \tag{9.5.3}$$

By (3.2.17) and (3.2.11) the even contraction of this continued fraction is

$$1 + \frac{2a_1 z|}{|1 - a_1 z} - \frac{a_2(1 - |a_1|^2)z|}{|a_1 + a_2 z} - \frac{a_1 a_3(1 - |a_2|^2)z|}{|a_2 + a_3 z} - \cdots \tag{9.5.4}$$

Let $A_n(z)$ and $B_n(z)$ denote the corresponding numerator and denominator polynomials of the continued fraction (9.5.3). If we rewrite the recursion formula (9.4.7) for $\varphi_n$ and $\varphi_n^*$ in the form

$$\varphi_n(z) = -\overline{a}_n \varphi_n^*(z) + (1 - |a_n|^2)z\varphi_{n-1}(z),$$

$$\tag{9.5.5}$$

$$\varphi_n^*(z) = -a_n z\varphi_{n-1}(z) + \varphi_{n-1}^*(z),$$

it follows from (3.2.2) that the partial denominators of the continued fraction in

(9.5.3) are given by

$$\varphi_n^*(z) = B_{2n}(z) \text{ and } \varphi_n(z) = B_{2n+1}(z). \tag{9.5.6}$$

The numerators $A_n(z)$ are related to the *associated Szegö polynomials with respect to the measure* $d\sigma$ defined by $\pi_0(z) = -1$, $\omega_0(z) = 1$, and for $n \geq 1$,

$$\pi_n(z) = \int_{-\pi}^{\pi} \frac{e^{i\theta} + z}{e^{i\theta} - z} \left[ \varphi_n(z) - \varphi_n(e^{i\theta}) \right] d\sigma(\theta),$$

$$\tag{9.5.7}$$

$$\omega_n(z) = \int_{-\pi}^{\pi} \frac{e^{i\theta} + z}{e^{i\theta} - z} \left[ \varphi_n^*(z) - \left( \frac{z}{e^{i\theta}} \right)^n \varphi_n^*(e^{i\theta}) \right] d\sigma(\theta).$$

For these polynomials we have the following lemma, which appears as Theorem 4.1 in Jones, Njåstad, and Thron (1989).

**Lemma 9.5.1** *If $\pi_n$ and $\omega_n$ are given by (9.5.7) then*
1. *for each $n \geq 0$, $\pi_n(z)$ is a polynomial of exact degree $n$, $\omega_n(z)$ is a polynomial of degree at most $n$, and*

$$\omega_n(z) = -\pi_n^*(z); \tag{9.5.8}$$

2. *for $n \geq 2, 0 \leq k \leq n - 1$,*

$$\pi_n(z) = \int_{-\pi}^{\pi} \frac{e^{i\theta} + z}{e^{i\theta} - z} \left[ \varphi_n(z) - \left( \frac{z}{e^{i\theta}} \right)^k \varphi_n(e^{i\theta}) \right] d\sigma(\theta), \tag{9.5.9}$$

*and for $n \geq 1, 1 \leq k \leq n$,*

$$\omega_n(z) = \int_{-\pi}^{\pi} \frac{e^{i\theta} + z}{e^{i\theta} - z} \left[ \varphi_n^*(z) - \left( \frac{z}{e^{i\theta}} \right)^k \varphi_n^*(e^{i\theta}) \right] d\sigma(\theta); \tag{9.5.10}$$

3. *If $A_l(z)$ denotes the lth partial numerator of the continued fraction (9.5.3), then for $n \geq 0$,*

$$\omega_n(z) = A_{2n}(z), \quad \pi_n(z) = A_{2n+1}(z). \tag{9.5.11}$$

*Proof* (1) The statements regarding the degrees of $\pi_n$ and $\omega_n$ follow, since $\varphi_n$ is a polynomial of exact degree $n$ and $\varphi_n^*$ is a polynomial of degree at most $n$.

From (9.2.4),

$$\pi_n^*(z) = z^n \overline{\pi_n(1/\bar{z})} = z^n \int_{-\pi}^{\pi} \frac{z + e^{i\theta}}{z - e^{i\theta}} \left[ \overline{\varphi}_n(z^{-1}) - \overline{\varphi}_n(e^{-i\theta}) \right] d\sigma(\theta)$$

$$= \int_{-\pi}^{\pi} \frac{z + e^{i\theta}}{z - e^{i\theta}} \left[ \varphi_n^*(z) - \left( \frac{z}{e^{i\theta}} \right)^n \varphi_n^*(e^{i\theta}) \right] d\sigma(\theta) = -w_n(z).$$

(2) For the statement regarding $\pi_n$, note that the case $k = 0$ is simply (9.5.7). For $1 \leq k \leq n - 1$ and $n \geq 2$, let $\rho_n(z)$ denote the right-hand side of (9.5.9), then

$$\rho_n(t) - \pi_n(z) = \int_{-\pi}^{\pi} \frac{e^{i\theta} + z}{e^{i\theta} - z} \left[ 1 - \left( \frac{z}{e^{i\theta}} \right)^k \right] \varphi_n(e^{i\theta}) d\sigma(\theta)$$

$$= \int_{-\pi}^{\pi} (e^{i\theta} + z) \left[ \sum_{m=0}^{k-1} (e^{i\theta})^{m-k} z^{k-1-m} \right] \varphi_n(e^{i\theta}) d\sigma(\theta),$$

which is zero by the orthogonality conditions of $\varphi_n$. The result for $w_n(z)$ is shown in a similar manner.

(3) This follows by taking $k = 1$ in both (9.5.9) and (9.5.10) and showing that $\pi_n(z)$ and $w_n(z)$, $n \geq 0$ satisfy the same recursive equations and boundary conditions as the partial numerators $A_{2n}(z)$ and $A_{2n+1}(z)$. ■

**Theorem 9.5.2** *The even convergents $A_{2n}(z)/B_{2n}(z)$ of the continued fraction (9.5.3) converge uniformly for $|z| < r < 1$ to $F(z)$ defined in (9.5.1).*

*Proof* Using (9.5.10) with $k = n$ and (9.5.11), we have

$$A_{2n}(z) = w_n(z) = \int_{-\pi}^{\pi} \frac{e^{i\theta} + z}{e^{i\theta} - z} \left[ \varphi_n^*(z) - z^n \overline{\varphi_n(e^{i\theta})} \right] d\sigma(\theta)$$

$$= \varphi_n^*(z) F(z) - z^n \int_{-\pi}^{\pi} \frac{e^{i\theta} + z}{e^{i\theta} - z} \overline{\varphi_n(e^{i\theta})} d\sigma(\theta).$$

Then from (9.5.6) it follows that

$$\left| \frac{A_{2n}(z)}{B_{2n}(z)} - F(z) \right| \leq \frac{|z|^n}{|\varphi_n^*(z)|} \int_{-\pi}^{\pi} \left| \frac{e^{i\theta} + z}{e^{i\theta} - z} \right| |\varphi_n(e^{i\theta})| d\sigma(\theta).$$

Finally the Christoffel-Darboux equation in the proof of Theorem 9.4.1 shows that

$$1 \leq \sum_{i=0}^{n} \frac{|\varphi_i(z)|^2}{k_i} = \frac{|\varphi_n^*(z)|^2 - |z|^2 |\varphi_n(z)|^2}{k_n(1 - |z|^2)} \leq \frac{|\varphi_n^*(z)|^2}{k_n(1 - |z|^2)}, \tag{9.5.12}$$

or equivalently

$$|\varphi_n^*(z)|^2 \geq k_n(1 - |z|^2).$$

Using this and Schwarz's inequality, we obtain that for $|z| \leq r < 1$,

$$\left| \frac{A_{2n}(z)}{B_{2n}(z)} - F(z) \right| \leq \frac{r^n}{\sqrt{k_n}(1 - r^2)^{1/2}} \frac{1+r}{1-r} \left[ \int_{-\pi}^{\pi} |\varphi_n(e^{i\theta})|^2 d\sigma(\theta) \right]^{1/2}$$

$$= \frac{r^n(1+r)}{(1 - r^2)^{1/2}(1-r)},$$

and the desired result follows. ∎

To invert the transform $F(z)$ defined in (9.5.1) the real part of $F(z)$ for $|z| < 1$ is considered. For $0 < r < 1$,

$$\frac{1}{2\pi} \mathcal{R}(F(re^{i\theta_0})) = \frac{1}{2\pi} \int_{-\pi}^{\pi} \frac{1 - r^2}{1 - 2r \cos (\theta - \theta_0) + r^2} d\sigma(\theta) \qquad (9.5.13)$$

$$= \int_{-\pi}^{\pi} K_r(\theta - \theta_0) d\sigma(\theta),$$

where

$$K_r(\theta) = \frac{1 - r^2}{2\pi(1 - 2r \cos \theta + r^2)} \qquad (9.5.14)$$

is the classical *Poisson kernel*. Viewed as a "density" in $\theta$, it is known that $K_r(\theta)$ integrates to one and converges weakly to the Dirac measure at $0$ as $r \to 1$.

The kernel $K_r(\theta)$ in (9.5.14) and $v^{-1}K(yv^{-1})$ in (3.6.8) have similar properties. The following theorem is the analogue of Theorem 3.6.7, and the proof is omitted. Theorems 3.6.4, 3.6.5, and 3.6.6 can be transferred in a similar manner.

**Theorem 9.5.3** *If $\mathcal{R}(F(re^{i\theta}))$ has an extension to $\mathcal{R}(F(e^{i\theta}))$ as $r \to 1$, which is continuous, then $\sigma$ has a density given by*

$$\frac{1}{2\pi} \mathcal{R}(F(e^{i\theta})). \qquad (9.5.15)$$

An important consequence of Theorem 9.5.3 is the following useful result:

**Theorem 9.5.4** *If $\sigma$ has canonical moments $|a_i| < 1$ for $i = 1, \ldots, m$ and $a_i = 0$ for $i > m$, then $\sigma$ has a continuous density given by*

$$\sigma'(\theta) = \frac{1}{2\pi} \frac{\Delta_m/\Delta_{m-1}}{|\varphi_m^*(e^{i\theta})|^2}. \qquad (9.5.16)$$

*Proof*   Using the recursive relations for $\varphi_n^*$ in (9.5.5) and (9.5.7), we find that for $\ell \geq 0$,

$$\varphi_{m+\ell}^*(z) = \varphi_m^*(z),$$

$$\omega_{m+\ell}(z) = \omega_m(z),$$

and hence by Theorem 9.5.2, (9.5.11), and (9.5.6),

$$F(z) = \lim_{k \to \infty} \frac{A_{2k}(z)}{B_{2k}(z)} = \frac{\omega_m(z)}{\varphi_m^*(z)}, \qquad |z| \leq r < 1.$$

This yields for the real part of the transform $F$,

$$\mathcal{R}(F(e^{i\theta})) = \frac{1}{|\varphi_m^*(e^{i\theta})|^2} \left( \frac{\omega_m(e^{i\theta})\overline{\varphi_m^*(e^{i\theta})} + \overline{\omega_m(e^{i\theta})}\varphi_m^*(e^{i\theta})}{2} \right)$$

From (9.2.4), (9.5.6), and Lemma 9.5.1 we have

$$\overline{\varphi_m^*(e^{i\theta})} = e^{-im\theta} B_{2m+1}(e^{i\theta}),$$

$$\overline{\omega_m(e^{i\theta})} = -e^{-im\theta} A_{2m+1}(e^{i\theta}),$$

so by Lemma 9.5.1 (part 3) and (9.5.6),

$$\mathcal{R}(F(e^{i\theta})) = \frac{e^{-im\theta}}{2|\varphi_m^*(e^{i\theta})|^2} [A_{2m}(e^{i\theta})B_{2m+1}(e^{i\theta}) - A_{2m+1}(e^{i\theta})B_{2m}(e^{i\theta})].$$

Observing that $A_n(z)$, $B_n(z)$ are the partial numerator and denominator of the continued fraction (9.5.3), this expression can be reduced by the determinant formula (3.2.12) to

$$\mathcal{R}(F(e^{i\theta})) = \frac{e^{-im\theta}}{2|\varphi_m^*(e^{i\theta})|^2} \left[ (-1)^{2m+1}(-2)e^{im\theta} \prod_{i=1}^{m}(1 - |a_i|^2) \right].$$

Because $\Delta_m > 0$, $\varphi_m^*(z)$ has no zeros in the region $|z| \leq 1$ (see Theorem 9.4.1), and formula (9.5.16) then follows from (9.5.15), Lemma 9.3.4, and Theorem 9.5.3.    ∎

**EXAMPLE 9.5.5**   If we take $a_1$ real, $|a_1| < 1$, and $a_i = 0$ for $i \geq 2$, then the corresponding distribution $\sigma$ has density

$$\sigma'(\theta) = \frac{1}{2\pi} \frac{1 - |a_1|^2}{1 - 2a_1 \cos \theta + a_1^2}. \tag{9.5.17}$$

This follows from Theorem 9.5.4, since $\varphi_1^*(z) = 1 - a_1 z$.

## 9.6   STATIONARY RANDOM PROCESSES

The canonical moments on the circle defined in Section 9.3 have appeared previously in various places, including approximation theory on the circle and in the theory of *discrete time stationary processes*.

Let $(\Omega, \mathcal{A}, P)$ denote an abstract probability space; that is, $\Omega$ is an arbitrary set with a $\sigma$-field of subsets on which is defined a probability measure $P$. A discrete time process is a sequence $\{X_n(\omega)\}_{n=-\infty}^{\infty}$ of $\mathcal{A}$-measurable complex valued functions defined on $\Omega$. The operator $E$ will denote "expectation" with respect to the measure $P$ defined by

$$Ef = \int_{\Omega} f(\omega)dP(\omega).$$

**Definition 9.6.1**   *A discrete time process $\{X_n(\omega)\}_{n=-\infty}^{\infty}$ is called stationary (or weakly stationary) if $E|X_n|^2 < \infty$ for all n, and $EX_n$ and*

$$\rho(h) = E(X_{n+h}\overline{X}_n) - EX_{n+h}E\overline{X}_n \qquad h = 0, \pm 1, \pm 2, \ldots, \tag{9.6.1}$$

*are independent of n. The function $\rho(h)$ is called the covariance function. We will assume that $EX_n = 0$ and $\rho(0) = 1$.*

**Definition 9.6.2**   *A measure $\sigma$ on the interval $[-\pi, \pi)$ is called the spectral measure of the stationary process $\{X_n(\omega)\}_{n=-\infty}^{\infty}$ if the covariance function can be represented as*

$$\rho(h) = \int_{-\pi}^{\pi} e^{ih\theta} d\sigma(\theta), \qquad h = 0, \pm 1, \pm 2, \ldots .. \tag{9.6.2}$$

The theory in the previous sections shows that a spectral measure $\sigma$ always exists. More precisely, the covariance function $\rho(h)$ has the property that

$$\sum_{i,j=0}^{n} d_i \overline{d}_j \rho(i-j) = E\left| \sum_{i=0}^{n} d_i X_i \right|^2 \geq 0 \tag{9.6.3}$$

for all complex $d_0, \ldots, d_n$. The same property holds for $\gamma_i = \overline{\rho(i)}$; therefore $\gamma_1, \gamma_2, \ldots$ can be viewed as a moment sequence, and the representation (9.6.2) is an immediate consequence of Theorem 9.2.7. As before the spectral measure of a stationary process is unique. The characterization of a nonnegative definitive function $\rho$ on the integers by the spectral representation (9.6.2) is called *Herglotz's theorem*.

Associated with the process $\{X_n\}$ is a Hilbert space of functions on $\Omega$. Starting with the usual $L_2$-space of functions on $(\Omega, \mathcal{A}, P)$ with inner product

$$(g, h) = Eg\overline{h},$$

the subspace $L_2(X)$ is defined as the closure of the linear space spanned by the process functions $X_n(\omega)$, $n = 0, \pm 1, \pm 2, \ldots$. The corresponding norm is denoted by $|Z|^2_{L_2(X)} = E|Z|^2$.

A related space is the $L_2$-space of complex-valued periodic functions on $[-\pi, \pi)$ that are square integrable with respect to the measure $d\sigma$. This space and the corresponding norm are denoted by $L_2(\sigma)$ and $|f|^2_{L_2(\sigma)} = \int_{-\pi}^{\pi} |f(\theta)|^2 d\sigma(\theta)$, respectively. Note that this space is the closure in $L_2(\sigma)$ of the subspace spanned by the doubly infinite sequence $e^{ik\theta}$, $k = 0, \pm 1, \ldots$.

Now define a transformation $S$ between $L_2(X)$ and $L_2(\sigma)$ by setting

$$S(e^{ik\theta}) = X_k \tag{9.6.4}$$

and extending it linearly so that

$$S\left(\sum_{k=0}^{n} d_k e^{ik\theta}\right) = \sum_{k=0}^{n} d_k X_k.$$

Then, using the spectral representation (9.6.2), it follows that

$$\left|\sum_{k=0}^{n} d_k X_k\right|^2_{L_2(X)} = E\left|\sum_{k=0}^{n} d_k X_k\right|^2$$

$$= \sum_{k,l=0}^{n} d_k \bar{d}_l \rho(k - l) = \sum_{k,l=0}^{n} d_k \bar{d}_l \int_{-\pi}^{\pi} e^{i\theta(k-l)} d\sigma(\theta)$$

$$= \int_{-\pi}^{\pi} \left|\sum_{k=0}^{n} d_k e^{ik\theta}\right|^2 d\sigma(\theta) = \left|\sum_{k=0}^{n} d_k e^{ik\theta}\right|^2_{L_2(\sigma)}. \tag{9.6.5}$$

Similarly, inner products of corresponding pairs of elements in $L_2(X)$ and $L_2(\sigma)$ are equal. The mapping $S$ may be extended to the whole of $L_2(\sigma)$ in the usual manner, and it defines an isometry between $L_2(X)$ and $L_2(\sigma)$. In this way any linear problem in $L_2(\sigma)$ can be viewed as a corresponding problem in $L_2(X)$. The following will illustrate this point:

In the stochastic setting one sometimes has available observed values of $X_1, \ldots, X_m$ and wishes to predict or estimate $X_{m+1}$ or $X_0$. Estimating $X_{m+1}$ is called a *forward prediction*, while estimating $X_0$ is called a *backward prediction*. The prediction is done using a linear combination of $X_1, \ldots, X_m$. Let $b_{1m}, \ldots, b_{mm}$ denote the values minimizing the norm

$$E\left|X_0 + \sum_{k=1}^{m} b_{km} X_k\right|^2 = \int_{-\pi}^{\pi} \left|1 + \sum_{k=1}^{m} b_{km} e^{ik\theta}\right|^2 d\sigma(\theta) \tag{9.6.6}$$

and $a_{0m}, \ldots, a_{m-1,m}$ the values minimizing

$$E\left|X_m + \sum_{k=0}^{m-1} a_{km} X_k\right|^2 = \int_{-\pi}^{\pi} \left|e^{im\theta} + \sum_{k=0}^{m-1} a_{km} e^{ik\theta}\right|^2 d\sigma(\theta). \tag{9.6.7}$$

From the above discussion and Equations (9.4.8) and (9.4.9) it follows that

$$z^m + \sum_{k=0}^{m-1} a_{km} z^k = \varphi_m(z),$$

$$\tag{9.6.8}$$

$$1 + \sum_{k=1}^{m} b_{km} z^k = \varphi_m^*(z),$$

and the "best" predictors of $X_m$ from $X_0, \ldots, X_{m-1}$ and $X_0$ from $X_1, \ldots, X_m$ are given by

$$\hat{X}_m = -\sum_{k=0}^{m-1} a_{km} X_k$$

and $\tag{9.6.9}$

$$\hat{X}_0 = -\sum_{k=1}^{m} b_{km} X_k,$$

where $a_{km}$ and $b_{km}$ can be obtained from the coefficients of the polynomials $\varphi_m(z)$ and $\varphi_m^*(z)$. Observing (9.2.4), it follows that the two sets of coefficients are related by

$$b_{m-k,m} = \overline{a}_{km}, \qquad k = 0, 1, \ldots, m-1. \tag{9.6.10}$$

If $a_m(\sigma)$ denotes the $m$th canonical moment of the spectral measure $\sigma$, then from (9.4.5) we obtain

$$a_m(\sigma) = -\overline{\varphi_m(0)} = -\overline{a}_{0m} = -b_{mm}. \tag{9.6.11}$$

The coefficients $a_{0m}$ and $b_{mm}$ are sometimes referred to as *reflection coefficients*. Because of (9.6.14) below $a_m(\sigma)$ is also called the *partial autocorrelation function*.

Finally by Lemma 9.3.4 the $m$-step prediction error in either direction is

$$E\left|X_m + \sum_{k=0}^{m-1} a_{km}X_k\right|^2 = E\left|X_0 + \sum_{k=1}^{m} b_{km}X_k\right|^2$$

$$= \int_{-\pi}^{\pi} |\varphi_m(e^{i\theta})|^2 d\sigma(\theta) \qquad (9.6.12)$$

$$= \int_{-\pi}^{\pi} |\varphi_m^*(e^{i\theta})|^2 d\sigma(\theta)$$

$$= \frac{\Delta_m}{\Delta_{m-1}} = \prod_{k=1}^{m}(1 - |a_k|^2).$$

In (9.6.11) we showed that the coefficient $-b_{mm}$ was equal to the canonical moment $a_m(\sigma)$. This coefficient has an alternative interpretation as the correlation between the forward and backward residuals in predicting $X_0$ and $X_m$ linearly from $X_1, \ldots, X_{m-1}$. Note that this is now an $m - 1$ step prediction rather than an $m$ step prediction. Let

$$U_m = X_m + \sum_{k=1}^{m-1} a_{k-1,m-1}X_k$$

and

$$V_m = X_0 + \sum_{k=1}^{m-1} b_{k,m-1}X_k.$$

The correlation between $U_m$ and $V_m$ is defined as

$$\text{corr}(V_m, U_m) = \frac{E(V_m m)}{|V_m|_{L_2(X)}|U_m|_{L_2(X)}} \qquad (9.6.13)$$

and is called the *partial autocorrelation function*. It is fairly straightforward to show that

$$a_m(\sigma) = \text{corr}(V_m, U_m). \qquad (9.6.14)$$

By the discussion in (9.6.12) the denominator in (9.6.13) is $\Delta_{m-1}/\Delta_{m-2}$. Observing the isometry between $L_2(X)$ and $L_2(\sigma)$, the numerator can be written as

$$\int_{-\pi}^{\pi} \varphi_{m-1}^*(e^{i\theta}) \overline{e^{i\theta}\varphi_{m-1}(e^{i\theta})} d\sigma(\theta). \qquad (9.6.15)$$

Equation (9.6.14) can then be verified by using (9.4.6) and the orthogonality relations (9.4.4) for $\varphi_{m-1}^*$.

An important problem is to be able to calculate the prediction equations (9.6.9) from the covariance function $\rho(h)$ and the inverse problem of calculating the covariance function from the prediction equations or the reflection coefficients in (9.6.11). This is the same as exhibiting more carefully the one-to-one map between the canonical moments $a_k = a_k(\sigma), k = 1, \ldots, m$ and the trigonometric moments $\gamma_k = \gamma_k(\sigma), k = 1, \ldots, m$. Essential use is made of the orthogonal equations (9.4.2) and (9.4.4) for $\varphi_n$ and $\varphi_n^*$ and the recursion equations (9.4.7). A careful discussion of many of these algorithms or procedures is given in Strobach (1990).

We first consider the problem of calculating the sequence of canonical moments from the sequence of trigonometric moments. Integrating the first equation in (9.4.7) with respect to $d\sigma$ and observing (9.4.2) yields

$$\bar{a}_{m+1} = \frac{\int_{-\pi}^{\pi} e^{i\theta} \varphi_m(e^{i\theta}) d\sigma(\theta)}{\int_{-\pi}^{\pi} \varphi_m^*(e^{i\theta}) d\sigma(\theta)} = \frac{\sum_{k=0}^{m} a_{km} \bar{\gamma}_{k+1}}{\sum_{k=0}^{m} b_{km} \bar{\gamma}_k} \tag{9.6.16}$$

where $a_{mm} = b_{0m} = 1$. Thus starting with $\gamma_0 = 1$ and $\varphi_0 = \varphi_0^* = 1$, we can calculate $\bar{a}_1 = \bar{\gamma}_1$. Using $\bar{a}_1$ in (9.4.7) then gives $\varphi_1(z) = z - \bar{a}_1$ and $\varphi_1^* = 1 - a_1 z$, which gives rise to $\bar{a}_2$ by (9.6.16), namely

$$a_2 = \frac{\gamma_2 - a_1 \gamma_1}{1 - \bar{a}_1 \gamma_1} = \frac{\gamma_2 - \gamma_1^2}{1 - |\gamma_1|^2}.$$

This in term gives the polynomials $\varphi_2$ and $\varphi_2^*$ and so on. On the one hand, (9.6.16) allows a recursive computation of the canonical moments on the circle from the trigonometric moments. On the other hand, the algorithm also gives the coefficients $a_{km}, b_{km}$ of the polynomials $\varphi_m$ and $\varphi_m^*$, which are used for the forward and backward prediction in discrete stationary processes. Note that the denominator in (9.6.16) can be simplified using the fact that

$$\int_{-\pi}^{\pi} \varphi_m^*(e^{i\theta}) d\sigma(\theta) = \frac{\Delta_m}{\Delta_{m-1}} = \prod_{i=1}^{m}(1 - |a_i|^2).$$

Observing $\gamma_i = \bar{\rho(i)}$ and (9.6.11), it follows that the coefficients in the forward prediction can be obtained recursively from the autocovariance function as

$$a_{0,m+1} = \frac{-\rho(m+1) - \sum_{k=0}^{m-1} a_{km} \rho(k+1)}{\prod_{k=1}^{m}(1 - |a_{0i}|^2)}. \tag{9.6.17}$$

Comparing the coefficients in (9.4.7) and observing (9.6.8), (9.6.10), and (9.6.11) yields

$$a_{km} = a_{k-1,m-1} + a_{0m} \bar{a}_{m-k-1,m-1}, \qquad k = 1, \ldots, m-1. \tag{9.6.18}$$

Equations (9.6.17) and (9.6.18) allow a recursive calculation of the coefficients $a_{km}$ used in the forward prediction (9.6.5). This algorithm is known as the Levinson or *Durbin-Levinson algorithm*.

To calculate the regular moments $\{\gamma_k\}$ from the canonical moments $\{a_k\}$, we use the fact that

$$0 = \int_{-\pi}^{\pi} \varphi_m(e^{i\theta})d\sigma(\theta) = \gamma_{-m} + \sum_{k=0}^{m-1} a_{km}\gamma_{-k}$$

or equivalently

$$\gamma_{-m} = -\sum_{k=0}^{m-1} a_{km}\gamma_{-k}. \qquad (9.6.19)$$

Thus starting with $a_1$, (9.4.7) gives $\varphi_1(z) = z - \overline{a_1}$, and $\gamma_1 = \overline{\gamma}_{-1}$ is calculated using (9.6.19), namely $\gamma_1 = a_1$. The second moment $a_2$ gives $\varphi_2$, and $\varphi_2$ together with $\gamma_1$ gives $\gamma_2$. Thus the moments $\gamma_k$ can be calculated successively using (9.4.7) and (9.6.19).

## 9.7  LIMITING PREDICTION AND MAXIMUM ENTROPY

An interesting dichotomy appears in the prediction problems discussed in Section 9.5. It is shown in (9.6.12) that the $m$-step prediction of $X_1$ using $X_0, X_{-1}, \ldots, X_{-m+1}$ has a prediction error

$$\mu_m = \frac{\Delta_m}{\Delta_{m-1}} \qquad (9.7.1)$$

$$= \prod_{i=1}^{m}(1 - |a_i|^2).$$

Note that this is also the best $L_2(\sigma)$ approximation of $e^{i\theta}$ or $e^{-im\theta}$ by the functions

$$1, e^{-i\theta}, e^{-2i\theta}, \ldots, e^{-(m-1)i\theta}. \qquad (9.7.2)$$

Since $|a_i| \leq 1$, then $\mu_m \geq \mu_{m+1}$ and $\mu_m$ must therefore converge to some value $\mu = \lim_{m\to\infty} \mu_m$. Two obvious cases arise, namely when $\mu = 0$ and $\mu > 0$. These two cases correspond to $\sum_{i=1}^{\infty} |a_i|^2 = \infty$ and $\sum_{i=1}^{\infty} |a_i|^2 < \infty$. When $\mu = 0$ the process is called *deterministic*, since $X_1$ can be predicted "perfectly" given all of the past history $X_k, k \leq 0$. This can further be iterated in the sense that $X_2$ can be "predicted" from $X_1, X_0, X_{-1}. \ldots$ and hence from $X_0, X_{-1}, \ldots$. Continuing this process, any "future" value $X_h, h \geq 1$ can be predicted from $X_k, k \leq 0$ in the sense that for each $h \geq 1$ there is a linear combination of $X_k, k \leq 0$ which is arbitrarily close to $X_h$ in the norm of $L_2(X)$.

In the space $L_2(\sigma)$ this result says that the functions

$$\{e^{-ik\theta}|k \geq 0\}$$

are complete in $L_2(\sigma)$. Recall that the full system of functions

$$\{e^{ik\theta}|-\infty < k < \infty\}$$

forms a complete set in $L_2(\sigma)$. If $\mu = 0$, each $e^{ih\theta}$, $h \geq 1$ can be approximated using the functions $e^{-ik\theta}$, $k \geq 0$, and it follows that the set $\{e^{-ik\theta}|k \geq 0\}$ is also complete in $L_2(\sigma)$. In the case $\mu > 0$, the process $\{X_n\}$ is called *nondeterministic*, and the set $\{e^{-ik\theta}|k \geq 0\}$ is not complete in $L_2(\sigma)$.

The condition that $\mu = 0$ or $\mu > 0$ can be equivalently stated in terms of the spectral measure $\sigma$. Let $\sigma = \sigma_a + \sigma_s$, where $\sigma_a$ corresponds to the absolutely continuous part of $\sigma$ and $\sigma_s$ contains the singular and/or discrete part. The condition $\mu = 0$ or $\mu > 0$ depends on the behavior of the integral

$$\int_{-\pi}^{\pi} \log \sigma_a'(\theta)d\theta, \tag{9.7.3}$$

where $\sigma_a'$ denotes the density of the absolute continuous component $\sigma_a$. Since $\log \sigma_a'(x) \leq \sigma_a'(x)$, it follows that

$$-\infty \leq \int_{-\pi}^{\pi} \log \sigma_a'(\theta)d\theta \leq 1.$$

**Theorem 9.7.1** *The prediction error is given by*

$$\mu = 2\pi \exp\left\{\frac{1}{2\pi}\int_{-\pi}^{\pi} \log \sigma_a'(\theta)d\theta\right\}. \tag{9.7.4}$$

*Here $\mu = 0$ if $\int_{-\pi}^{\pi} \log \sigma_a'(\theta)d\theta = -\infty$.*

*Proof* A complete proof of this theorem is given in Grenander and Szegö (1958). ∎

A simple special case of Theorem 9.7.1, however, can be easily proved here. This is the situation described in Theorem 9.5.4, where $\sigma_m$ has density

$$\sigma_m'(\theta) = \frac{1}{2\pi}\frac{\Delta_m}{\Delta_{m-1}}|\varphi_m^*(e^{i\theta})|^{-2} \tag{9.7.5}$$

and corresponding canonical moments

$$a_1, \ldots, a_m, 0, 0, \ldots. \tag{9.7.6}$$

For these canonical moments the minimum prediction error is

$$\mu = \mu_m = \frac{\Delta_m}{\Delta_{m-1}} = \prod_{i=1}^{m}(1 - |a_i|^2).$$

$(9.7.7)$

**Lemma 9.7.2**   *If $\sigma'_m(\theta)$ is given by (9.7.5), then*

$$2\pi \exp\left\{\frac{1}{2\pi}\int_{-\pi}^{\pi} \log\, \sigma'_m(\theta)d\theta\right\} = \frac{\Delta_m}{\Delta_{m-1}}.$$

$(9.7.8)$

*Proof*   Recall that all the zeros of $\varphi^*_m$ are located in the region $|z| > 1$ so that $\varphi^*_m$ is analytic for $|z| < 1$. In this case $\log |\varphi^*_m(z)|$ is harmonic in $|z| < 1$ and since $\log(\varphi^*_m(z))$ is continuous on $|z| \leq 1$

$$\frac{1}{2\pi}\int_{-\pi}^{\pi} \log\, |\varphi^*_m(e^{i\theta})|d\theta = \log\, |\varphi^*_m(0)|.$$

However, this is zero since $\varphi^*_m(0) = 1$, and equation (9.7.8) then follows from (9.7.5).    ∎

A full proof of Theorem 9.7.1 uses Lemma 9.7.2 as an important intermediary step. For general $\sigma$ the proof requires a separate argument regarding the singular and discrete part of $\sigma$ and an argument that approximates $\sigma'_a$ by the inverse of a nonnegative trigonometric polynomial.

A consequence of Theorem 9.7.1 involves stationary processes for which the correlation structure $\rho(h)$ in (9.6.1) is specified only for a finite set of values, $h = 0, 1, \ldots, m$. Thus, if one knows or can estimate the covariances between $X_i$ values that are less than $m$ time intervals apart, how should one choose $\rho(k)$, $k \geq m$ so that the process is fully defined? One such possibility is to ensure that no further additional information is added to the process in the sense that we should require, for the $m$-step prediction errors, that

$$\mu_{m+k} = \mu_m \qquad \text{for } k \geq 1.$$

$(9.7.9)$

By (9.7.1) this corresponds to requiring that the canonical moments of the spectral measure $\sigma$ are given by (9.7.6).

The expression

$$E(\sigma) = \int_{-\pi}^{\pi} \log\, \sigma'_a(\theta)d\theta$$

is sometimes used as a measure of *entropy* in $\sigma$ or in the process $\{X_n\}$. If $\sigma$ has no absolutely continuous part, then $E(\sigma) = -\infty$. The above analysis answers the question of finding the maximum of $E(\sigma)$ over those $\sigma$ satisfying

$$\gamma_k = \int_{-\pi}^{\pi} e^{-ik\theta}d\sigma(\theta), \qquad k = 1, 2, \ldots, m.$$

where $\gamma_1, \gamma_2, \ldots, \gamma_m$ are prescribed. By Lemma 9.7.2 and Theorem 9.7.1 the required $\sigma$ has density (9.7.5).

The entropy result in stochastic processes was explored by Burg (1967) and is known as Burg's maximal entropy spectrum theorem. See also Childers (1978), Choi and Cover (1984), and Cover and Thomas (1991) for other references.

## 9.8   EXPERIMENTAL DESIGN FOR TRIGONOMETRIC REGRESSION

In Chapters 5 and 6 the theory of canonical moments for the powers $1, x, \ldots, x^m$ on an interval $[a, b]$ was applied to investigate some problems in experimental design for ordinary polynomial regression. In this section a brief discussion of some analogous results will be given for the trigonometric functions.

For our response function $\theta^T f(x)$ in (5.2.1) a linear combination of the functions

$$f^T(\theta) = (1, \ \cos\theta, \ \sin\theta, \ldots, \ \cos m\theta, \ \sin m\theta) \tag{9.8.1}$$

for $\theta \in [-\pi, \pi)$ will be used instead of the polynomials in (5.2.16). In this case the number of functions is now $k = 2m + 1$, and our basic space $\mathcal{X}$ is the circle or $\mathcal{X} = [-\pi, \pi)$. In Chapter 5 the basic Hankel matrix

$$M(\xi) = (c_{i+j}(\xi))_{i,j=0}^m,$$

where $c_\ell = \int_a^b x^\ell d\xi(x)$, was used extensively. The "moment" matrix for the system of functions in (9.8.1) is given by the $(2m + 1) \times (2m + 1)$ size matrix

$$M_f(\sigma) = \int_{-\pi}^{\pi} f(\theta)f(\theta)^T d\sigma(\theta). \tag{9.8.2}$$

For certain questions it is easier to work with the $2m + 1$ functions

$$g(\theta) = (e^{-im\theta}, e^{-i(m-1)\theta}, \ldots, 1, e^{i\theta}, \ldots, e^{im\theta}). \tag{9.8.3}$$

The functions in $f$ are simple linear combinations of the functions in $g$, so $f = Sg$ where $S$ is a nonsingular square matrix of size $2m + 1$. The matrix in (9.8.2) can then be written as

$$M_f(\sigma) = SM_g(\sigma)S^T.$$
$$= ST_{2m}(\sigma)JS^T$$
$$= ST_{2m}(\sigma)S^*,$$

where, $J$ has ones down the diagonal from the upper right to the lower left and zero elsewhere, and $S^*$ denotes complex conjugate of $S$. The matrix $T_{2m}(\sigma)$ is the Toeplitz matrix of size $(2m+1) \times (2m+1)$ from (9.2.6).

**Theorem 9.8.1** *A design or measure $\sigma$ is D-optimal for the system (9.8.1) on $[-\pi, \pi)$ if and only if $a_i(\sigma) = 0 = \gamma_i(\sigma)$, $i = 1, 2, \ldots, 2m$.*

*Proof* By Corollary 9.3.5 the determinant of $M_f(\sigma)$ is given by

$$|M_f(\sigma)| = |S|^2 \Delta_{2m}(\sigma)$$

$$= |S|^2 \prod_{i=1}^{2m} (1 - |a_i(\sigma)|^2)^{2m-i+1}. \qquad (9.8.4)$$

This is clearly maximized if and only if $a_i(\sigma) = 0$, $i = 1, 2, \ldots, 2m$. In view of (9.6.17) this can be shown, by induction, to be equivalent to $\gamma_i(\sigma) = 0$, $i = 1, \ldots, 2m$. ∎

The D-optimal design for the trigonometric functions on $[-\pi, \pi)$ is not unique as in the ordinary polynomial case in Chapter 5. The uniform measure $(2\pi)^{-1} d\theta$ is D-optimal, since $\gamma_i = 0$ for all $i$. Any measure with equal weight on at least $2m + 1$ equally spaced points on the circle can be shown to have $\gamma_i = 0$ for $i = 1, 2, \ldots, 2m$, so these designs are all D-optimal. The class of all D-optimal designs is quite large and contains many "nonuniform" type designs. It is clear that any D-optimal design for trigonometric regression of order $m$ in (9.8.1) will also be D-optimal for any lower order.

    If the experimenter wants to estimate only the highest $s$ pairs of coefficients, then a $D_s$-optimal design may be used. This optimality criterion was discussed more fully in Section 5.6. Let $M_f(\sigma; 2m) := M_f(\sigma)$ denote the moment matrix in (9.8.2) where the index $2m$ corresponds to the order of the system in (9.8.1). By (5.6.1) the determination of the $D_{2s}$-optimal design is equivalent to maximizing

$$\frac{|M_f(\sigma; 2m)|}{|M_f(\sigma; 2(m-s))|}.$$

Since this is proportional to $\Delta_{2m}/\Delta_{2(m-s)}$, it follows from Corollary 9.3.5 that the corresponding class of $D_s$-optimal designs is the same as the class of D-optimal designs in Theorem 9.8.1.

    The $D_s$-optimal designs for estimating the coefficients of the sine or cosine terms separately (and some further results) can be worked out by reverting to results for the ordinary polynomials. Let $M_s(\sigma)$ and $M_c(\sigma)$ denote the "moment" matrices obtained by using the sine and cosine terms

$$(\sin \theta, \sin 2\theta, \ldots, \sin m\theta)$$

and

$$(1, \cos\theta, \cos 2\theta, \ldots, \cos m\theta),$$

respectively. The $D_{m+1}$-optimal designs for estimating the coefficients of the cosine terms would maximize

$$\phi_c(\sigma) = \frac{|M_f(\sigma)|}{|M_s(\sigma)|}. \tag{9.8.5}$$

Similarly the $D_m$-optimal design for estimating the coefficients of the sine terms would maximize

$$\phi_s(\sigma) = \frac{|M_f(\sigma)|}{|M_c(\sigma)|}. \tag{9.8.6}$$

To accomplish this, we first demonstrate that a design symmetric about zero will maximize $\phi_c(\sigma)$ or $\phi_s(\sigma)$. Note that if $d\mu(\theta) = d\sigma(-\theta)$, then $\mu$ and $\sigma$ have the same determinant for $|M_f|$, $|M_c|$, and $|M_s|$. In this case we have $\phi_c(\mu) = \phi_c(\sigma)$ and $\phi_s(\mu) = \phi_s(\sigma)$. By Theorem 5.4.4 $\phi_c(\sigma)$ and $\phi_s(\sigma)$ are concave in $\sigma$ and the symmetric design $\rho(\theta) = 2^{-1}(\sigma(\theta) + \sigma(-\theta))$ will have a larger value for $\phi_c$ and $\phi_s$ than $\sigma$. Now for a symmetric design any terms in $M_f(\sigma)$ involving a product of a sine and cosine will vanish, so the determinant of $M_f(\sigma)$ splits into two parts

$$|M_f(\sigma)| = |M_c(\sigma)||M_s(\sigma)|.$$

Thus the $D_{m+1}$-optimal design for estimating the coefficient of the cosine terms will maximize $|M_c(\sigma)|$ and the $D_m$-optimal design for the sine terms will maximize $|M_s(\sigma)|$.

The two determinants $|M_c(\sigma)|$ and $|M_s(\sigma)|$ can be maximized over the class of symmetric measures by reverting to $[-1, 1]$ and using the Chebyshev polynomials of the first and second kind. Thus, if $x = \cos\theta$, then $\cos k\theta = T_k(x)$ and $\sin(k+1)\theta = U_k(x)\sin\theta$, where $T_k(x)$ and $U_k(x)$ are the Chebyshev polynomials introduced in Example 2.1.1.

Since the highest coefficient in $T_k(x)$ is $2^{k-1}$, it follows from (5.5.8) that

$$|M_c(\sigma)| = 2^{(m-1)m}|M_h(\xi)|$$

$$= 2^{2m^2} \prod_{i=1}^{m} (\zeta_{2i-1}\zeta_{2i})^{m+1-i},$$

where $h(x) = (1, x, \ldots, x^m)^T$, $\xi$ is the "projection" of $\sigma$ onto $[-1, 1]$ via

$x = \cos\theta$ and $p_1, p_2, \ldots$ denote the canonical moments of $\xi$. Thus the $D_{m+1}$-optimal design for estimating the coefficient of the cosine terms in the trigonometric regression (9.8.1) can be obtained from the $D$-optimal design for the polynomial regression of degree $m$ on $[-1, 1]$. It is shown in the proof of Theorem 5.5.3 that the canonical moments for this are given by $p_{2i-1} = \frac{1}{2}$, $i = 1, \ldots, m$, and

$$p_{2i} = \frac{m-i+1}{2m-2i+1}, \qquad i = 1, \ldots, m.$$

A similar result using

$$(\sin\theta, \ldots, \sin m\theta) = \pm(1-x^2)^{1/2}(1, U_1(x), \ldots, U_{m-1}(x))$$

$(\theta = \arccos x)$ shows that

$$M_s(\sigma) = 2^{2m^2} \prod_{i=1}^{m} (\gamma_{2i-1}(\xi)\gamma_{2i}(\xi))^{m-i+1}.$$

This is maximized by the measure $\xi$ with canonical moments $p_{2i-1} = \frac{1}{2}$, $i = 1, \ldots, m$, and

$$p_{2i} = \frac{m-i}{2m-2i+1}, \qquad i = 1, \ldots, m.$$

**Theorem 9.8.2**

1. *The $D_{m+1}$-optimal design for estimating the coefficients of $(1, \cos\theta, \ldots, \cos m\theta)$ in the full trigonometric model has canonical moments*

$$a_{2i-1} = 0, \qquad i = 1, \ldots, m,$$

$$a_{2i} = \frac{1}{2m-2i+1}, \qquad i = 1, \ldots, m.$$

2. *The $D_m$-optimal design for estimating the coefficients of $(\sin\theta, \ldots, \sin m\theta)$ in the full trigonometric model has canonical moments*

$$a_{2i-1} = 0, \qquad i = 1, \ldots, m,$$

$$a_{2i} = \frac{-1}{2m-2i+1}, \qquad i = 1, \ldots, m.$$

*Proof*  The result follows from the preceding discussion observing Remark 9.3.3, which shows that the canonical moments of $\sigma$ and its projection $\xi$ onto

$[-1, 1]$ are related by

$$a_i = 2p_i - 1.$$ ∎

**EXAMPLE 9.8.3**  For $m = 1$, we estimate the coefficients of 1, $\cos \theta$, in the model with 1, $\cos \theta$, $\sin \theta$, with a design having $a_1 = 0$ and $a_2 = 1$. This has equal mass at $\theta = 0$ and $\theta = -\pi$. The coefficient of the single term $\sin \theta$ is best estimated by a design with $a_1 = 0$ and $a_2 = -1$, which has equal mass on $\pm \pi/2$. If $m = 2$, the set 1, $\cos \theta$, $\cos 2\theta$ is estimated using a design with $a_1 = a_3 = 0$, $a_2 = \frac{1}{3}$, and $a_4 = 1$. This corresponds to masses $\frac{1}{3}$ on $\theta = 0$ and $-\pi$ and $\frac{1}{6}$ on $\theta = \pm \pi/2$. For $\sin \theta$, $\sin 2\theta$ we use $a_1 = a_3 = 0$, $a_2 = -\frac{1}{3}$, and $a_4 = -1$ which has equal mass on the four points corresponding to $\cos \theta = \pm 1/\sqrt{3}$. The trigonometric regression is a good example of a situation where $D_s$-optimal designs are not very reasonable in the sense that they are not very good for anything except their intended purpose.

The remaining results of this section discuss two extensions of Theorem 9.8.2 that can be proved by similar arguments. If the relative importance of the sine and cosine terms can be quantified, then the designs from the following theorem might be useful:

**Theorem 9.8.4**  *The design $\sigma$ maximizing $|M_c(\sigma)|^\alpha |M_s(\sigma)|^\beta$, where $0 \le \alpha \le 1$ and $\alpha + \beta = 1$, has canonical moments*

$$a_{2i-1} = 0,$$

$$a_{2i} = \frac{2\alpha - 1}{2m - 2i + 1}, \qquad i = 1, 2, \ldots, m.$$

A simple extension of this involves putting a "prior" $\alpha_k$ and $\beta_k$, $k = 1, 2, \ldots, m$ on the various models

$$(1, \ \cos \theta, \ \ldots, \ \cos k\theta), \tag{9.8.7}$$

$$(\sin \theta, \ \ldots, \ \sin k\theta), \tag{9.8.8}$$

where $\alpha_k \ge 0$, $\beta_k \ge 0$ and $\sum_{k=1}^{m} (\alpha_k + \beta_k) = 1$. In this case $\beta_k$, for example, is the prior probability or belief that the true model is in the space spanned by $(\sin \theta, \ \ldots, \ \sin k\theta)$. Following the discussion in Section 6.4 a reasonable criterion might be to maximize the geometric mean

$$\phi(\sigma) = \sum_{k=1}^{m} \alpha_k \log |M_{ck}(\sigma)| + \sum_{k=1}^{m} \beta_k \log |M_{sk}(\sigma)|, \tag{9.8.9}$$

where $M_{ck}(\sigma)$ and $M_{sk}(\sigma)$ are the information matrices involving the terms in (9.8.7) and (9.8.8), respectively.

**Theorem 9.8.5**   *The design maximizing $\phi(\sigma)$ in (9.8.9) has canonical moments*

$$a_{2i-1} = \frac{1}{2}, \qquad i = 1, \ldots, m,$$

$$a_{2i} = \frac{\displaystyle\sum_{j=1}^{m-i+1} (\alpha_{j+i-1} - \beta_{j+i-1})}{\displaystyle\sum_{j=1}^{m-i+1} (2j-1)(\alpha_{j+i-1} + \beta_{j+i-1})}, \qquad i = 1, 2, \ldots, m.$$

# CHAPTER 10

# Further Applications

## 10.1 INTRODUCTION

There are many other interesting applications of canonical moments in the literature in addition to those presented in Chapters 4 to 8. As mentioned in the preface, Morris Skibinsky has a number of papers describing the usefulness of canonical moments and T. S. Lau also has some interesting applications. These are all listed in the bibliography. In this chapter two particular applications are described, one related to Bayesian analysis for the binomial distribution by Skibinsky (1968) and the other concerning a limit theorem in the moment space $M$ by Chang, Kemperman, and Studden (1993).

## 10.2 BAYESIAN BINOMIAL ESTIMATION

In Sections 10.2 to 10.5 some problems from Skibinsky (1968) for the binomial distribution are described. The general formulation is to consider a pair of random variables $(X, \Theta)$, $\Theta$ being distributed on $[0,1]$ and given $\Theta = \theta$, $X$ having a binomial distribution

$$P_\theta(X = j) = P(X = j | \Theta = \theta) = b(j; n, \theta)$$

$$= \binom{n}{j} \theta^j (1 - \theta)^{n-j} \qquad j = 0, 1, \ldots, n$$

(10.2.1)

The integer $n$ will be fixed. The "marginal" distribution of $\Theta$ is a measure $\mu$ on the interval $[0,1]$, which is called the *prior distribution*. From a statistical viewpoint one observes $X = j$ but not $\Theta$, and on the basis of $X$ one wants to estimate $\Theta$. Any function $\delta(i)$, $i = 0, 1, \ldots$, $n$ with values in $[0,1]$ is called an estimate of $\Theta$. If $X = j$ is observed, one estimates $\Theta$ by $\delta(j)$. The quality of the estimate $\delta$ is sometimes measured by the *mean squared error*

$$r_n(\delta, \theta) = E_{X|\theta}(\delta(X) - \theta)^2$$

for $\Theta = \theta$ or by the *risk*

$$R_n(\delta, \mu) = E(\delta(X) - \Theta)^2 \qquad (10.2.2)$$

$$= \int_0^1 r_n(\delta, \theta) d\mu(\theta)$$

$$= \int_0^1 \sum_{j=0}^n (\delta(j) - \theta)^2 \binom{n}{j} \theta^j (1 - \theta)^{n-j} d\mu(\theta).$$

The estimate $\delta$ minimizing the risk $R_n(\delta, \mu)$ for a given prior $\mu$ is called a *Bayes estimate* relative to $\mu$ or corresponding to $\mu$ and will be denoted by $\delta_n(i) = \delta_n(i, \mu)$. The *Bayes envelope* $R_n(\mu)$ is defined by

$$R_n(\mu) = \inf_\delta R_n(\delta, \mu) = R_n(\delta_n, \mu).$$

The quantities

$$\xi_{nj} = \xi_{nj}(\mu) = \int_0^1 \theta^j (1 - \theta)^{n-j} d\mu$$

play an important role here and in the theory of moments.

**Lemma 10.2.1** *The estimate*

$$\delta_n(j, \mu) = \begin{cases} \dfrac{\xi_{n+1,j+1}}{\xi_{nj}}, & \xi_{nj} > 0, \\ 0 & \xi_{nj} = 0, \end{cases} \qquad (10.2.3)$$

*is a Bayes estimate. The Bayes estimate is unique if $\xi_{nj} > 0$ for all $j = 0, \ldots, n$ and the Bayes envelope is given by*

$$R_n(\mu) = c_2(\mu) - \sum_{j=0}^n \binom{n}{j} \xi_{nj} \delta_n^2(j, \mu).$$

*Proof* Observing (10.2.2), the risk $R_n(\delta, \mu)$ may be written as

$$R_n(\delta, \mu) = c_2(\mu) + \sum_{j=0}^n \binom{n}{j} [\delta^2(j)\xi_{nj} - 2\delta(j)\xi_{n+1,j+1}].$$

This is clearly minimized by $\delta_n$ given in the statement of the theorem, and it is unique if $\xi_{nj} > 0$ for all $j$. ∎

The conditional distribution of $\Theta$ given $X = j$, $j = 0, 1, \ldots, n$, is called the *posterior distribution*, and it takes the form

$$d\mu_j(\theta) = \frac{\theta^j(1-\theta)^{n-j}d\mu(\theta)}{\xi_{nj}} \tag{10.2.4}$$

provided that $\xi_{nj} > 0$; in this case $\delta_n(j,\mu)$ is simply the mean of the posterior distribution $\mu_j$. Observing (10.2.2) and (10.2.4), the Bayes envelope may be written as

$$R_n(\mu) = \sum_{j=0}^{n} \xi_{nj} \binom{n}{j} \int_0^1 (\theta - \delta_n(j,\mu))^2 d\mu_j(\theta) \tag{10.2.5}$$

$$= \sum_{j=0}^{n} p_1^{(j)} q_1^{(j)} p_2^{(j)} P(X=j),$$

where $p_i^{(j)}$ are the canonical moments of posterior distribution $\mu_j$.

In moment theory the quantities $\xi_{nj}$ enter the picture if the moment space $M_n$ is enclosed in another space $S_n$, which is the convex hull of the $n+1$ points $x^{(k)} = (x_1^{(k)}, \ldots, x_n^{(k)}) \in R^{n \times n}$, $k = 0, 1, \ldots, n$ given by

$$x_i^{(k)} = \frac{\binom{k}{i}}{\binom{n}{i}}, \qquad i = 1, \ldots, n.$$

Using the identity

$$\binom{n}{k}\binom{k}{i} = \binom{n-i}{k-i}\binom{n}{i}, \qquad i, k = 0, 1, \ldots, n,$$

it follows that

$$\theta^i = \sum_{k=0}^{n} \binom{n}{k} \theta^k (1-\theta)^{n-k} \frac{\binom{k}{i}}{\binom{n}{i}}, \tag{10.2.6}$$

which implies that $M_n \subset S_n$. Integrating (10.2.6) with respect to $\mu$ also gives

$$c_n(\mu) = \sum_{k=0}^{n} \binom{n}{k} \xi_{nk} x^{(k)}, \tag{10.2.7}$$

where $c_n(\mu) = (c_1(\mu), \ldots, c_n(\mu))$ denotes the vector of the first $n$ moments of the prior $\mu$.

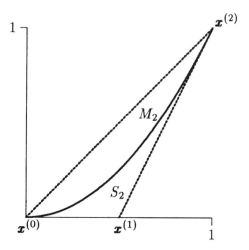

**Figure 10.1**   The moment space $M_2$ included by the set $S_2$.

Clearly

$$P(X = i) = \int_0^1 \binom{n}{i} \theta^i (1 - \theta)^{n-i} d\mu(\theta) = \binom{n}{i} \xi_{ni}, \qquad (10.2.8)$$

$$\binom{n}{k} \geq 0, \text{ and } \sum_{k=0}^n \binom{n}{k} \xi_{nk} = 1,$$

so

$$\binom{n}{k} \xi_{nk}, \qquad k = 0, 1, \ldots, n,$$

give the "coordinates" of the representation of $(c_1, \ldots, c_n)$ by the extreme points $x^{(k)}$ generating $S_n$.

The uniqueness statement, that $\xi_{nj} > 0$ for all $j$, appearing in Lemma 10.2.1 is equivalent to

$$(c_1(\mu), \ldots, c_n(\mu)) \in \text{Int } S_n, \qquad (10.2.9)$$

which also holds if $(c_1(\mu_1), \ldots, c_n(\mu)) \in \text{Int } M_n$. Note also that if $X$ has a binomial distribution given $\Theta = \theta$, then (10.2.6) may be interpreted as

$$E_{X|\theta} \frac{\binom{X}{i}}{\binom{n}{i}} = \theta^i, \qquad (10.2.10)$$

so

$$\frac{\binom{X}{i}}{\binom{n}{i}}$$

may be regarded as an estimate of $\theta^i$.

## 10.3  CANONICAL MOMENT EXPRESSIONS

From our interest in canonical moments it is appropriate that we write many of the expressions in Section 10.2 in terms of canonical moments. To accomplish this, it is helpful to recall a number of results from Chapter 1. The first of these is that if $\mu$ has canonical moments $p_1, p_2, \ldots$, then by Theorem 1.3.3, the "reversed" measure $d_\mu^{(r)}(\theta) = d\mu(1 - \theta)$ has canonical moments

$$p_{2i-1}^{(r)} = 1 - p_{2i-1} = q_{2i-1}, \tag{10.3.1}$$

$$p_{2i}^{(r)} = p_{2i},$$

for $i \geq 1$ whenever the $p_i$ are defined. If the dependence of $\xi_{ni}$ on $p_1, p_2, \ldots$ is denoted by $\xi_{ni} = \xi_{ni}(p_1, p_2, \ldots)$, then from the definition of $\xi_{ni}$ it follows that

$$\xi_{ni}(p_1, p_2, \ldots) = \xi_{n,n-i}(q_1, p_2, q_3, \ldots). \tag{10.3.2}$$

For example, if

$$\xi_{22} = \int_0^1 \theta^2 d\mu(\theta) = p_1(p_1 + q_1 p_2),$$

then

$$\xi_{20} = \int_0^1 (1 - \theta)^2 d\mu(\theta) = q_1(q_1 + p_1 p_2).$$

The other useful items are Equation (1.5.12) in Theorem 1.5.4 and Equation (1.5.13) which is an alternate form of the result in Theorem 1.5.5. These describe the canonical moments of $(1 - \theta)d\mu(\theta)$ and $\theta d\mu(\theta)$ in terms of the canonical moments of $\mu$. If $\mu$ has canonical moments $p_1, p_2, \ldots$ and the canonical moments of $\theta d\mu(\theta)/p_1$ and $(1 - \theta)d\mu(\theta)/q_1$ are denoted by $p_1', p_2', \ldots$ and

$p''$, $p_2''$, . . ., respectively, then from Equation (1.5.12),

$$p_1'' = p_1 q_2,$$
$$q_1'' p_2'' = p_2 q_3,$$
$$q_2'' p_3'' = p_3 q_4,$$
$$\vdots$$

(10.3.3)

and from Equation (1.5.13),

$$q_1' = q_1 q_2,$$
$$p_1' p_2' = p_2 p_3,$$
$$q_2' q_3' = q_3 q_4,$$
$$\vdots$$

(10.3.4)

Expressions for $P_n(X = i)$, $R_n(\mu)$ and $\delta_\mu$ can be easily written down for small values of $n$. It will be assumed that $(c_1(\mu), \ldots, c_n(\mu)) \in \text{Int } M_n$. The idea is to repeatedly apply the transformations (10.3.3) and (10.3.4). Let $\mu'$ and $\mu''$ denote the measures corresponding to $\theta d\mu / p_1$ and $(1 - \theta) d\mu / q_1$, respectively. Since

$$\delta_n(i, \mu) = \frac{\int_0^1 \theta \theta^i (1 - \theta)^{n-i} d\mu(\theta)}{\int_0^1 \theta^i (1 - \theta)^{n-i} d\mu(\theta)},$$

it follows that

$$\delta_n(i, \mu) = \delta_{n-1}(i, \mu''),$$
$$\delta_n(i, \mu) = \delta_{n-1}(i - 1, \mu').$$

(10.3.5)

The Bayes estimate for $n = 1$ can be obtained directly from (10.3.3), (10.3.4), and (10.3.5), giving

$$\delta_1(0, \mu) = \delta_0(0, \mu'') = p_1'' = p_1 q_2,$$           (10.3.6)
$$\delta_1(1, \mu) = \delta_0(0, \mu') = p_1' = 1 - q_1 q_2$$
$$= p_1 q_2 + p_2$$
$$= p_1 + q_1 p_2.$$

For later reference note that these values differ by $p_2$ or

$$\delta_1(1, \mu) = \delta_1(0, \mu) + p_2.$$

For $n = 2$, the equations in (10.3.5) and (10.3.6) can be used to give

$$\delta_2(1, \mu) = \delta_1(1, \mu'')$$
$$= p_1'' + q_1'' p_2''$$
$$= p_1 q_2 + p_2 q_3.$$

Also

$$\delta_2(0, \mu) = \delta_1(0, \mu'') = \delta_1(1, \mu'') - p_2''$$
$$= p_1 q_2 + p_2 q_3 - \frac{p_2 q_3}{1 - p_1 q_2}$$

and

$$\delta_2(2, \mu) = \delta_1(1, \mu') = \delta_1(0, \mu') + p_2' = \delta_2(1, \mu) + p_2'$$
$$= p_1 q_2 + p_2 q_3 + \frac{p_2 p_3}{1 - q_1 q_2}.$$

Note also that the values $\delta_n(i, \mu)$ have a certain symmetry in the sense that

$$\delta_n(i, \mu) = 1 - \delta_n(n - i, \mu^{(r)}).$$

Thus we could have obtained $\delta_2(2, \mu)$ from

$$\delta_2(2, \mu) = 1 - \delta_2(0, \mu^{(r)})$$
$$= 1 - \left[ q_1 q_2 + p_2 p_3 - \frac{p_2 p_3}{1 - q_1 q_2} \right]$$
$$= p_1 q_2 + p_2 q_3 + \frac{p_2 p_3}{1 - q_1 q_2}.$$

By judiciously exploiting the above, it can be shown that $\delta_3(i, \mu)$ is given by

$$p_1 q_2 + p_2 q_3 - \frac{p_2 q_3(1 - p_1 p_2 p_3 p_4)}{(1 - p_1 q_2)^2 + p_1 p_2 q_2 q_3}, \qquad i = 0,$$

$$p_1 q_2 + p_2 q_3 - \frac{p_2 p_3 q_3 q_4}{q_1 q_2 + p_2 p_3}, \qquad i = 1,$$

$$p_1 q_2 + p_2 q_3 + \frac{p_2 p_3 q_3 q_4}{p_1 q_2 + p_2 q_3}, \qquad i = 2,$$

$$p_1 q_2 + p_2 q_3 + \frac{p_2 p_3(1 - q_1 q_2 q_3 q_4)}{(1 - q_1 q_2)^2 + q_1 p_2 q_2 p_3}, \qquad i = 3.$$

The Bayes risk $R_n(\mu)$ also has an interesting property in that

$$R_n(\mu) = q_1 R_{n-1}(\mu'') + p_1 R_{n-1}(\mu'). \qquad (10.3.7)$$

The proof of (10.3.7) is fairly straightforward by observing that $p_1 \xi_{n-1,j-1}(\mu') = \xi_{nj}(\mu)$ and $q_1 \xi_{n-1,j}(\mu'') = \xi_{nj}(\mu)$. This gives for the right-hand side of (10.3.7),

$$c_2(\mu) - q_1 R_{n-1}(\mu'') - p_1 R_{n-1}(\mu')$$

$$= \sum_{j=0}^{n-1} \left\{ q_1 \binom{n-1}{j} \xi_{n-1,j}(\mu'') \delta_{n-1}^2(j,\mu'') + p_1 \binom{n-1}{j} \xi_{n-1,j}(\mu') \delta_{n-1}^2(j,\mu') \right\}$$

$$= \sum_{j=0}^{n-1} \binom{n-1}{j} \xi_{nj}(\mu) \delta_n^2(j,\mu) + \sum_{j=1}^{n} \binom{n-1}{j-1} \xi_{nj}(\mu) \delta_n^2(j,\mu)$$

$$= \sum_{j=0}^{n} \binom{n}{j} \xi_{nj}(\mu) \delta_n^2(j,\mu) = c_2(\mu) - R_n(\mu),$$

where we used Lemma 10.2.1 and (10.3.5) for the first and second equality.

For example, since $R_0(\mu) = p_1 q_1 p_2$, then

$$R_1(\mu) = q_1(p_1 q_2 p_2 q_3) + p_1(q_1 q_2 p_2 p_3)$$

$$= p_1 q_1 p_2 q_2 = r_2 \qquad (10.3.8)$$

and

$$R_2(\mu) = q_1 \left[ p_1 q_2 p_2 q_3 \left( 1 - \frac{p_2 q_3}{1 - p_1 q_2} \right) \right] + p_1 \left[ q_1 q_2 p_2 p_3 \left( 1 - \frac{p_2 p_3}{1 - q_1 q_2} \right) \right]$$

$$= r_2 \left[ 1 - p_2 \left( \frac{q_3^2}{1 - p_1 q_2} + \frac{p_3^2}{1 - q_1 q_2} \right) \right]. \qquad (10.3.9)$$

## 10.4   MINIMAX ESTIMATES IN $\mathcal{P}(c_n)$

In the previous sections Bayes estimates of $\Theta$ corresponding to $\mu$ were considered. If $\mu$ is unavailable, a Bayes estimate corresponding to a *noninformative prior* $\mu$ (possibly $\mu = $ Lebesgue measure or $\mu = $ arc-sine law) is sometimes used. An alternative method is to use some sort of *minimax estimate*. The usual minimax estimate is one that minimizes

$$\sup_{\mu} R_n(\mu, \delta).$$

If the supremum is taken over all prior distributions $\mu$, the solution is to use the Bayes rule corresponding to a Beta prior with $\alpha = \beta = (\sqrt{n}/2) - 1$ given by

$$\delta(i) = \frac{i + \sqrt{n}/2}{n + \sqrt{n}} \qquad i = 0, 1, \ldots, n.$$

More generally the supremum is taken over a suitable subclass $\mathcal{P}^*$ of all the measures $\mathcal{P}$. A prior $\mu^*$ is called a *least favorable prior* in $\mathcal{P}^*$ if

$$R_n(\mu^*) = \sup_{\mu \in \mathcal{P}^*} R_n(\mu).$$

From game theory considerations,

$$\inf_{\delta} \sup_{\mu \in \mathcal{P}^*} R_n(\delta, \mu) = \sup_{\mu \in \mathcal{P}^*} \inf_{\delta} R_n(\delta, \mu)$$

$$= \sup_{\mu \in \mathcal{P}^*} R_n(\mu),$$

and the minimax estimate in $\mathcal{P}^*$ is the Bayes rule $\delta_n^*$ corresponding to the least favorable distribution in $\mathcal{P}^*$. We will be concerned here with the situation where $\mathcal{P}^* = \mathcal{P}(c_n)$ is the set of those $\mu$ with a fixed set of first $n$ moments given by $c_n = (c_1, \ldots, c_n) \in M_n$. Further motivation for this will be given in Section 10.5. Note that the Bayes envelope $R_n(\mu)$ depends on the prior $\mu$ only through its first $n + 1$ moments, and therefore the Bayes estimate relative to $\mu$ is also called the Bayes rule corresponding to $(c_1, \ldots, c_{n+1}) \in M_{n+1}$. In this case the maximum of $R_n(\mu)$ over $\mathcal{P}(c_n)$ is simply over the last moment $c_{n+1}(\mu)$ and $R_n(\mu)$ is a quadratic function of $c_{n+1}$.

For $n = 1$ and $n = 2$ the minimax estimate can be obtained from (10.3.8) and (10.3.9). From (10.3.8)

$$R_1(\mu) = p_1 q_1 p_2 q_2,$$

so the least favorable measures in $\mathcal{P}(c_1)$ have $p_2 = q_2 = \frac{1}{2}$. By (10.3.6) the corresponding minimax estimates have $\delta_1(0, \mu^*) = p_1/2$ and $\delta_1(1, \mu^*) = 1 - q_1/2$. The case $n = 2$ can be obtained from (10.3.9); the least favorable $p_3$ is

$$p_3 = \frac{p_2 + q_2 p_1}{p_2 + 1}. \tag{10.4.1}$$

For general $n$ the dependence on $c_{n+1}$ can be isolated by first writing, for $j = 0, 1, \ldots, n$,

$$\xi_{n+1,j+1}(c_{n+1}) = \gamma_{nj}(c_n) + (-1)^{n+j} c_{n+1}, \tag{10.4.2}$$

where $\gamma_{nn}(c_n) = 0$ and

$$\gamma_{nj}(c_n) = \sum_{k=j+1}^{n} (-1)^{j+1+k} \binom{n-j}{n+1-k} c_k, \qquad j = 0, \ldots, n-1. \qquad (10.4.3)$$

Then, if $c_n \in \text{Int } M_n$, we obtain from Lemma 10.2.1,

$$R_n(\mu) = B_n(c_n) - A_n(c_n)(c_{n+1} - \hat{c}_{n+1})^2, \qquad (10.4.4)$$

where $c_k = c_k(\mu)$, $k = 1, \ldots, n+1$, $B_n(c_n)$ is a constant depending only on $c_n = (c_1, \ldots, c_n)$, and

$$A_n = A_n(c_n) = \sum_{j=0}^{n} \frac{\binom{n}{j}}{\xi_{nj}}, \qquad (10.4.5)$$

$$\hat{c}_{n+1} = A_n^{-1} \sum_{j=0}^{n-1} (-1)^{n+1+j} \binom{n}{j} \frac{\gamma_{nj}}{\xi_{nj}}.$$

The maximum over $\mathcal{P}(c_n)$ will occur whenever the $(n+1)$th moment is given by

$$c_{n+1}^* = c_{n+1}^*(c_n) = \begin{cases} c_{n+1}^-, & \hat{c}_{n+1} < c_{n+1}^-, \\ \hat{c}_{n+1}, & c_{n+1}^- \leq \hat{c}_{n+1} \leq c_{n+1}^+, \\ c_{n+1}^+, & c_{n+1}^+ < \hat{c}_{n+1}. \end{cases} \qquad (10.4.6)$$

Thus we have the following lemma:

**Lemma 10.4.1**   *For all* $c_n \in M_n$, *a measure* $\mu^* \in \mathcal{P}(c_n)$ *is least favorable if and only if*

$$c_{n+1}(\mu^*) = c_{n+1}^*(c_n),$$

*where* $c_{n+1}^*(c_n)$ *is defined by* (10.4.6). *The corresponding Bayes estimate* $\delta_n^*$ *is minimax in* $\mathcal{P}(c_n)$ *and is unique if and only if* $c_n \in \text{Int } S_n$.

The usual simple idea in proving minimaxity of an estimate $\delta_n^*$ is to verify that the risk $R_n(\delta_n^*, \mu)$ is constant over $\mu \in \mathcal{P}^*$, and it is of some interest to check when this is the case in our situation. It is evident that if $c_n \in \text{Int } M_n$, then the risk function $R_n(\delta_n^*, \mu)$ is constant over $\mu \in \mathcal{P}(c_n)$ if and only if $c_{n+1}^* = \hat{c}_{n+1}$. For the case $n = 1$ and $n = 2$ discussed briefly above, the $(n+1)$st canonical moment of the least favorable distribution is in [0,1]. For $n = 1$ the least favorable $p_2 = \frac{1}{2}$, so $\hat{c}_2 = (c_1 + c_1^2)/2$. For $n = 2$ the least favorable $p_3$ in (10.4.1) is also in [0,1]. For $n = 3$ Skibinsky exhibits cases where $\hat{c}_{n+1} < c_{n+1}^-$, which means that in this case the risk function is not constant.

In terms of canonical moments the expression for the risk $R_n(\mu)$ in (10.4.4) can be expressed, for $c_n \in \text{Int } M_n$, as

$$R_n(\mu) = B_n(c_n) - A_n(c_n)r_n^2(p_{n+1} - \hat{p}_{n+1})^2, \tag{10.4.7}$$

where $\hat{p}_{n+1} = (\hat{c}_{n+1} - c_{n+1}^-)/r_n$ and $r_n = c_{n+1}^+ - c_{n+1}^-$. Note that $\hat{p}_{n+1}$ is not necessarily in [0,1].

**Lemma 10.4.2**  *For $c_n \in \text{Int } M_n$, let $\delta_n^+$ and $\delta_n^-$ denote the Bayes estimates corresponding to $(c_n, c_{n+1}^+)$ and $(c_n, c_{n+1}^-)$. If $c_n \in \text{Int } M_n$, then*

$$\hat{p}_{n+1} = \frac{-h_n(\delta_n^-)}{A_n r_n}, \quad \hat{q}_{n+1} = 1 - \hat{p}_{n+1} = \frac{h_n(\delta_n^+)}{A_n r_n}, \tag{10.4.8}$$

*where*

$$h_n(\delta) = \sum_{j=0}^{n}(-1)^{n+j}\binom{n}{j}\delta(j)$$

*and $A_n = A_n(c_n)$ is defined by (10.4.5).*

*Proof*  The expression for $\hat{q}_{n+1}$ follows by first writing

$$\xi_{n+1,j+1}^+ = \xi_{n+1,j+1}(c_n, c_{n+1}^+) = \gamma_{nj}(c_n) + (-1)^{n+j}c_{n+1}^+,$$

so by (10.4.5),

$$h_n(\delta_n^+) = \sum_{j=0}^{n}(-1)^{n+j}\binom{n}{j}\xi_{n+1,j+1}^+/\xi_{nj} = A_n(c_{n+1}^+ - \hat{c}_{n+1}),$$

from which the result follows. The result for $\hat{p}_{n+1}$ is similar.  ∎

We might also note here that if $R_n^+$ and $R_n^-$ denote the expression in (10.4.7) with $p_{n+1}$ replaced by $p_{n+1}^+$ and $p_{n+1}^-$, respectively, then

$$R_n(\mu) = R_n^- + A_n r_n^2[\hat{p}_{n+1}^2 - (p_{n+1} - \hat{p}_{n+1})^2] \tag{10.4.9}$$

$$= R_n^+ + A_n r_n^2[\hat{q}_{n+1}^2 - (q_{n+1} - \hat{q}_{n+1})^2].$$

## 10.5  RISK  DIFFERENCE

The introduction of the minimax estimate in $\mathcal{P}^* = P(c_n)$ results mainly from an *empirical Bayes approach* developed by Robbins (1955, 1964). In this approach a sequence $(X_1, \Theta_1), (X_2, \Theta_2), \ldots$ of independent pairs with common distribution $(X, \Theta)$ is considered, where only $X_i$, $i = 1, \ldots, m$, are observed. It is desired to estimate $\Theta_{m+1}$ from $X_{m+1}$ using an adaptive estimate that may depend on $X_1, \ldots, X_m$. The values of $X_1, \ldots, X_m$ will hopefully provide information about $\mu$. In the present context $X$ depends on $\mu$ only through its first $n$ moments. Since the Bayes estimate $\delta_n(\cdot, \mu)$ depends on the first $(n+1)$ moments, one cannot expect to adaptively approximate the Bayes estimate $\delta_n(\cdot, \mu)$ from $X_1, \ldots, X_m$. From (10.2.10),

$$\hat{c}_k = \frac{1}{m} \sum_{i=1}^{m} \frac{\binom{X_i}{k}}{\binom{n}{k}}, \qquad k = 0, 1, \ldots, n,$$

will consistently estimate the moments $c_k$, $k = 0, 1, \ldots, n$. In this case, by observing $X_1, X_2, \ldots$, one can approximate the class $\mathcal{P}(c_n)$ to which $\mu$ belongs. For $\mu \in \mathcal{P}(c_n)$ and $c_n \in M_n$ we therefore define the difference between the minimax and Bayes risks:

$$W_n(\mu) = \sup_{\rho \in \mathcal{P}^*} R_n(\rho) - R_n(\mu), \tag{10.5.1}$$

where $\mathcal{P}^* = \mathcal{P}(c_n)$. Since the risk for the standard estimate $X/n$ is bounded above by $1/4n$, this is also a bound for $W_n(\mu)$, so $W_n(\mu)$ approaches zero as $n$ goes to infinity uniformly in $\mu$. The following theorem shows that it approaches zero exponentially fast:

**Theorem 10.5.1**  *For every $\mu \in \mathcal{P}^* = \mathcal{P}^*(c_n)$,*

$$W_n(\mu) \leq r_n 2^n \leq 2^{-n},$$

*where*

$$r_n = r_n(\mu) = \prod_{i=1}^{n} p_i(\mu) q_i(\mu).$$

*Proof*  If $c_n$ is a boundary point in the $n$th-moment space $M_n$, then $W_n(\mu) = 0$ by Theorem 1.2.5. Let $c_n \in \text{Int } M_n$. If $\delta_n^*$ denotes the minimax estimate in $\mathcal{P}(c_n)$, then

$$W_n(\mu) = \sum_{i=0}^{n} \binom{n}{i} \xi_{ni} [(\delta_n(i, \mu))^2 - (\delta_n^*(i, \mu^*))^2],$$

where $\mu^*$ denotes the least favorable distribution in $\mathcal{P}^*$. Assume that $c_{n+1} > c_{n+1}^* = c_{n+1}^*(\mu^*)$; the remaining case is treated exactly in the same way. By (10.4.2) and Lemma 10.2.1,

$$\delta_n(i, \mu) - \delta_n^*(i, \mu^*) = \frac{(-1)^{n+i}(c_{n+1} - c_{n+1}^*)}{\xi_{ni}},$$

so $(r_n = c_{n+1}^+ - c_{n+1}^-)$,

$$W_n(\mu) = (c_{n+1} - c_{n+1}^*) \sum_{i=0}^{n} (-1)^{n+i} \binom{n}{i} (\delta_n(i, \mu) + \delta_n^*(i, \mu))$$

$$\leq r_n \sum_{i=0}^{n} \binom{n}{i} [1 + (-1)^{n+i}] = r_n 2^n. \qquad \blacksquare$$

From (10.5.1) and (10.4.4) it follows that $W_n(\mu)$ has *a* maximum over $\mathcal{P}(c_n)$ for $\mu$ either $\mu_n^+$ or $\mu_n^-$. Using this and the expression in (10.4.9), it is possible to show the following:

**Lemma 10.5.2**   *If $c_n \in Int\ M_n$, then*

$$W_n(\mu) \leq \begin{cases} \frac{1}{4} A_n r_n^2 (1 + |\hat{p}_{n+1} - \hat{q}_{n+1}|)^2, & 0 \leq \hat{p}_{n+1} \leq 1, \\ \\ A_n r_n^2 |\hat{p}_{n+1} - \hat{q}_{n+1}| & \text{otherwise} \end{cases}$$

*This bound is attained for either $\mu_n^+$ or $\mu_n^-$.*

**REMARK 10.5.3**   It was conjectured by Robbins and proved by Joshi (1975) that not only does sup $W_n(\mu)$ go to zero but also that

$$\limsup_{n \to \infty} \frac{W_n(\mu)}{R_n(\mu)} = 0,$$

where the supremum is taken over all measures for which $W_n(\mu)$ and $R_n(\mu)$ do not both vanish. The proof of this fact will not be given here.

## 10.6   A LIMIT THEOREM IN $M_n$

We turn now to an entirely different problem regarding a limit theorem in the moment space $M$ or $M_n$ that is given in the paper by Chang, Kemperman, and Studden (1993).

The problem stems from an attempt to understand the shape or structure of the moment space $M_n = M_n[0, 1]$ by assigning a uniform measure over $M_n$ and then viewing $c_k$ or $\mathbf{c}_k = (c_1, \ldots, c_k)$ as random variables. Let $v_n$ be the uniform probability measure over $M_n$, that is, ordinary Lebesgue measure divided by the volume $V_n$ of $M_n$. Since $v_n$ assigns measure zero to the boundary of $M_n$, all moment points considered here will be in the interior of $M_n$.

It was shown in Example 1.4.12, in particular, Equation (1.4.8), that the volume $V_n$ of $M_n$ is given by

$$V_n = \text{vol } M_n = \prod_{k=1}^{n} B(k, k). \tag{10.6.1}$$

As $n$ approaches infinity, one can show that $\log V_n \approx -n^2 \log 2$, so $M_n$ is a very small part of the $n$-dimensional unit cube. For $n = 2$, $M_2$ is determined by $0 \le c_1^2 \le c_2 \le c_1 \le 1$, $V_2 = \frac{1}{6}$, and $dv_2 = 6 dc_1 dc_2$. With the probability measure on $M_n$ any function of $(c_1, \ldots, c_n)$ can be viewed as a random variable. For example, if $n = 2$, $c_1$ has density $6c_1(1 - c_1)$, $0 < c_1 < 1$, and $c_2$ has density $6(\sqrt{c_2} - c_2)$, $0 < c_2 < 1$. The corresponding means are $E_2(c_1) = \frac{1}{2}$ and $E_2(c_2) = \frac{2}{5}$. Here $E_n$ denotes expectation or integration with respect to the uniform distribution $v_n$ on the $n$th-moment space.

The "center" $(c_1^{\circ}, \ldots, c_n^{\circ})$ of the moment space $M_n$ where $p_i^{\circ} = \frac{1}{2}$, $i = 1, \ldots, n$, corresponds to the arc-sine measure $\mu^{\circ}$ on the interval $[0,1]$ with density

$$\frac{1}{\pi\sqrt{x(1 - x)}}, \qquad 0 < x < 1.$$

The moments of $\mu^{\circ}$ are given by

$$c_k^{\circ} = \frac{1}{2^{2k}} \binom{2k}{k} \approx \frac{1}{\sqrt{\pi k}} \qquad \text{as } k \to \infty. \tag{10.6.2}$$

The main result of this section shows that the random vector $(c_1, \ldots, c_k)$ is asymptotically $(n \to \infty)$ normal with mean $(c_1^{\circ}, \ldots, c_k^{\circ})$.

**Theorem 10.6.1**   *As $n \to \infty$, the distribution of*

$$\sqrt{n}[(c_1, \ldots, c_k) - (c_1^{\circ}, \ldots, c_k^{\circ})]$$

*relative to the uniform measure $v_n$ over $M_n$ converges to a multivariate normal distribution $\mathcal{N}(0, \Sigma_k)$ with mean zero and covariance $\Sigma_k = \frac{1}{2} A_k A_k'$, where $A_k$ is the lower triangular $k \times k$ matrix with elements*

$$a_{ij} = \begin{cases} 2^{-2i+1} \binom{2i}{i-j}, & 1 \le j \le i, \\ 0, & i < j. \end{cases} \tag{10.6.3}$$

The proof of Theorem 10.6.1 is conceptually simple, the idea being, of course, to switch to canonical moments. It was shown in Example 1.4.12 that the Jacobian of the mapping from $(c_1, \ldots, c_n)$ to $(p_1, \ldots, p_n)$ is given by

$$\frac{\partial(c_1, \ldots, c_n)}{\partial(p_1, \ldots, p_n)} = \prod_{i=1}^{n-1} (p_i q_i)^{n-i}. \tag{10.6.4}$$

Thus the uniform measure over $M_n$ converts to the canonical moments $p_1, \ldots, p_n$ being independent random variables, where $p_k$ has a symmetric Beta distribution $B(\alpha_k, \alpha_k)$, with parameter $\alpha_k = n - k$, $k = 1, \ldots, n$ (see Example 1.3.6). The symmetric Beta distribution $B(\alpha, \alpha)$ has mean $\frac{1}{2}$ and variance $1/(8\alpha + 12)$. Therefore $E_n(p_k) = \frac{1}{2}$ and

$$\text{Var}_n(p_k) = \frac{1}{4(2(n-k) + 3)}$$

$$= \frac{1}{8n} + O\left(\frac{1}{n^2}\right).$$

The limiting distribution is well known, and it is easy to show that if $n \to \infty$,

$$\sqrt{n}(p_k - \tfrac{1}{2}) \longrightarrow \mathcal{N}(0, \tfrac{1}{8}),$$

where $\mathcal{N}(\mu, \sigma^2)$ denotes a normal distribution with mean $\mu$ and variance $\sigma^2$. This is summarized as

**Lemma 10.6.2** *If $(c_1, \ldots, c_n)$ is a random vector with uniform distribution over $M_n$, then the random variables*

$$\sqrt{n}(2p_i - 1), \qquad i = 1, \ldots, k,$$

*are independent and if $n \to \infty$, their joint limiting distribution is $\mathcal{N}(0, \tfrac{1}{2} I_k)$.*

The proof of Theorem 10.6.1 involves transferring the result of Lemma 10.6.2 to the ordinary moments $c_k$. This is done via the following lemma:

**Lemma 10.6.3** *The first order Taylor expansion of $c_k = c_k(p_1, \ldots, p_k)$ around $(p_1^\circ, \ldots, p_k^\circ)$, $p_i^\circ \equiv \frac{1}{2}$, is given by*

$$c_k = c_k^\circ + \sum_{i=1}^{k} a_{ki}(2p_i - 1) + O\left(\sum_{i=1}^{k}(2p_i - 1)^2\right), \tag{10.6.5}$$

*where $a_{ki}$ are defined in (10.6.3).*

Theorem 10.6.1 is then a direct consequence of Lemmas 10.6.2 and 10.6.3. In the paper of Chang, Kemperman, and Studden (1993) two different proofs of the result in (10.6.5) are given. The first one is a direct induction proof using the recursive formula (2.4.7) and Theorem 2.4.4 expressing $c_k$ in terms of $S_{ij}$. The second proof uses an interesting random walk argument by way of the fact that $c_k = P_{00}^{2k}$ (see Theorem 8.5.5), where $P_{ij}^n$ denotes the transition probabilities for a random walk on the integers $0, 1, \ldots$ with reflecting barrier at zero and up and down one-step transition probabilities $q_i$ and $p_i$, respectively.

It was then noticed that the diagonalization of $\Sigma_k$ in Theorem 10.6.1 was intimately connected with the Chebyshev polynomials of the first kind. We will give here a possibly more direct proof using the Chebyshev polynomials.


## 10.7 MORE CHEBYSHEV POLYNOMIALS

Let $T_r(y)$, $y \in [-1, 1]$, $r \geq 0$, denote the Chebyshev polynomials of the first kind on $[-1, 1]$ which are orthogonal with respect to the arc-sine distribution on the interval $[-1, 1]$ (see Example 2.1.1). Also let $T_r^*(x) = T_r(2x - 1)$ denote the corresponding polynomial on $[0, 1]$. The expansion in (10.6.5) will be derived by expanding $x^k$ in terms of $T_1^*, \ldots, T_k^*$ and connecting these to the $p_i$ by using the canonical moments on the circle defined in Chapter 9. To expand the powers in terms of the Chebyshev polynomials, we use the following:

$$y^k = b_k^\circ + \sum_{r=1}^{k} b_{kr} T_r(y),$$

$$x^k = c_k^\circ + \sum_{r=1}^{k} a_{kr} T_r^*(x).$$

(10.7.1)

Here $c_k^\circ$ and $a_{kr}$ are defined in (10.6.2) and (10.6.3). For the first expression

$$b_k^\circ = \int_{-1}^{1} y^k \, d\nu^\circ(y),$$

where $\nu^\circ$ is the arc-sine distribution on $[-1, 1]$. It is easily seen that

$$b_k^\circ = \begin{cases} 0, & k \text{ odd}, \\ c_i^\circ, & k = 2i. \end{cases}$$

(10.7.2)

The $b_{kr}$ are given by

$$b_{k,k-2j} = \begin{cases} \dfrac{1}{2^{k-1}} \dbinom{k}{j}, & j = 0, 1, \ldots, \left[\dfrac{k}{2}\right], \\ 0 & \text{otherwise.} \end{cases}$$

(10.7.3)

The trigonometric representation of the Chebyshev polynomials of the first kind in (2.1.5) gives $T_n(T_m(x)) = T_{nm}(x)$, and it follows that

$$T_k^*(x^2) = T_k(2x^2 - 1)$$
$$= T_k(T_2(x))$$
$$= T_{2k}(x).$$

This implies that

$$a_{ij} = b_{2i,2j}. \tag{10.7.4}$$

The first expression in (10.7.1) follows by writing $y = \cos\theta$ and expanding $y^n = (\cos\theta)^n$ after writing $\cos\theta$ in exponential form. The second follows in a similar manner using $y = 2x - 1 = (\cos\theta/2)^2$.

For a given measure $\mu$ on the interval $[0, 1]$, let $\nu$ denote the corresponding measure induced by the transformation $y = 2x - 1$ on $[-1, 1]$. Integrating the expressions in (10.7.1) with respect to $\mu$ on $[0, 1]$ and the corresponding $\nu$ on $[-1, 1]$ gives

$$c_k = c_k^{\circ} + \sum_{r=1}^{k} a_{kr} \int_0^1 T_r^*(x) d\mu(x),$$

$$b_k = b_k^{\circ} + \sum_{r=1}^{k} b_{kr} \int_{-1}^1 T_r(y) d\nu(y). \tag{10.7.5}$$

Let

$$\gamma_r = \int_{-1}^1 T_r(y) d\nu(y) = \int_0^1 T_r^*(x) d\mu(x). \tag{10.7.6}$$

The quantity $\gamma_r$ is associated with the canonical moments on the circle defined in Chapter 9. For a measure $\sigma(\theta)$ on $[-\pi, \pi)$ or on the unit circle that is symmetric about zero, we have

$$\gamma_k = \int_{-\pi}^{\pi} e^{-ik\theta} d\sigma(\theta)$$

$$= \int_{-\pi}^{\pi} \cos k\theta d\sigma(\theta)$$

$$= \int_{-1}^1 T_k(y) d\nu(y),$$

where $\nu$ is the measure induced from $\sigma$ by the mapping $y = \cos\theta$. Recall from (9.3.8) that the canonical moments for the circle $a_k$ of a symmetric measure $\sigma$ are real and related to the canonical moments $p_k$ of the induced measure $\nu$ on the interval $[-1, 1]$ by

$$a_k = 2p_k - 1.$$

We will assume in the discussion below that the $a_k$ are all real. The measure on the circle corresponding to the arc-sine distribution is the distribution with $a_i = 0$, $i \geq 1$. By Example 9.3.2 this is the uniform measure on the circle.

**Lemma 10.7.1** *The first-order Taylor expansion of* $\gamma_k = \gamma_k(a_1, \ldots, a_k)$ *about* $a_i \equiv 0$ *is given by*

$$\gamma_k = a_k + O\left(\sum_{i=1}^{k} |a_i|^2\right). \tag{10.7.7}$$

*Proof* By (9.3.7) the $k$th canonical moments $a_k$ for a symmetric $\sigma$ is given by

$$a_k = \frac{(-1)^{k-1}}{\Delta_{k-1}} \begin{vmatrix} \gamma_1 & \gamma_2 & \cdots & \gamma_k \\ \gamma_0 & \gamma_1 & \cdots & \gamma_{k-1} \\ \vdots & \vdots & & \vdots \\ \gamma_{-k+2} & \gamma_{-k+3} & \cdots & \gamma_1 \end{vmatrix},$$

where $\Delta_k$ is the determinant of the Toeplitz matrix defined in (9.2.6). Using Lemma 9.3.4, we have

$$\gamma_k = a_k \prod_{i=1}^{k-1} (1 - |a_i|^2) + (-1)^k \frac{R_{k-1}(0)}{\Delta_{k-2}}, \tag{10.7.8}$$

where $R_{k-1}(0)$ is defined in (9.3.4) by

$$R_{k-1}(0) = \begin{vmatrix} \gamma_1 & \cdots & \gamma_{k-1} & 0 \\ \gamma_0 & \cdots & \gamma_{k-2} & \gamma_{k-1} \\ \vdots & & \vdots & \vdots \\ \gamma_{-k+2} & \cdots & \gamma_0 & \gamma_1 \end{vmatrix}.$$

Since $\gamma_i$ depends only on $(a_1, \ldots, a_i)$, we have

$$\left.\frac{\partial \gamma_i}{\partial a_i}\right|_0 = 1,$$

where the "zero" denotes evaluation at $a_j \equiv 0$. By Lemma 9.3.4,

$$\Delta_{k-2} = \prod_{i=1}^{k-2} (1 - |a_i|^2)^{k-i-1},$$

and it suffices to show that

$$\frac{\partial R_{k-1}(0)}{\partial a_r}\bigg|_0 = 0, \qquad r \leq k - 1.$$

However, this follows by differentiating $R_{k-1}(0)$ columnwise (or rowwise), since

$$\gamma_i^\circ = \int_{-1}^1 T_i(y)dv^\circ(y) = 0 \qquad \text{for } i \geq 1. \qquad \blacksquare$$

Lemma 10.6.3 is a direct consequence of Lemma 10.7.1. Using the second part of (10.7.5), it follows also for the interval $[-1,1]$ and the uniform measure on the $n$th-moment space $M_n(-1,1)$ that the distribution of

$$\sqrt{n}\,[(b_1, \ldots, b_k) - (b_1^\circ, \ldots, b_k^\circ)] \qquad (10.7.9)$$

converges to a $\mathcal{N}(0, \frac{1}{2}\Gamma_k)$, where $\Gamma_k = B_k B_k'$ and $B_k$ is lower triangular with elements $b_{ij}$ given in (10.7.3).

Note that $B_k$ is the upper left "section" of $B_{k+1}$, and a similar remark holds for $A_k$. Let $A$ and $B$ be the corresponding infinite lower triangular matrices, $\Gamma = BB'$ and $\Sigma = AA'$. The elements $\lambda_{ij}$ and $\sigma_{ij}$ of $\Gamma$ and $\Sigma$ have the following simple structure:

**Lemma 10.7.2** *If $c_i^\circ$ and $b_i^\circ$ are as given in (10.6.2) and (10.7.2), then*

$$\lambda_{ij} = b_{i+j}^\circ - b_i^\circ b_j^\circ$$

*and*

$$\sigma_{ij} = c_{i+j}^\circ - c_i^\circ c_j^\circ.$$

*Proof* Only the result for $\sigma_{ij}$ is proved. By (10.7.4) this would give the result for $\lambda_{ij}$ where $i$ and $j$ are both even. The case where $i$ and $j$ are both odd is similar. Now

$$\sigma_{ij} = \frac{1}{2} \sum_{r=1}^{\min(i,j)} a_{ir} a_{jr}$$

$$= 2^{-2j-2i+1} \sum_{r=1}^{\min(i,j)} \binom{2i}{i-r}\binom{2j}{j-r}$$

$$= -c_i^\circ c_j^\circ + \sum_{r=-\min(i,j)}^{r=\min(i,j)} 2^{-2i}\binom{2i}{i-r} 2^{-2j}\binom{2j}{j-r}$$

$$= -c_i^\circ c_j^\circ + c_{i+j}^\circ.$$

Here the last summation is obtained by considering the coefficient of $z^{i+j}$ in the expansion of $((1+z)/2)^{2i}((1+z)/2)^{2j}$. ∎

Using Lemma 10.7.2 and the asymptotics for $c_k^\circ$ in (10.6.2), one can get some idea of the size of the correlations between the random variables $c_i$ and $c_j$ in the limiting distribution. More specifically, let

$$\rho_{ij} = \frac{\sigma_{ij}}{\sqrt{\sigma_{ii}\sigma_{jj}}}.$$

Then it can be shown that

1. if $s$ is fixed $\rho_{s,s+r} \longrightarrow 0$ as $r \to \infty$;
2. if $r$ is fixed $\rho_{s,s+r} \longrightarrow 1$ as $s \to \infty$;
3. more generally, if $i, j \longrightarrow \infty$ and $i/j \to K$, then

$$\rho_{ij} \longrightarrow \left(\frac{4K}{(K+1)^2}\right)^{1/4}.$$

# Bibliography

Al-Salam, W., Allaway, W. R., and Askey, R. (1984). Sieved ultraspherical polynomials. *Transactions of the American Mathematical Society*, **284**, 39–55.

Abramowitz, M., and Stegun, I. A. (1965). *Handbook of Mathematical Functions.* Dover, New York.

Achieser, N. I. (1956). *Theory of Approximation.* Ungar, New York.

Achieser, N. I. (1965). *The Classical Moment Problem and Some Related Questions in Analysis.* Hafner, New York, NY.

Anderson, T. W. (1962). The choice of the degree of a polynomial regression as a multiple decision problem. *Annals of Mathematical Statistics*, **33**, 255–265.

Anderson, W. J. (1991). *Continuous-Time Markov Chains.* Springer-Verlag, New York.

Askey, R. and Ismail, M. E. H. (1984). Recurrence relations, continued fractions, and orthogonal polynomials. American Mathematical Society Memoirs, **49**, AMS, Providence, RI.

Atkinson, A. C., and Cox, D. R. (1974). Planning experiments for discriminating between models. *Journal of the Royal Statistical Society, Ser. B*, **36**, 321–334.

Bingham, N. H. (1991). Fluctuation theory for the Ehrenfest urn. *Advances Applied Probability*, **23**, 598–611.

Bordes, G., and Roehner, G. (1983). Application of Stieltjes theory for $S$-fractions to birth and death processes. *Advances Applied Probability*, **15**, 507–530.

Bowman, K. O., and Shenton, L. R. (1989). *Continued Fractions in Statistical Applications.* Marcel Dekker, New York.

Brezinski, C. (1980). *Padé-Type Approximation and General Orthogonal Polynomials.* Birkhäuser, Boston.

Brezinski, C. (1991). *History of Continued Fractions and Padé Approximants.* Springer Series in Computational Mathematics, **12**. Springer-Verlag, Berlin.

Burg, J. P. (1967). Maximum entropy spectral analysis. Proc. 37th Meeting of the Society of Exploration Geophysicists. Reprinted in *Modern Spectrum Analysis*, D. G. Childers, ed. IEEE Press, New York, 1978, pp. 34–41.

Carleman, T. (1922). Sur le probléme des moments. *Comptes Rendus*, **174**, 1680–1682.

Chang, F. C., Kemperman, J. H. B., and Studden, W. J. (1993). A normal limit theorem for moment sequences. *Annals of Probability*, **21**, 1295–1309.

Chebyshev, P. L. (1859). Sur les questions de minima qui se retachent à la représentation approximative de fonctions. *Mémoires de l'Académie Impériale des Sciences de St-Pétersbourg. Sixième Série Sciences Mathématiques et Physiques*, **7**, 199–291.

Chihara, T. S. (1978). *Introduction to Orthogonal Polynomials.* Gordon and Breach, New York.

Childers, D. G. (1978). *Modern Spectrum Analysis.* IEEE Press, New York.

Choi, B. S., and Cover, T. M. (1984). An information-theoretic proof of Burg's maximum entropy spectrum. *Proc. IEEE*, **72**, 1094–1095.

Cover, T. M., and Thomas, J. (1991). *Elements of Information Theory.* Wiley, New York.

DasGupta, A., Mukhopadhyay, S., and Studden, W. J. (1992). Compromise designs in heteroscedastic linear models. *Journal of Statistical Planning and Inference*, **32**, 362–384.

Dehesa, J. S., Van Assche, W., and Yáñez, R. J. (1997). Information entropy of classical orthogonal polynomials and their application to harmonic oscillator and Coulomb potentials. To appear in: *Methods and Applications in Analysis.*

Dette, H. (1990). A generalization of $D$- and $D_1$-optimal designs in polynomial regression. *Annals of Statistics*, **18**, 1784–1805.

Dette, H. (1991). A note on robust designs for polynomial regression. *Journal of Statistical Planning and Inference*, **28**, 223–232.

Dette, H. (1992). Optimal designs for a class of polynomials of odd or even degree. *Annals of Statistics*, **20**, 238–259.

Dette, H. (1993a). On a mixture of $D$- and $D_1$-optimality in polynomial regression. *Journal of Statistical Planning and Inference*, **35**, 233–249.

Dette, H. (1993b). New Identities for orthogonal polynomials on a compact interval. *Journal of Mathematical Analysis and Its Applications*, **179**, 547–573.

Dette, H. (1994a). Extremal properties of ultraspherical polynomials. *Journal of Approximation Theory*, **76**, 246–273.

Dette, H. (1994b). Robust designs for multivariate polynomial regression on the $d$-cube. *Journal of Statistical Planning and Inference*, **38**, 105–124.

Dette, H. (1994c). Discrimination designs for polynomial regression on a compact interval. *Annals of Statistics*, **22**, 890–904.

Dette, H. (1994d). On a generalization of the Ehrenfest urn model. *Journal of Applied Probability*, **31**, 930–939.

Dette, H. (1995a). Optimal designs for identifying the degree of a polynomial regression. *Annals of Statistics*, **23**, 1248–1267.

Dette, H. (1995b). New bounds for Hahn- and Krawtchouk polynomials. *SIAM Journal on Mathematical Analysis*, **26**, 1647–1659.

Dette, H. (1995c). A note on some peculiar extremal phenomena of the Chebyshev polynomials. *Proc. Edinburgh Mathematical Society*, **38**, 343–355.

Dette, H. (1995d). On the minimum of the Christoffel function. *Journal of Computational and Applied Mathematics*, **65**, 85–96.

Dette, H. (1995e). Characterizations for generalized Hermite and sieved ultraspherical polynomials. *Transactions of the American Mathematical Society*, **346**, 691–712.

Dette, H. (1996). On the generating functions of a random walk on the nonnegative integers. *Journal of Applied Probability*, **33**, 1033–1052.

Dette, H., Fill, J., Pitman, J., and Studden, W. J. (1997). Wall and Siegmund duality relations for birth and death chains with reflecting barrier. Technical Report 94–27. Department of Statistics, Purdue University. *Journal of Theoretical Probability*, **10**, 349–374.

Dette, H., and Röder, I. (1995). Designs for multivariate regression with missing terms. *Scandinavian Journal of Statistics*, **23**, 195–203.

Dette, H., and Röder, I. (1997). Optimal discrimination designs for multi-factor models. To appear in: *Annals of Statistics*, June.

Dette, H., and Studden, W. J. (1992). On a new characterization of the classical orthogonal polynomials. *Journal of Approximation Theory*, **71**, 3–17.

Dette, H., and Studden, W. J. (1995a). Some new asymptotic properties for the zeros of Jacobi, Laguerre and Hermite polynomials. *Constructive Approximation*, **11**, 227–238.

Dette, H., and Studden, W. J. (1995b). Optimal designs for polynomial regression when the degree of the polynomial is not known. *Statistica Sinica*, **5**, 459–474.

Dette, H., and Studden, W. J. (1996). How many random walks correspond to a given set of return probabilities. *Stochastic Processes and Their Applications*, **64**, 17–30.

Dette, H., and Wong, W. K. (1995). On $G$-efficiency calculation for polynomial models. *Annals of Statistics*, **23**, 2081–2101.

Eberlein, P. J. (1964). A two parametric test matrix. *Mathematics of Computation*, **18**, 296–298.

Ehrenfest, P., and Ehrenfest, T. (1907). Über zwei bekannte Einwände gegen das Boltzmannsche H-Theorem. *Physikalische Zeitschrift*, **8**, 311–314.

Euler, L. (1767). De Fractionibus Continuis Observationes. *Commentarii Academiae scientarum Petropolitanae*, **11**, 32–81. *Opera Omnia, Ser.* 1, Vol. **14**, B.G. Teubner, Lipzig (1925), 291–349.

Farrell, R. H., Kiefer, J. C., and Walbran, A. (1967). Optimum multivariate designs. *Proc. Fifth Berkeley Symposium*, vol. **1** eds. L. M. LeCam and J. Neyman, University of California Press, Berkeley, 113–138.

Favard, J. (1935). Sur les polynomes de Tchebicheff. *Comptes Rendus de l'Académie des Sciences*, Paris, **200**, 2052–2053.

Fedorov, V. V. (1972). *Theory of Optimal Experiments*. Academic Press, New York.

Feller, W. (1966). *An Introduction to Probability Theory and Its Applications*, vol. **2**. Wiley, New York.

Feller, W. (1968). *An Introduction to Probability Theory and Its Applications*, vol. **1**, 3rd ed. Wiley, New York.

Flajolet, P. (1980). Combinatorical aspects of continued fractions. *Discrete Mathematics*, **32**, 125–162.

Freud, G. (1971). *Orthogonal polynomials*. Pergamon, Oxford.

Gaffke, N. (1987). Further characterizations of design optimality and admissibility. *Annals of Statistics*, **15**, 942–957.

Gantmacher, F. R. (1959). *The Theory of Matrices*, vols. **1, 2**, Chelsea Publishing, New York.

Gauss, C. F. (1813). Disquisitiones gererales circa seriem infinitam

$$1 + \frac{\alpha\beta}{1 \cdot \gamma} + \frac{\alpha(\alpha+1)\beta(\beta+1)}{1 \cdot 2 \cdot \gamma(\gamma+1)} + \frac{\alpha(\alpha+1)(\alpha+2)\beta(\beta+1)(\beta+2)}{1 \cdot 2 \cdot 3 \cdot \gamma(\gamma+1)(\gamma+2)} + etc.$$

*Commentationes Societatis Regiae Scientiarum Gottingensis Recentiones*, vol. **2**, 1–46; Werke, vol. **3**, Königliche Gesellschaft der Wissenschaften, Göttingen (1876), 123–162.

Gerl, P. (1983). Continued fraction methods for random walks on $N$ and on trees. *Probability measures on groups VII, Proc. Oberwolfach*. H. Heyer, ed. Springer-Verlag, New York, 131–146.

Geronimus, L. Ya. (1961). *Orthogonal Polynomials*. Consultants Bureau, New York.

Gohberg, I., and Landau, H. J. (1994). Prediction and the inverse of Toeplitz matrices. In *Approximation and Computation*: A Festschrift in Honor of Walter Gautschi. Proc. Purdue Symposium, Lafayette, IN, December 1993. International Series in Numerical Mathematics, **119**. Birkhäuser, Boston, 219–230.

Good, I. J. (1958). Random motion and analytic continued fractions. *Proc. Cambridge Philosophical Society*, **54**, 43–47.

Grenander, U., and Szegö, G. (1958). *Toeplitz Forms and Their Applications*. University of California Press, Berkeley.

Guest, P. G. (1958). The spacing of observations in polynomial regression. *Annals of Mathematical Statistics*, **29**, 294–299.

Hahn, W. (1949). Über Orthogonalpolynome, die q-Differenzengleichungen genügen. *Mathematische Nachrichten*, **2**, 4–34.

Henrici, P. (1974). *Applied and Computational Complex Analysis*, vol. **1**. Wiley, New York.

Henrici, P. (1977). *Applied and Computational Complex Analysis*, vol. **2**, *Special Functions, Integral Transforms, Asymptotics and Continued Fractions*. Wiley, New York.

Hoel, P. G. (1958). Efficiency problems in polynomial estimation. *Annals of Mathematical Statistics*, **29**, 1134–1145.

Jacobi, C. G. J. (1859). Untersuchungen über die Differentialgleichung der hypergeometrischen Reihe. *Journal für die reine und angewandte Mathematik*, **56**, 149–165. *Gesammelte Werke*, **6**, 184–202.

Jacobsen, L. (1990). Orthogonal polynomials, chain sequences, three-term recurrence relations and continued fractions. *Proc. Conference on Computational Methods and Function Theory, Valparaiso 1989*, E. St. Ruscheweyh, B. Saff, L. C. Calinas, and R. S. Varga, eds. Lecture Notes in Mathematics, **1435**, Springer-Verlag, Berlin, 89–101.

Johnson, N. L., and Kotz, S. (1970). *Continuous Univariate Distributions*. Houghton Mifflin, Boston.

Johnson, N. L., and Kotz, S. (1977). *Urn Models and Their Applications*. Wiley, New-York.

Johnson, N. L., Kotz, S., and Kemp, A. W. (1992). *Univariate Discrete Distributions*. Wiley, New York.

Jones, W. B., Njåstad, O., and Thron, W. J. (1986a). Schur fractions, Perron-Carathéodory fractions and Szegö polynomials: A survey. *Analytic Theory of Continued Fractions II*, W. J. Thron, ed. Lecture Notes in Mathematics, **1199**. Springer-Verlag, Berlin, 127–158.

Jones, W. B., Njåstad, O., and Thron, W. J. (1986b). Continued fractions associated with trigonometric and other strong moment problems. *Constructive Approximation*, **2**, 197–211.

Jones, W. B., Njåstad, O., and Thron, W. J. (1989). Moment theory, orthogonal polynomials, quadrature, and continued fractions associated with the unit circle. *Bulletin of the London Mathematical Society*, **21**, 113–152.

Jones, W. B., and Steinhardt, A. (1982). Digital filters and continued fractions. *Analytic Theory of Continued Fractions*, W. B. Jones, W. J. Thron, and H. Waadeland, eds. Lecture Notes in Mathematics, **932**. Springer-Verlag, Berlin, 129–151.

Jones, W. B., and Thron, W. J. (1980). *Continued Fractions: Analytic Theory and Applications*. Encyclopedia of Mathematics and Its Applications, **11**. Addison-Wesley, Reading, MA. Now distributed by Cambridge University Press, New York.

Jones, W. B., and Thron, W. J. (1982). Survey of continued fraction methods for solving moment problems and related topics. *Analytic Theory of Continued Fractions*, Lecture Notes in Mathematics, **932**. W. B. Jones, W. J. Thron, and H. Waadeland, eds. Springer-Verlag, Berlin, 4–37.

Joshi, V. M. (1975). On the minimax estimation of a random probability with known first *N* moments. *Annals Statistics*, **3**, 680–687.

Karlin, S. (1963). Representations theorems for positive functions. *Journal of Mathematics and Mechanics*, **12**, 599–618.

Karlin, S., and Shapley, L. S. (1953). Geometry of Moment Spaces, *American Mathematical Society Memoirs*, **12**, AMS, Providence, RI.

Karlin S., and Studden, W. J. (1966a). *Tchebycheff Systems: With Applications in Analysis and Statistics*. Wiley Interscience, New York.

Karlin S., and Studden, W. J. (1966b). Optimal experimental designs. *Annals of Mathematical Statistics*, **37**, 783–815.

Karlin, S., and McGregor, J. (1957a). The differential equations of birth and death processes and the Stieltjes moment problem. *Transactions of the American Mathematical Society*, **85**, 489–546.

Karlin, S., and McGregor, J. (1957b). The classification of birth and death processes. *Transactions of the American Mathematical Society*, **86**, 366–400.

Karlin, S., and McGregor, J. (1958a). Many server queuing processes with Poisson input and exponential service times. *Pacific Journal of Mathematics*, **8**, 87–118.

Karlin, S., and McGregor, J. (1958b). Linear growth birth and death processes. *Journal of Mathematics and Mechanics*, **7**, 643–662.

Karlin, S., and McGregor, J. (1959a). Random walks. *Illinois Journal of Mathematics*, **3**, 66–81.

Karlin, S., and McGregor, J. (1959b). A characterization of birth and death processes. *Proc. National Academy of Sciences*, **45**, 375–379.

Karlin, S., and McGregor, J. (1961). The Hahn polynomials, formulas and an application. *Scripta Mathematica*, **26**, 33–46.

Karlin, S., and McGregor, J. (1962). On a genetics model of Moran. *Proc. Cambridge Philosophical Society*, **58**, 299–311.

Karlin, S., and McGregor, J. (1965). Ehrenfest urn models. *Journal of Applied Probability*, **2**, 352–376.

Kemperman, J. H. B. (1961). *The Passage Problem for a Stationary Markov Chain*. University of Chicago Press, Chicago.

Kendall, D. G. (1958). Integral representations for Markov transition probabilities. *Bulletin of the American Mathematical Society*, **64**, 358–362.

Kersting, G. (1974). Strong ratio limit property and *R*-recurrence of reversible Markov chains. *Zeitschrift für Wahrscheinlichkeitstheorie und Verwandte Gebiete*, **30**, 343–356.

Khovanskii, A. N. (1963). *The Application of Continued Fractions and Their Generalizations to Problems in Approximation Theory*, trans. by Peter Wynn. P. Noordhoff, Groningen, The Netherlands.

Kiefer, J. C. (1985). *Collected Papers*, L. D. Brown, I. Olkin, J. Sacks, and H. P. Wynn, eds. Springer-Verlag, New York.

Kiefer, J. C., and Studden, W. J. (1976). Optimal designs for large degree polynomial regression. *Annals of Statistics*, **4**, 1113–1123.

Kiefer, J. C., and Wolfowitz, J. (1960). The equivalence of two extremum problems. *Canadian Journal of Mathematics*, **12**, 363–366.

Kozek, A. (1982). Towards a calculus of admissibility. *Annals Statistics*, **10**, 825–837.

Krafft, O., and Schaefer, M. (1993). Mean passage times for triangular transition matrices and a two parameter Ehrenfest urn model. *Journal of Applied Probability*, **30**, 964–970.

Krafft, O., and Schaefer, M. (1995). Exact Elfving-minimax designs for quadratic regression. *Statistica Sinica*, **5**, 475–485.

Krawtchouk, M. (1929). Sur une généralisation des polynomés d' Hermite. *Comptes Rendus de l' Académie des Sciences, Paris*, **189**, 620–622.

Krein, M. G. (1951). The ideas of P. L. Chebyshev and A. A. Markov in the theory of limiting values of integrals and their further development. *American Mathematical Society Translations*, Ser. 2, **12**, 1–22.

Krein, M. G., and Nudelman, A. A. (1977). *The Markov Moment Problem and Extremal Problems*. Translations of Mathematical Monographs, **50**. AMS, Providence, RI.

Läuter, E. (1974a). Experimental design for a class of models. *Mathematische Operationsforschung und Statistik*, **5**, 379–398.

Läuter, E. (1974b). Optimal multipurpose designs for regression models. *Mathematische Operationsforschung und Statistik*, **7**, 51–68.

Landau, H. J. (editor). (1987). *Moments in Mathematics*. Proc. *Symposia in Applied Mathematics*, **37**. AMS, Providence, RI.

Lau, T. S. (1983). Theory of canonical moments and its application in polynomial regression I and II. Technical Reports 83–23, 83–24, Department of Statistics, Purdue University, Lafayette, IN.

Lau, T. S. (1988). $D$-optimal design on the unit $q$-ball. *Journal of Statistical Planning and Inference*, **19**, 299–315.

Lau, T. S. (1989). On moment inequalities. *Statistics and Probability Letters* **8**, 9–16.

Lau, T. S. (1991). On dependent screening tests. *Biometrics*, **47**, 77–86.

Lau, T. S. (1992). The reliability of exchangeable binary systems. *Statistics and Probability Letters*, **13**, 153–158.

Lau, T. S. (1993). Higher-order kappa-type statistics for a dichotomous attribute in multiple ratings. *Biometrics*, **49**, 535–542.

Lau, T. S., and Studden, W. J. (1985). Optimal designs for trigonometric and polynomial regression using canonical moments. *Annals of Statistics*, **13**, 383–394.

Lau, T. S., and Studden, W. J. (1988). On an extremal problem of Fejér. *Journal of Approximation Theory*, **53**, 184–194.

Lee, C. M. S. (1988). $D$-optimal designs for polynomial regression when the lower degree parameters are more important. *Utilitas Mathematica*, **34**, 53–63.

Legendre, A. M. (1785). Recherches sur l'attraction des sphéroides homogènes. *Mèm. Math. Phys.*, prèsentés à l'Acad. Sciences, **10**, 411–434.

Lim, Y. B., and Studden, W. J. (1988). Efficient $D_s$-optimal design for multivariate polynomial regression on the $q$-cube. *Annals of Statistics*, **16**, 1225–1240.

Lim, Y. B., Wynn, H. P., and Studden, W. J. (1988). A note on approximate $D$-optimal designs for $G \times 2^k$. *Fourth Purdue Symposium on Statistical Decision Theory and Related Topics*, Lafayette, IN., vol. **2**. S. S. Gupta and J. O. Berger, eds., 351–362, Springer-Verlag, New York.

Lorentzen, L., and Waadeland, H. (1992). *Continued Fractions with Applications*. North-Holland, Amsterdam.

McNeil, D. R. (1970). Integral functionals of birth and death processes and related limiting distribution. *Annals of Mathematical Statistics*, **41**, 480–485.

Murphy, J. A., and O'Donohoe, M. R. (1975). Some properties of continued fractions with applications in Markov processes. *Journal of the Institute of Mathematics and its Applications*, **16**, 57–71.

Natanson, T. P. (1955). *Konstruktive Funktionentheorie*. Akademie Verlag, Berlin.

Nevai, P. G. (1979). *Orthogonal Polynomials*. American Mathematical Society Memoirs, **213**. AMS, Providence, RI.

Palacios, J. L. (1993). Fluctuation theory for the Ehrenfest urn via electronic network. *Advances Applied Probability*, **25**, 472–476.

Palacios, J. L. (1996). A note on expected hitting times for birth and death chains. *Statistics and Probability Letters*, **30**, 119–125.

Perron, O. (1954a). *Die Lehre von den Kettenbrüchen*, vol. 1, Teubner, Stuttgart.

Perron, O. (1954b). *Die Lehre von den Kettenbrüchen*, vol. 2. Teubner, Stuttgart.

Pruitt, W. E. (1963). Bilateral birth and death processes. *Transactions of the American Mathematical Society*, **107**, 508–525.

Pukelsheim, F. (1993). *Optimal Design of Experiments*. Wiley, New York.

Pukelsheim, F., and Rieder, S. (1992). Efficient rounding of approximate designs. *Biometrika*, **79**, 763–770.

Rahman, M. (1978). A positive kernel for Hahn Eberlein polynomials. *SIAM Journal on Mathematical Analysis*, **9**, 891–905.

Rivlin, T. J. (1990). *Chebyshev Polynomials*. Wiley, New York.

Robbins, H. (1955). An empirical Bayes approach to statistics. *Proc. of the Third Berkeley Symposium on Mathematical Statistics and Probability* **1**, 157–164, University of California Press, Berkeley, CA.

Robbins, H. (1964). The empirical Bayes approach to statistical decision problems. *Annals of Mathematical Statistics*, **35**, 1–20.

Rodrigues, O. (1816). Mémoire sur l'attraction des spheroides. Paris, Ecole Polytechnique Correspondance **3**, 1814–1816, 361–385.

Royden, H. L. (1968). *Real Analysis*, Macmillan, New York.

Schur, I. (1917, 1918). Über Potenzreihen, die im Innern des Einheitskreises beschränkt sind. *Journal für die reine und angewandte Mathematik*, **147**, **148**.

Selberg, A. (1944). Bemerkninger om et Multipelt Integral. *Norsk Matematisk Tidskrift*, **26**, 71–78.

Shohat, J. A., and Tamarkin, J. D. (1943). *The Problem of Moments*. Mathematical Surveys, **1**. AMS. Providence, RI.

Silvey, S. D. (1980). *Optimal Design*. Chapman and Hall, London.

Skibinsky, M. (1967). The range of the $(n + 1)$th moment for distributions on $[0, 1]$. *Journal of Applied Probability*, **4**, 543–552.

Skibinsky, M. (1968a). Extreme $n$th moments on $[0, 1]$ and the inverse of a moment space map. *Journal of Applied Probability*, **5**, 693–701.

Skibinsky, M. (1968b). Minimax estimation of a random probability whose first $N$ moments are known. *Annals of Mathematical Statistics*, **39**, 492–501.

Skibinsky, M. (1969). Some striking properties of Binomial and Beta moments. *Annals of Mathematical Statistics*, **40**, 1753–1764.

Skibinsky, M. (1976). Sharp upper bounds for probability on an interval when the first three moments are known. *Annals of Statistics*, **4**, 187–213.

Skibinsky, M. (1986). Principal representations and canonical moment sequences for distributions on an interval. *Journal of Mathematical Analysis and Its Applications*, **120**, 95–120.

Skibinsky, M., and Rukhin, A. L. (1989). Admissible estimators of Binomial probability and the inverse Bayes rule map. *Annals of the Institute of Statistical Mathematics*, **41**, 699–716.

Skibinsky, M., and Rukhin, A. L. (1991). Bayes estimators whose range has convex hull of zero prior probability. *Journal of Theoretical Probability*, **4**, 465–472.

Smith, K. (1918). On the standard deviations of adjusted and interpolated values of an observed polynomial function and its constants and the guidance they give towards a proper choice of the distribution of observations. *Biometrika*, **12**, 1–85.

Spitzer, F. L., and Stone, C. J. (1960). A class of Toeplitz forms and their application to probability theory. *Illinois Journal of Mathematics*, **4**, 253–277.

Spruill, M. G. (1990). Good designs for testing the degree of a polynomial mean. *Sankhyā, The Indian Journal of Statistics, Ser. B*, **52**, 67–74.

Stahl, H., and Totik, V. (1992). *General Orthogonal Polynomials*. Cambridge University Press, Cambridge.

Stieltjes, T. J. (1884). Quelques recherches sur la théorie des quadratures dites mecaniques. *Annales scientifiques de l'école normale supériorieure, Paris*, **1**, 409–426. *Ouevres*, **1**, 377–396.

Stieltjes, T. J. (1886). Recherches sur quelques séries semi-convergentes. *Annales scientifiques de l'école normale supériorieure, Paris*, **3**, 201–258. *Ouevres*, **2**, 2–59.

Stieltjes, T. J. (1889). Sur la réducion en fraction continue d'une série précéden suivant les pouissances descendants d'une variable. *Annales de la faculté des sciences de Toulouse pour les sciences mathématiques et les sciences physiques*, **3**, 1–17. *Oeuvres*, **2**, 184–200.

Stieltjes, T. J. (1890). Sur quelques intégrales développment en fractions continues. *Quarterly Journal of Pure and Applied Mathematics*, **24**, 370–382. *Oeuvres*, **2**, 378–394.

Stieltjes, T. J. (1894). Recherches sur les fractions continues. *Annales de la faculté des sciences de Toulouse pour les sciences mathématiques et les sciences physiques*, **8**, J, 1–122; **9**, A, 1–47; *Oeuvres*, **2**, 402–566. Also published in *Memoires Présentés par divers savants à l'Académie de sciences de l'Institut National de France*, **33**, 1–196.

Strobach, P. (1990). *Linear Prediction Theory*. Springer-Verlag, New York.

Studden, W. J. (1980a). $D_s$-optimal designs for polynomial regression using continued fractions. *Annals of Statistics*, **8**, 1132–1141.

Studden, W. J. (1980b). On a problem of Chebyshev. *Journal of Approximation Theory*, **29**, 253–260.

Studden, W. J. (1982a). Optimal designs for weighted polynomial regression using canonical moments. *Third Purdue Symposium on Decision Theory and Related Topics*, vol. **2**, S. S. Gupta, J. O. Berger eds. 335–350.

Studden, W. J. (1982b). Some robust-type $D$-optimal designs in polynomial regression. *Journal of the American Statistical Association*, **77**, 916–921.

Studden, W. J. (1989). Note on some $\phi_p$-optimal designs for polynomial regression. *Annals of Statistics*, **17**, 618–623.

Szegö, G. (1975). *Orthogonal Polynomials*. American Mathematical Society Colloquium Publications, vol. **23**, AMS, Providence, RI.

Van Assche, W. (1987). *Asymptotics for Orthogonal Polynomials*. Lecture Notes in Mathematics, **1265**. Springer-Verlag, Berlin.

Van Doorn, E. A. (1981). The transient state problem for a queuing model. *Journal of Applied Probability*, **18**, 499–506.

Van Doorn, E. A., and Schrijner, P. (1993). Random walk polynomials and random walk measures. *Journal of Computational and Applied Mathematics*, **49**, 289–296.

Van Doorn, E. A., and Schrijner, P. (1995). Ratio limits and quasi-limiting distributions for random walks. *Journal of Mathematical Analysis and Its Applications*, **190**, 263–284.

Wall, H. S. (1948). *Analytic Theory of Continued Fractions*. Van Nostrand, New York.

Weber, M., and Erdélyi, A. (1952). On a finite difference analogue of Rodrigues-formula. *American Mathematical Monthly*, **59**, 163–168.

Whitehurst, T. A. (1982). An application of orthogonal polynomials to random walks. *Pacific Journal of Mathematics*, **99**, 205–213.

Wilson, M. W. (1970). On the Hahn polynomials. *SIAM Journal on Mathematical Analysis*, **1**, 131–139.

Woess, W. (1985). Random walks and periodic continued fractions. *Advances Applied Probability*, **17**, 67–84.

Wynn, P. (1964). On Some Recent Developments in the Theory and Application of Continued Fractions. *SIAM Journal on Numerical Analysis*, Ser. B, **1**, 177–197.

Zaremba, S. K. (1975). Some properties of polynomials orthogonal over the set $1, 2, \ldots, N$, *Annali di Matematica Pura ed Applicate.*, **105**, 333–345.

# Index of Symbols

The number following each symbol is the page number where the symbol is introduced.

**320**

# Author Index

# Subject Index

# WILEY SERIES IN PROBABILITY AND STATISTICS

*Now available in a lower priced paperback edition in the Wiley Classics Library.

*Now available in a lower priced paperback edition in the Wiley Classics Library.

*Now available in a lower priced paperback edition in the Wiley Classics Library.

*Now available in a lower priced paperback edition in the Wiley Classics Library.

*Now available in a lower priced paperback edition in the Wiley Classics Library.